Advances in Unconventional Reservoirs and Enhanced Oil Recovery

Advances in Unconventional Reservoirs and Enhanced Oil Recovery

Guest Editors

Saipeng Huang
Jun Liu
Ang Li

MDPI

Basel • Beijing • Wuhan • Barcelona • Belgrade • Novi Sad • Cluj • Manchester

Guest Editors

Saipeng Huang
School of Resources and Safety Engineering
Chongqing University
Chongqing
China

Jun Liu
Institute of New Energy and Low-Carbon Technology
Sichuan University
Chengdu
China

Ang Li
College of Earth Sciences
Jilin University
Changchun
China

Editorial Office
MDPI AG
Grosspeteranlage 5
4052 Basel, Switzerland

This is a reprint of the Special Issue, published open access by the journal *Energies* (ISSN 1996-1073), freely accessible at: https://www.mdpi.com/journal/energies/special_issues/6G8C35W3RJ.

For citation purposes, cite each article independently as indicated on the article page online and as indicated below:

Lastname, A.A.; Lastname, B.B. Article Title. *Journal Name* **Year**, *Volume Number*, Page Range.

ISBN 978-3-7258-7404-0 (Hbk)
ISBN 978-3-7258-7405-7 (PDF)
https://doi.org/10.3390/books978-3-7258-7405-7

Contents

About the Editors

Saipeng Huang

Saipeng Huang, of Chongqing University, has been awarded one project from the National Natural Science Foundation of China, one project from the New Chongqing Talent Project, one project from the China Postdoctoral Science Foundation, one project from the Postdoctoral International Exchange Program (Academic Exchange Project), and one project from the Heilongjiang Postdoctoral Science Foundation. He primarily focuses on "in situ stress evaluation in unconventional oil and gas reservoirs", "hydraulic fracturing fracture propagation mechanisms", "multi-scale control mechanisms of fractures in shale oil/gas reservoirs," and "reservoir numerical simulation based on thermal–hydraulic–mechanical coupling". His notable strengths have been developed in areas such as detailed zonation of horizontal in situ stress in reservoirs, finite element simulation of in situ stress distribution characteristics, simulation of fracture initiation and propagation mechanisms, and detailed characterization of shale fractures. His interests are shale gas and oil; CBM; FEM; nature fracture propagation; CCUS; hydraulic fracturing; and in situ stress evaluation in deep reservoirs.

Jun Liu

Jun Liu serves as an Associate Professor and doctoral supervisor at Sichuan University, China, and has a petroleum geology background in the realm of unconventional gas exploration and exploitation. His interests are reservoir geology, CO_2 enhanced gas recovery, and CO_2 geological sequestration in unconventional hydrocarbon reservoirs, like shale and coal measures.

Ang Li

Ang Li is currently an Associate Professor at the School of Earth Sciences, Jilin University, with a main research focus on unconventional oil and gas geology and evaluation. Ang Li is currently engaged in research on the formation mechanism and oil and gas accumulation mechanism of volcanic reservoirs in ancient buried hills, as well as organic geochemistry of shale oil and gas.

energies

MDPI

Article

Surfactant–Polymer Formulation for Chemical Flooding in Oil Reservoirs

Dmitriy Podoprigora, Mikhail Rogachev and Roman Byazrov *

Department of Oil and Gas Fields Development and Operation, Empress Catherine II Saint Petersburg Mining University, 2, 21st Line, 199106 St. Petersburg, Russia; podoprigora-1990@mail.ru (D.P.); rogatchev@mail.ru (M.R.)
* Correspondence: byazrow97@gmail.com

Abstract: A significant part of oil fields has reached a late stage of development, where technologies aimed at increasing the oil recovery factor are becoming particularly relevant. One such technology is surfactant–polymer flooding. To implement this technology, it is necessary to select a chemical formulation that retains its properties under reservoir conditions and enhances the efficiency of water flooding. This work presents a laboratory evaluation of various polymer and surfactant samples to develop an effective chemical formulation. The results demonstrate that anionic surfactants based on sodium laureth sulphate and betaine significantly reduce interfacial tension at the oil–water interface of the target reservoir. Furthermore, when combined with a partially hydrolysed polymer, the sodium laureth sulphate-based surfactant increases the capillary number by 4500 times (reducing interfacial tension from 32.77 mN/m to 0.065 mN/m and increasing the viscosity of the injected agent from 0.5 mPa·s to 4.36 mPa·s). Based on core flooding studies, it can be concluded that the proposed surfactant–polymer composition increases the oil displacement factor from the core sample by 0.15–0.24, depending on the injection volume. The selected formulation can be recommended for application in water flooding at the target reservoir.

Keywords: enhanced oil recovery; polymer; surfactant; chemical flooding; surfactant–polymer flooding

1. Introduction

Today, a significant portion of the world's oil fields has entered a late stage of development, characterised by a high water cut, a low recovery factor relative to initial recoverable reserves, and a slow pace of production [1–4]. One such object, located in the West Siberian Oil and Gas Basin in Russia, is represented by heterogeneous terrigenous reservoirs associated with the Alym and Cherkashin formations.

The reservoirs under consideration are characterised by a high degree of heterogeneity and lithological variability both laterally and vertically. Typically, the overall clay content and compartmentalisation of the section increase from the crest to the flanks of the reservoirs. A distinctive feature is the alternation of massive sandstone intervals with thin interbedded sand–clay sequences within the section. The average effective thickness is 3.7 m, ranging from 1.7 to 8.3 m, while permeability varies from 0.008 to 0.43 μm^2. The oils in the AV reservoirs are low-viscosity, with an average viscosity of 2 mPa·s, ranging from 1 to 4.4 mPa·s. The clay content in the productive formation is moderate, up to 15%. The salinity of formation water varies between 19 and 32 g/L. The average water cut of the produced fluids is 90%, ranging from 70% to 97%.

The AV_2 development target, which includes the oil-bearing $AV_{1/3}$ and AV_2 reservoirs, is the primary focus for the subsoil user in terms of production volumes and available potential reserves. Although these reservoirs are hydrodynamically connected by more than 70% and share a common oil–water contact (OWC), they differ significantly in terms of structural characteristics and reservoir properties. The ratio of permeability values for these reservoirs varies from 3 to 8 across different fields.

The development of the vast majority of such reservoirs is carried out with reservoir pressure maintenance through continuous water injection into the formation via injection wells. However, the efficiency of traditional water flooding in heterogeneous terrigenous reservoirs, in terms of oil recovery factor, is still considered unsatisfactory [1].

For increasing the level of extraction in mature oil fields, various enhanced oil recovery (EOR) technologies can be applied [2,5]. Some of these technologies aim to improve the displacement and sweeping efficiency of water, primarily by adding various chemical agents [6,7]. Such methods are commonly referred to as chemical EOR methods, among which the following main types can be distinguished: polymer flooding, surfactant–polymer (SP) flooding, and alkaline-surfactant-polymer (ASP) flooding.

The addition of thickening agents (polymers) to increase the viscosity of the displacing fluid can be implemented in two scenarios [8]:

- When there is an unfavourable mobility ratio between oil and water;
- When the mobility ratio is favourable but the oil viscosity is low in reservoirs with a certain degree of heterogeneity.

In the first case, inefficient displacement at the macroscopic level is observed, leading to early water breakthrough followed by a prolonged period of oil production with increasing water cut [9,10]. This situation is well illustrated by the phenomenon of "viscous fingering", which occurs due to the viscosity contrast between oil and water, resulting in the formation of unswept zones [11]. This issue can also arise when injecting insufficiently viscous polymer into reservoirs containing heavy oil. The above is clearly described by the mobility ratio formula, which, under ideal conditions, should be equal to (1):

$$M = \frac{\lambda_w}{\lambda_o} = \frac{k_{w(S_{or})} \cdot \mu_o}{k_{o(S_{wi})} \cdot \mu_w} \tag{1}$$

λ_w—mobility of water (dimensionless);
λ_o—mobility of oil (dimensionless);
$k_{w(S_{or})}$—relative permeability to water at residual oil saturation (dimensionless);
$k_{o(S_{wi})}$—relative permeability to oil at irreducible water saturation (dimensionless);
μ_o—viscosity of oil, mPa·s;
μ_w—viscosity of water, mPa·s.

In the second case, during conventional water flooding with a favourable mobility ratio, the presence of high-permeability channels or significant reservoir compartmentalisation can lead to a reduction in sweep efficiency, both areal (lateral) and vertical [12,13]. The presence of high-permeability layers can also result in premature water breakthrough. Under such conditions, increasing the viscosity through polymer flooding can significantly improve sweep efficiency [14].

In the case of polymer flooding, it is not so much the compartmentalisation of the reservoir that needs to be considered, but rather the degree of permeability contrast between its different parts. With a small permeability contrast, polymer flooding can have a favorable effect by balancing the displacement profile between low-permeability and high-permeability zones [15]. However, in cases of high heterogeneity (where the ratio of maximum to minimum permeability is 20 times or more), the polyacrylamide (PAM) solution flows will predominantly occur in the high-permeability zone, which is typically already highly depleted. In such cases, technologies aimed at completely plugging high-permeability intervals are necessary [16].

Additionally, the value of absolute permeability must be taken into account. The lower limit, which is approximately 0.1–0.2 μm^2 (the exact value depends on the properties of the specific polymer), is determined by the ability of PAM molecules to pass through small pores [17–19]. At lower permeability values, the size of PAM molecules becomes comparable to the diameter of the pore channels, which can lead to pore plugging.

It is also important to note that, due to various microscopic effects associated with the viscoelastic properties of polymer solutions, polymer flooding not only improves the sweep efficiency, but also enhances the oil displacement factor. This refers to the additional displacement of a portion of oil contained in dead-end pores, residual oil in already swept pore channels of varying permeability, and oil held by capillary forces [20].

Some studies [21] suggest that the greatest improvement in oil displacement by polymer solutions is attributed to the mechanism of elastic turbulence. This involves the chaotic redistribution of flows in the boundary layer [22], causing micro-level pressure fluctuations that, in turn, lead to variations in shear rates and changes in the mobility of oil and water in zones with increased heterogeneity.

The authors [23] note that a polymer solution with viscoelastic properties displaces a greater amount of oil than a Newtonian fluid of the same viscosity. Studies [24,25] associate the reduction in residual oil saturation with the molecular weight of the polymer and the relaxation time of the molecules during flow in a porous medium.

The pivotal parameter governing the efficacy of polymer flooding implementation is the viscosity of the injected solution. The primary cause of viscosity loss stems from chain scission, i.e., the rupture of polymer backbone structures. It is imperative to distinguish between chain scission and degradation processes, as they entail fundamentally different physicochemical mechanisms with distinct implications for polymer performance under reservoir conditions [26].

Degradation refers to the deterioration of material properties—such as polymer solubility or solution viscosity loss—without the cleavage of macromolecular chains. This phenomenon arises from alterations in macromolecular conformation, changes in polymer coil dimensions, modifications to the chemical activity of functional groups, and their interactions with polyvalent salt ions present in aqueous media.

In contrast, chain destruction entails the rupture and fragmentation of macromolecules into smaller segments, thereby reducing the polymer's molecular weight. This process invariably diminishes polymer solution viscosity. Destruction mechanisms may be thermal, chemical, mechanical, or biological in nature, each inducing distinct pathways of molecular breakdown [27].

Elevated reservoir temperatures accelerate the hydrolysis of polyacrylamide. At 50 °C, PAM hydrolysis proceeds at a negligible rate, with solutions maintaining prolonged stability irrespective of brine salinity or ionic composition [28]. At temperatures exceeding 70 °C, amide group hydrolysis occurs at a pronounced rate, rendering the polymer susceptible to polymineral aggression. Consequently, under such thermal conditions, PAM application remains viable only when formation water contains calcium ions (Ca^{2+}) at concentrations ≤ 200 ppm [29].

Chemical or oxidative degradation predominantly proceeds via free-radical chain reactions. The active centre forms in the polymer backbone through the abstraction of a tertiary hydrogen atom. Subsequent oxygen molecule attachment generates a peroxide radical, which may either abstract another tertiary hydrogen or induce chain destruction [30]. In polymer flooding applications, such free radicals typically form through redox reactions between dissolved oxygen and either iron (II) species or hydrogen sulphide (H_2S). The resulting chain scission events reduce the polymer's hydrodynamic volume, ultimately leading to viscosity degradation [31].

The mechanical degradation of polymers occurs when macromolecules are subjected to excessive shear rates or abrupt pressure differentials within pipes, chokes, orifices, or pumping equipment. This induces molecular chain rupture, consequently diminishing polymer viscosity [32,33].

Biological degradation can occur in both partially hydrolysed polyacrylamides and polysaccharides, though with a higher probability in the latter case. The governing factors include bacterial strains present in solution, pressure, temperature, salinity, and coexisting chemical species. Biodegradation proceeds more rapidly in low-salinity environments at moderate temperatures [33].

The degradation challenges posed by high salinity and temperature can be mitigated through acrylamide copolymerisation with hydrolysis-resistant monomers, such as acrylamide-tert-butyl sulfonate (ATBS) [34].

ATBS is synthesised via the Ritter reaction utilising acrylonitrile and isobutylene in the presence of sulphuric acid and water. This monomer has been extensively investigated to address amide group stability issues under elevated temperature conditions.

Surfactants can enhance the oil recovery factor through the following two mechanisms:

(1) Reduction ininterfacial tension (IFT) at the oil–water interface;
(2) Alteration of rock wettability.

Interfacial tension can be defined as the force acting between molecules at the interface of two phases [35], specifically at the oil–water interface.

Reducing interfacial tension contributes to an increase in the capillary number, which is a dimensionless ratio of viscous forces to local capillary forces:

$$N_{ca} = \frac{F_V}{F_C} = \frac{\mu v}{\sigma \cos \cos \theta} \tag{2}$$

F_V—viscous forces;
F_C—capillary forces;
μ—viscosity of the displacing fluid, mPa·s;
v—velocity of the displacing fluid, m/s;
σ—interfacial tension (surface tension), mN/m;
θ—contact angle (wettability angle), degrees.

Experimental data show that as the capillary number increases, residual oil saturation decreases [36].

It is important to note that the chemical substances mentioned above can also be used in other field operations unrelated to enhanced oil recovery (EOR) technologies. For example, surfactants can be employed in well-killing operations [37–39] or in technologies aimed at preventing or removing asphalt–resin–paraffin deposits (ARPD) [40–42]. In turn, polymers can be used in the preparation of drilling fluids [43,44], for sand control production [45,46], reducing hydraulic resistance in pipes, and other operations [47].

Surfactants have the ability to dissolve in both water and oil, thereby reducing the interfacial tension at the oil–water interface to sufficiently low values. This enables a high capillary number, which helps overcome capillary forces and extract additional oil [48].

The first results of laboratory and field tests of ionogenic surfactants used as additives in water flooding were published in the USA in the 1940s–1950s. There were more than 30 surfactant injection (water-soluble and oil-soluble surfactants) pilots on different fields in Russia, but the first one was carried out in 1964 on Arlanskoye field. Aqueous solutions of OP-10 were used [49].

To date, over 700 chemical flooding projects have been implemented across diverse geological and petrophysical environments. Given this study's specific focus on surfactant–polymer (SP) flooding technology, Table 1 summarises key operational parameters and performance outcomes from the most thoroughly documented case studies in the literature.

Table 1. Key characteristics of fields and brief results of the SP flooding implementation.

№	Field, Country	Technology	Reservoir Type	Temperature, °C	Oil Viscosity, mPa·s	Salinity, g/L	Permeability, mD	Porosity	Results
1	Romashkino Russia [50]	SP	Sandstone	25	30	240	1300	0.23	Pilot is being carried out
2	Kharyaga Russia [51]	SP	Carbonate	62	0.8–1.1	200	300	0.09	Pilot preparation
3	Daqing China [52]	SP	Sandstone	52	12	6	1400	0.26	Increase in oil production from 0.2 mt/y to 4.06 mt/y
4	Zhongyuan China [53]	SP	Not given	80–90	Not given	120	716	Not given	Increase in recovery factor by 13.7%
5	Jilin China [54]	SP	Not given	55	Not given	14	163	Not given	Increase in recovery factor by 14.8%
6	Liaohe China [55]	SP	Not given	55	Not given	3.5	285.9	Not given	Increase in recovery factor by 15.4%
7	Changqing China [56]	SP	Not given	51	Not given	12–26	67	Not given	Increase in recovery factor by 15.1%
8	Dagang China [57]	SP	Not given	53	Not given	13.45	675	Not given	Increase in recovery factor by 13.0%
9	Algyo Hungary [58]	SP	Sandstone	98	0.64	0.15	70	0.23	Not given

As evidenced by Table 1, surfactant–polymer (SP) flooding implementation demonstrates enhanced ultimate recovery factors, confirming the method's significant potential and operational viability.

However, the implementation of surfactant–polymer (SP) flooding technology requires an individualised approach to develop an effective chemical formulation [59] that is compatible with reservoir fluids and rock, maintains its key properties under reservoir conditions, and maximises the potential for oil recovery. This article proposes a laboratory program for the selection and evaluation of a formulation for chemical flooding, and presents the results of research on the development of a SP composition for the AV reservoir.

2. Materials and Methods

2.1. Experimental Materials and Equipment

2.1.1. Polymer and Surfactant Samples

As part of the research, samples of partially hydrolysed polyacrylamide (HPAM) with a hydrolysis degree of 20–30%, a molecular weight ranging from 7.3 to 20.3 million Da, and an acrylamide-tert-butyl sulfonic acid content in the range of 0–15% were considered. Preliminary screening was primarily governed by the target formation's petrophysical properties. Given the elevated reservoir temperatures and compromised rock–fluid characteristics, conventional polymers risk property degradation or formation damage. However, deploying thermally stable low-molecular-weight polymers necessitates higher capital expenditures. Consequently, the aforementioned molecular weight and ATBS content ranges were established to probabilistically optimise polymer selection.

For surfactant pre-screening, the literature-based prediction of optimal surfactant–crude oil interactions proves exceptionally challenging under specific reservoir conditions. Therefore, we evaluated commercially available surfactants with documented field applications in chemical flooding projects. The surfactants evaluated included the following:

1. Anionic surfactants based on sodium alkylbenzene sulfonate, diethanolamide of lauric acid, sodium laureth sulphate, dodecylbenzene sulfonate, and alpha-olefin sulfonate;
2. Amphoteric surfactants based on betaine and oleyl betaine.

Detailed information about the chemical reagents used is provided in Tables 2 and 3.

Table 2. Chemical properties of the examined polymers.

Polymer	Intrinsic Viscosity, dL/g	ATBS Content, %%	Hydrolysis Degree, %	Molecular Weight, 10^6 Da
FP 3230S	16.21	0	30	7.3
FP 3330S	18.3	0	30	9.3
FP 3630S	27.62	0	30	20.3
FP 5205 SH	20.61	5	20	12.7
FP 5205 VHM	23.32	5	20	16.1
FP 5115 SH	19.84	15	20	11.8

Table 3. Chemical properties of the examined surfactants.

Surfactant	Type	Chemical Formula
Sodium Alkylbenzene Sulfonate	Anionic	$C_{12}H_{25}C_6H_4SO_3Na$
Sodium Alkylbenzene Sulfonate	Anionic	$C_{16}H_{33}C6H_4SO_3Na$
Diethanolamide of Lauric Acid	Anionic	$C_{16}H_{33}NO_3$
Sodium Laureth Sulfate	Anionic	$CH_3(CH_2)_{11}(CH2CH_2O)2_SO_3Na$
Dodecylbenzene Sulfonate	Anionic	$CH_3(CH_2)_{10}CH_2\text{-}SO_3Na$
Alpha-Olefin Sulfonate	Anionic	$C_{14}H_{27}SO_3Na$
Betaine	Amphoteric	$C_5H_{11}NO_2$
Oleyl Betaine	Amphoteric	$C_{22}H_{43}NO_2$

2.1.2. Brine Model

The solutions were prepared using a reservoir water model. The chemical composition of the brine used for preparing the working solutions is presented in Table 4.

Table 4. Parameters of brine.

Parameter	Value
Ca^{2+}, mg/L	940.0
Mg^{2+}, mg/L	96.0
$Na^{++}K^{+}$, mg/L	7753.1
Cl^{-}, mg/L	13,736.88
HCO_3^{-}, mg/L	268.4
SO_2^{-}, mg/L	1.26

2.1.3. Core Samples and Oil Samples

For the core flooding studies, extracted, dried, cylindrical core samples from the target reservoir and oil collected from the field's wells were used.

Cylindrical cores were drilled from reservoir rock samples. The plugs underwent hydrocarbon extraction using an azeotropic alcohol–benzene mixture (1:3 vol ratio) in a Soxhlet apparatus, followed by salt removal via distilled water rinsing. Cleaned samples were oven-dried to constant mass at 105 ± 2 °C and stored in desiccators containing anhydrous calcium chloride.

Petrophysical characterisation employed an automated system for porosity and permeability measurements by Geologica. Following property determination, cores were vacuum-saturated with crude oil. Initial oil volume was quantified through gravimetric analysis pre- and post-saturation. Saturated cores were then loaded into core holders and conditioned at test temperature (57.0 ± 0.5 °C) for 24 h.

Table 5 provides information on the core samples used.

Table 5. Information on the core samples used.

№	Sample Length, cm	Diameter of the Sample, cm	Porosity, %	Permeability, μm^2	Mass of the Dry Sample, g	The Mass of the Sample Saturated with Oil, g	Oil Volume, mL	Sample Volume, cm^3
Core samples for core flooding studies of polymer solutions								
1	3.42	2.98	22.43	0.193	47.45	52.11	5.35	23.85
2	3.37	3.01	21.48	0.172	48.05	52.54	5.15	23.98
3	3.18	2.99	23.24	0.191	45.25	49.77	5.19	22.33
4	3.24	2.97	23.43	0.189	44.9	49.48	5.26	22.45
Core samples for core flooding studies surfactant polymer solutions								
1	3.37	3.02	22.12	0.075	49.15	53.8	5.34	24.14
2	3.17	3.00	24.5	0.116	48.94	53.72	5.49	22.41
3	3.07	3.01	25.63	0.069	48.86	53.74	5.6	21.85

2.1.4. Experimental Equipment

The rheological properties of the solutions were studied using a Brookfield DV2T rotational viscometer (Brookfield Ametek, Middleboro, MA, USA) equipped with a UL adapter for measuring low viscosity. To accurately evaluate the interfacial tension at the oil–surfactant solution interface, studies were conducted using the KRÜSS SDT 100 tensiometer (KRÜSS Scientific, Hamburg, Germany) with the spinning drop method. Core flooding studies were performed using an automated unit, the RPS-812 (CoretestSystemsCorporation, Morgan Hill, CA, USA).

2.2. Experimental Methods

The research program for polymer solutions, surfactant solutions, and surfactant–polymer compositions is presented in Figure 1.

Studies on Polymer Solution Properties

Determination of polymer dissolution time; evaluation of rheological properties of solutions: viscosity curves as a function of shear rate and polymer concentration; measurement of static polymer adsorption; assessment of polymer solution stability; determination of polymer displacement characteristics

Studies on Surfactant Solution Properties

Evaluation of surfactant solubility in injected water; phase behavior of surfactant solutions with oil; dependence of IFT at the surfactant solution–oil interface on surfactant concentration; measurement of static surfactant adsorption; assessment of the influence of temperature and system salinity on IFT at the surfactant solution–oil interface.

Studies on Surfactant–Polymer Formulation Properties

Influence of surfactant and polymer concentration on the key properties of the chemical composition; core flooding studies

Figure 1. Laboratory tests program (developed by the authors).

2.2.1. Requirements for Polymers and Experimental Methodology

The main criteria for ranking different polymer grades are as follows:

1. Compatibility with the reservoir water (absence of lumps or precipitates during solution preparation). The methodology is based on the periodic visual assessment of polymer powder dissolution during solution preparation. As the polymer dissolves, the solution becomes homogeneous, and no polymer particles should remain. The acceptable dissolution time should not exceed 240 min.
2. Rheological properties, specifically achieving maximum solution viscosity at minimum concentration. The methodology involves measuring the viscosity of polymer solutions at various spindle rotation speeds (shear rates) using a rotational viscometer.
3. Resistance to mechanical degradation. The mechanical stability coefficient is calculated as the ratio of the viscosity of the degraded solution to the original solution. The mechanical stability coefficient should be above 0.85. The methodology involves measuring the viscosity of the polymer solution before and after vigorous stirring using a paddle mixer at high rotation speeds (2000 rpm and above).
4. Thermal stability of polymers. This is determined as the ratio of the viscosity of the solution after heating and prolonged exposure to the viscosity of the original solution at reservoir temperature. The thermal stability coefficient should be above 0.85.
5. Compatibility of polymer molecular weight with the reservoir properties. Given the poor reservoir properties (low-permeability zones), the polymer formulation must ensure an appropriate flow level (no excessive plugging). The key factor influencing the free flow of the polymer solution is its molecular weight. According to the information presented in [8], an empirical correlation between the number–average molecular weight and the absolute permeability of the rock has been proposed and is presented in Table 6.

Table 6. Empirical correlation between number and average molecular weight and absolute rock permeability.

Average Molecular Weight, Million Da	Absolute Permeability, μm^2
>20	1
18–20	0.75
15–18	0.5
12–15	0.35
8–12	0.2
5–8	0.1
1–5	0.01

It is important to note that Table 6 should be used with caution, as chemical compositions and salinity can significantly affect the hydrodynamic volume of the polymer, sometimes enabling its propagation in lower-permeability layers. During the study, the most suitable polymer formulations will be tested in core flooding experiments to verify the compatibility of the molecular weight with the flow properties of the rock.

2.2.2. Requirements for the Surfactant

Based on the review of scientific publications and studies on this topic, the following key criteria/requirements for surfactants have been formulated [60]:

1. Good solubility in water under standard (20 °C) and reservoir conditions. The methodology is based on a visual assessment of the solution's appearance during mixing. Some surfactants form cloudy solutions when dissolved, and in some cases, salting out or precipitation of the surfactant may occur (Figure 2). The criterion for surfactant solubility is the preservation of solution transparency for at least 3 days.

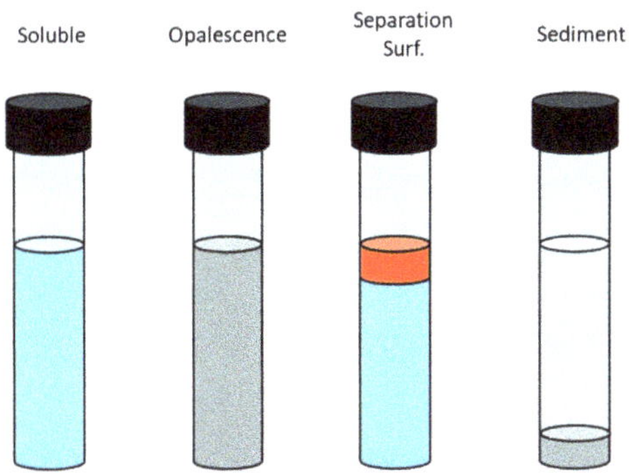

Figure 2. Solubility surfactant in water.

2. Ensuring a consistently low level of IFT for dispersing trapped oil and washing residual oil films from the rock. Two approaches are used to evaluate the reduction in IFT. The first method is based on assessing the phase behavior of surfactant solutions and oil (determining Winsor types), a rapid method. The second method involves a quantitative assessment of IFT in mN/m using the spinning drop method, a more accurate evaluation method.
3. Low level of adsorption on the rock. The methodology involves keeping surfactant solutions in contact with the rock (with periodic stirring) for 2 days in a thermal chamber at reservoir temperature (57 °C). After this period, the solutions are separated from the rock by centrifugation, and the residual surfactant concentration in the solution is measured. In this case, concentrations are determined indirectly, based on the dependence on interfacial tension.
4. Stability of the surfactant solution across various temperature and salinity ranges. The methodology involves measuring IFT using the spinning drop method at different temperatures and salinities.

Next, the compatibility of surfactants and polymers was evaluated. This stage involves assessing solubility in the water model and evaluating the influence of component concentrations on the key properties of the chemical composition (viscosity, interfacial tension). In the final stage, core flooding studies on the developed composition were conducted. Core flooding studies were performed in accordance with OST 39–195-86 "Oil. Method for determining the oil displacement coefficient by water under laboratory conditions" [61].

3. Results and Discussions

3.1. Polymer Studies

3.1.1. Evaluation of Polymer Solubility

Polymer solutions with concentrations of 5000 ppm were prepared at a water temperature of 20 °C using an overhead stirrer at a rotation speed of 300 rpm. The results of the study are presented in Table 7.

Table 7. Dissolution time of polymers.

Polymer	Dissolution Time, min
FP 3230S	60
FP 3330S	90
FP 3630S	160
FP 5205 SH	120
FP 5205 VHM	140
FP 5115 SH	110

As can be seen, all the polymer samples under consideration met the solubility criterion: the dissolution time was less than 240 min.

3.1.2. Rheological Curves for Polymer Samples

The results of the studies are presented in Figures 3–8. Figure 9 shows the dependence of the effective viscosity of the polymer solution on concentration (at a shear rate of 7.34 s^{-1}).

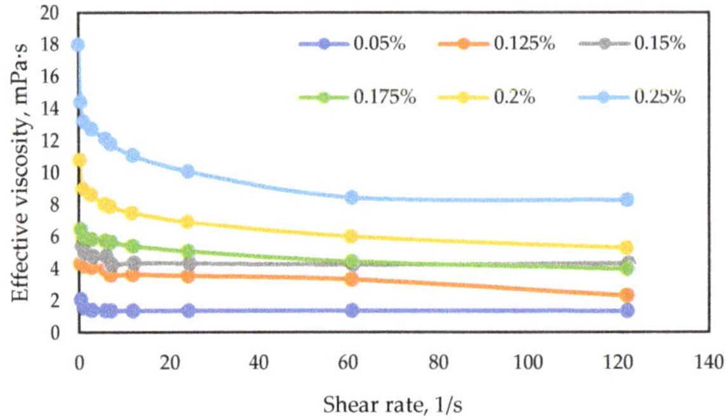

Figure 3. Dependence of effective viscosity on shear rate at various mass concentrations of polymer FP 3230S.

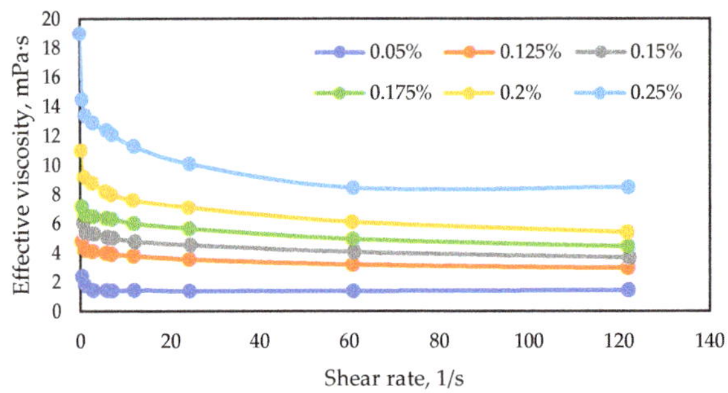

Figure 4. Dependence of effective viscosity on shear rate at various mass concentrations of polymer FP 3330S.

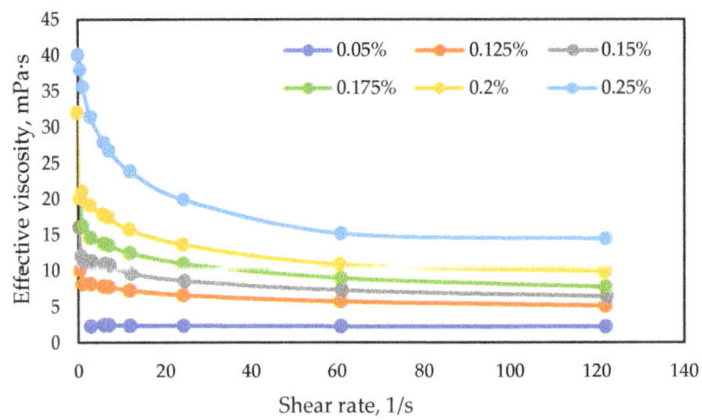

Figure 5. Dependence of effective viscosity on shear rate at various mass concentrations of polymer FP 3630S.

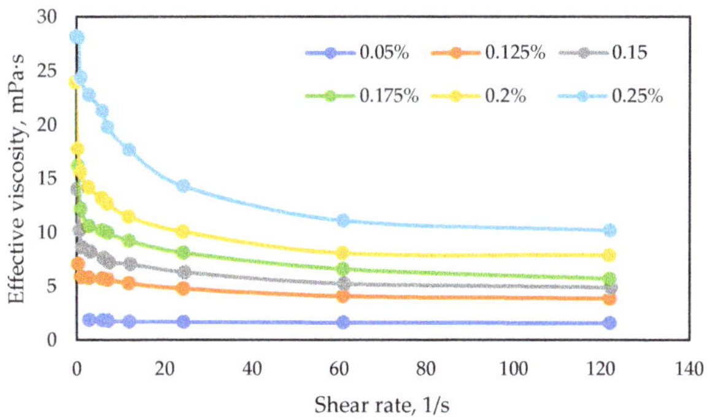

Figure 6. Dependence of effective viscosity on shear rate at various mass concentrations of polymer FP 5205 SH.

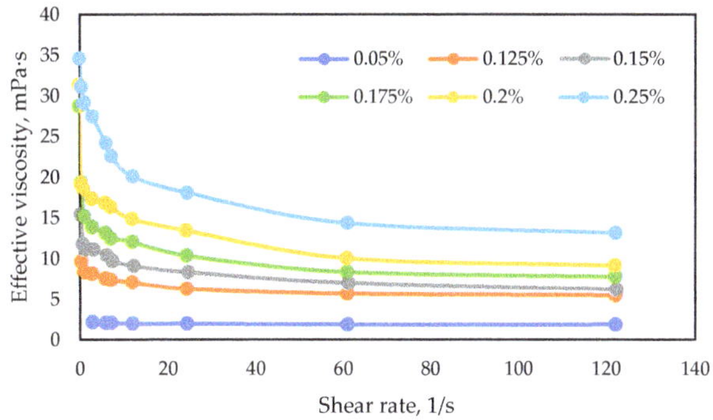

Figure 7. Dependence of effective viscosity on shear rate at various mass concentrations of polymer FP 5205 VHM.

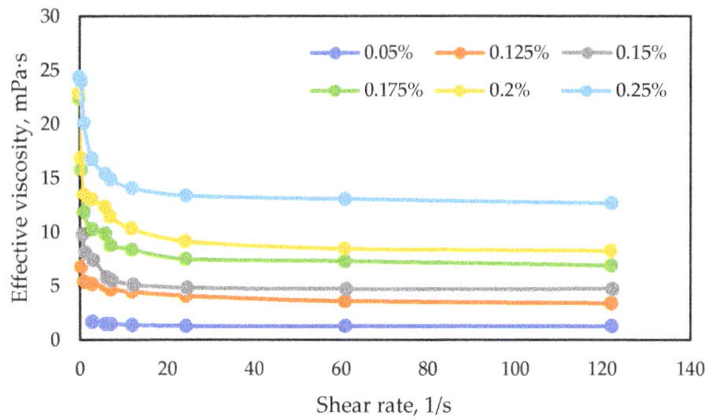

Figure 8. Dependence of effective viscosity on shear rate at various mass concentrations of polymer FP 5115 SH.

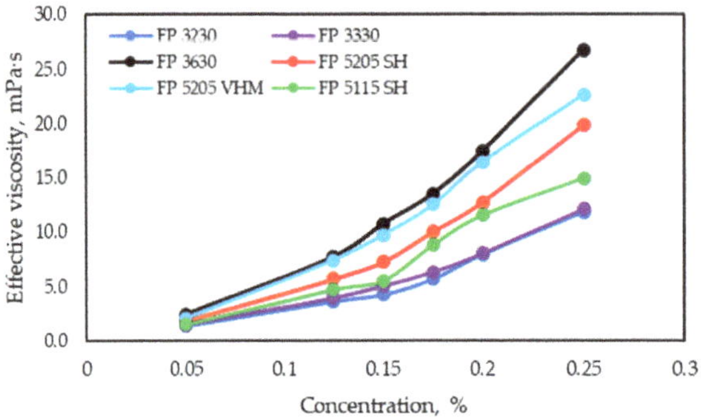

Figure 9. Dependence of effective viscosity of polymer solution on polymer concentration.

According to the obtained rheological curves, all prepared polymer solutions exhibited mainly pseudoplastic behavior in the shear rate range from 0.6 to 122.3 s^{-1}. It is important to note that studies of polymer solutions using highly sensitive viscometers have shown that at low shear rates, Newtonian flow predominates. After the first critical value, the apparent viscosity decreases (pseudoplastic behavior), and upon reaching the second critical shear rate value, the flow becomes dilatant (increase in apparent viscosity) [25,62]. Once the maximum possible stretching of polymer chains is achieved at high shear rates, molecular degradation occurs, leading to a reduction in the apparent viscosity of the solution [63]. Determining the dependence of polymer solution viscosity on shear rate allows for the further modeling of its movement in the injection system and reservoir, enabling the selection of the most suitable composition for specific conditions [64].

Given the range of oil viscosity for the target reservoir (1 to 4.4 mPa·s), the viscosity of the polymer solution should exceed this value. When selecting the concentration, the viscosity corresponding to the second Newtonian plateau (viscosity at high shear rates) must be considered. Thus, the target concentrations for the polymers are as follows: FP 3230 S—0.23 wt.%; FP 3330 S—0.23 wt.%; FP 3630 S—0.15 wt.%; FP 5205 SH—0.175 wt.%; FP 5205 VHM—0.15 wt.%; FP 5115 SH—0.18 wt.%.

3.1.3. Studies on the Mechanical and Thermal Stability of Polymers

During the implementation of polymer flooding technology, mechanical degradation of the polymer occurs during the injection process (in pump units, perforation intervals, and near-wellbore zones) due to shear stresses [32,33]. The degree of mechanical degradation depends on the intensity and duration of shear exposure and must be considered when designing the technology. Polymer solutions with working concentrations were prepared using model water. The viscosity of the initial solutions was measured at a temperature of 57 °C and a shear rate of 7.34 s^{-1}. Then, the solutions (500 cm^3) were stirred at a mixer speed of 2000 rpm for 10 min, allowed to settle until air bubbles dissipated, and their viscosity was measured after mechanical degradation. The results of the studies are presented in Table 8.

Table 8. Results of the study of polymer solution stability/resistance to mechanical degradation.

Parameter	FP 3230	FP 3330	FP 3630S	FP 5205 SH	FP 5205 VHM	FP 5115 SH
Initial Concentration, wt.%	0.23	0.23	0.15	0.175	0.15	0.18
Viscosity of Initial Solution, mPa·s	9.9	9.8	10.3	10.2	10.1	9.7
Viscosity of Degraded Solution, mPa·s	8.9	8.7	8.8	9.3	9.1	9.1
Mechanical Stability Coefficient	0.90	0.89	0.85	0.91	0.90	0.94

Polyacrylamide polymers under reservoir conditions can undergo thermo-oxidative degradation due to high temperatures and reactive components present in the water, rock, and the polymer reagent itself. This leads to a reduction in solution viscosity and a deterioration of their technological properties [31]. The thermal stability of the polymers was evaluated based on the dynamics of viscosity changes under the influence of temperature. The resistance coefficient to thermo-oxidative degradation was calculated as the ratio of the viscosity after a specified exposure time to the viscosity of the initial solutions at a fixed shear rate (7.34 s^{-1}).

The results of the studies are presented in Table 9 and Figure 10.

Table 9. Results of the study of polymer solution stability/resistance to thermo-oxidative degradation.

Parameter	FP 3230	FP 3330	FP 3630S	FP 5205 SH	FP 5205 VHM	FP 5115 SH
Viscosity of Initial Solution, mPa·s	9.80	9.70	10.30	10.20	10.10	9.70
Viscosity of Degraded Solution, mPa·s	8.70	8.20	8.00	9.60	9.40	9.40
Thermal Stability Coefficient	0.89	0.84	0.78	0.94	0.93	0.97

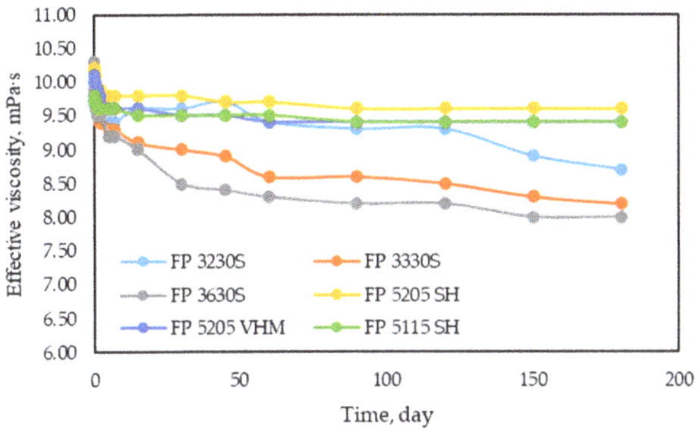

Figure 10. Dependence of polymer solution viscosity on exposure time at 57 °C (shear rate 7.34 s^{-1}).

As can be seen from the obtained results, polymer solutions containing ATBS (acrylamide-tert-butyl sulfonic acid) demonstrate better resistance to thermo-oxidative degradation. The viscosity loss coefficient for the aforementioned polymer solutions does not exceed 10%.

3.1.4. Evaluation of Polymer Solutions Flowability in Porous Media

Table 10 provides the main information about the polymer solution samples used in the experiments.

Table 10. Main information on polymer solution samples for core flooding experiments.

№	Polymer Sample	Concentration, wt.%	Effective Viscosity (7.34 s-1.57 °C), mPa·s
1	FP 3630 S	0.15	9.8
2	FP 5205 VHM	0.15	9.9
3	FP 5205 SH	0.175	10.1
4	FP 3230 S	0.23	9.9

Figures 11–14 illustrate the dynamics of pressure drop changes depending on the pore volume injected for the following cases.

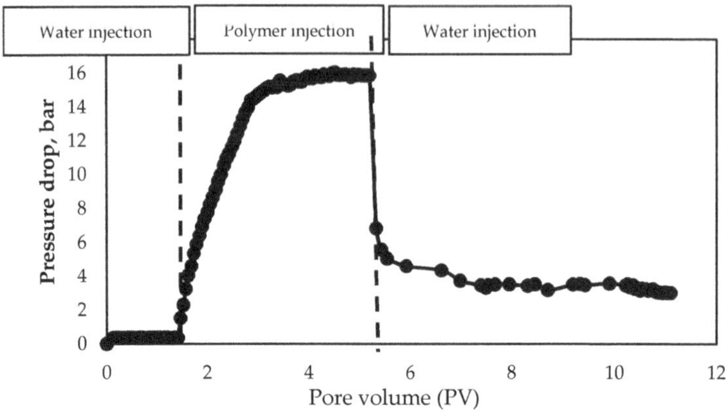

Figure 11. Dynamics of pressure drop change vs. pore volume injected (Experiment No. 1).

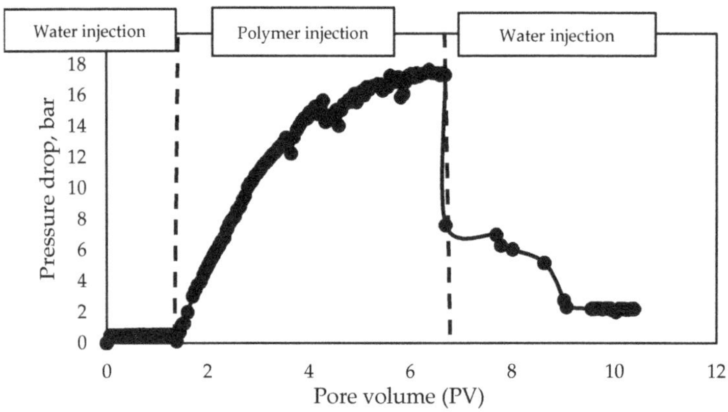

Figure 12. Dynamics of pressure drop change vs. pore volume injected (Experiment No. 2).

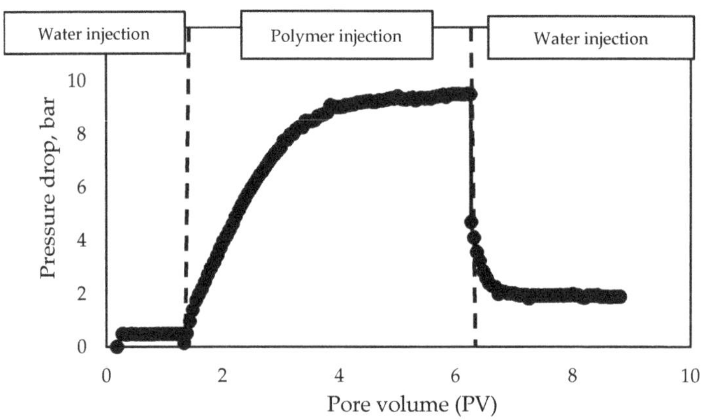

Figure 13. Dynamics of pressure drop change vs. pore volume injected (Experiment No. 3).

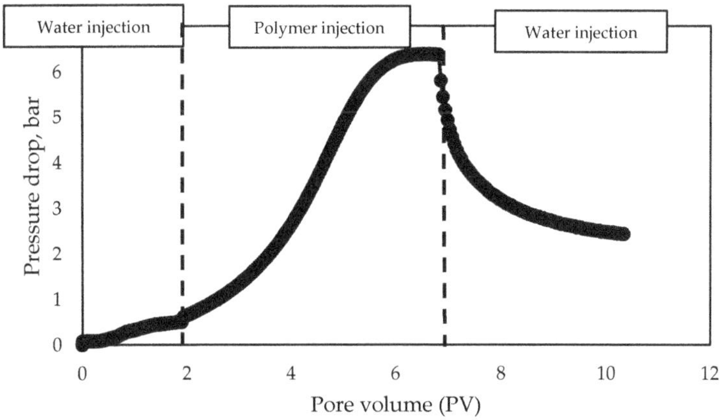

Figure 14. Dynamics of pressure drop change vs. pore volume injected (Experiment No. 4).

After all preparatory procedures and placing the core in the core holder, the first stage of the core flooding experiment involved simulating water flooding of the core by injecting water at a specified flow rate until oil displacement from the core holder ceased.

In the next stage of the core flooding study, a polyacrylamide solution prepared using the model injection water was injected, with simultaneous sampling at the outlet of the model until pressure stabilization. After sampling, the polymer solution was continuously injected until pressure stabilization was achieved at each flow rate. To assess mechanical degradation, polymer samples were collected after pressure stabilization at each flow rate.

In the final stage of the experiment, at the minimum rate, the polymer solution was displaced by injecting water while collecting samples. Thus, the core flooding experiment can be divided into three main stages:

1. Water injection before polymer solution injection;
2. Polymer solution injection;
3. Water injection after polymer solution injection.

Water, polymer, and surfactant–polymer solutions were injected at a constant rate of 1 mL/min. The injection rate selection was primarily guided by capillary number. While the hydrodynamic modelling of reservoir conditions suggested an optimal rate of 0.1 mL/min, laboratory implementation at this rate would require prohibitively long experimental durations.

Employing this parameter for capillary number calculation yields a dimensionless value of 10^{-7}. When recalculated at the experimental injection rate of 1 mL/min, the resulting capillary number magnitude decreases to 10^{-8}. This order-of-magnitude variation remains methodologically acceptable as it conserves the capillary-dominated displacement regime ($Nc \leq 10^{-6}$), thereby preserving the fundamental fluid dynamics governing the process.

Table 11 presents the main results of the core flooding experiments.

Figure 15 shows the dependence of the RF resistance factor on the value of the molecular weight of the polymer for the same values of the effective viscosity of 10 mPa s (shear rate 7.34 s^{-1}, temperature 57 °C).

The resistance factor is an indicator that reflects how many times the mobility of the displacing agent has changed when switching from a water injection to a polymer solution injection. Thus, the resistance factor is determined by two main mechanisms: an increase in the viscosity of the displacing agent and a decrease in relative permeability.

Table 11. Results of the core flooding tests with polymer solutions.

Parameter	Exp. 1	Exp. 2	Exp. 3	Exp. 4
Core Sample Length and Diameter, cm	3.2/2.98	3.37/3.01	3.18/2.99	3.24/2.97
Core Sample Porosity, %	22.43	21.48	23.23	23.43
Core Sample Permeability, μm^2	0.193	0.172	0.191	0.190
Polymer Sample	FP 3630S	FP 5205 VHM	FP 5205 SH	FP 3230 S
Concentration, wt.%	0.15	0.15	0.175	0.23
Effective Viscosity, mPa·s	9.8	9.9	10.1	9.9
Displaced Oil Volume (by Water), mL	2.6	2.65	2.70	2.9
Displacement Coefficient (by Water)	0.49	0.51	0.52	0.55
Displaced Oil Volume (with Polymer), mL	2.9	3.00	3.25	3.35
Displacement Coefficient (by Polymer)	0.54	0.58	0.63	0.64
Increase in Displacement Coefficient	0.06	0.07	0.11	0.09
Resistance Factor (RF)	43.91	33.02	19.51	12.73
Residual Resistance Factor (RRF)	8.35	4.17	3.88	4.86

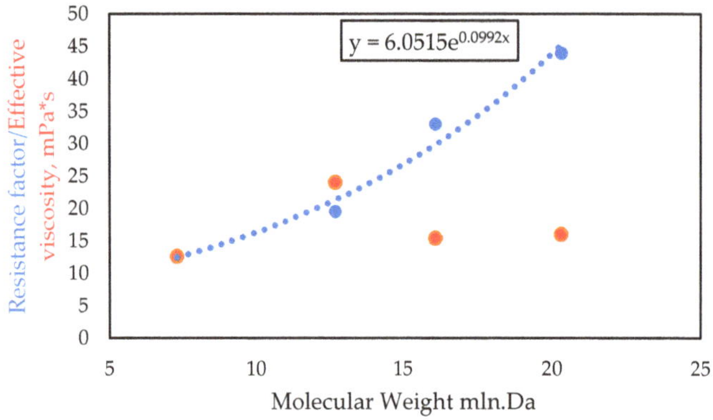

Figure 15. Dependence of RF on polymer molecular weight for an effective viscosity of 10 mPa·s (shear rate: 7.34 s^{-1}, temperature: 57 °C).

Additionally, in Figure 15, the red dots indicate the values of the maximum effective viscosity measured during the studies using a rotational viscometer for each of the considered polymers. A significant excess of the resistance factor over this value may indicate excessive retention of the polymer on the rock.

Polymer retention in the porous medium is caused by the following three mechanisms: mechanical entrapment, hydrodynamic retention, and adsorption, as illustrated in Figure 16.

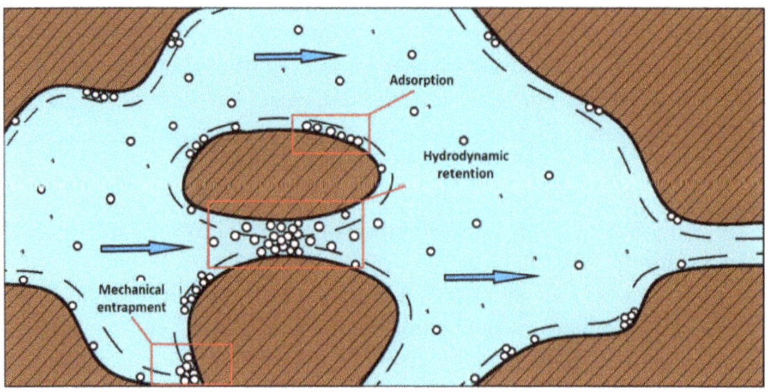

Figure 16. Polymer retention in pore volume [65].

The mechanical entrapment of polymers is caused by the large size of their molecules relative to the pore size. Therefore, when selecting a polymer, attention must be paid to the molecular size and the flow properties of the reservoir to avoid channel plugging. Typically, the influence of this mechanism in rocks (except for low-permeability ones) is small, and in most practical cases, it can be neglected by properly selecting the polymer composition [33,66].

Hydrodynamic retention is the least studied mechanism of polymer retention and depends on the injection rate. As the injection rate increases, hydrodynamic retention also increases. However, it is worth noting that this component does not significantly affect the overall level of polymer retention in the pore volume and is not a critical factor for polymer flooding in field conditions.

Adsorption refers to the interaction between polymer molecules and the solid surface. This interaction involves the bonding of polymer molecules with elements of the rock, typically those with low free energy [67].

Thus, polymers with a resistance factor significantly exceeding the maximum effective viscosity value are not suitable candidates for use in the target reservoir from the perspective of flowability in porous media.

On the other hand, the increase in the displacement factor for various polymers ranges from 0.06 to 0.11. Moreover, there is a slight increase in the oil displacement coefficient with a decrease in molecular weight. The highest increase in the additional displacement factor is observed for the FP 5205 SH sample.

Based on the obtained results from the polymer studies, the FP 5205 SH sample will be considered for further testing.

3.2. The Study of Surfactant Properties

3.2.1. Evaluation of Surfactant Solubility

The first stage of laboratory studies involved assessing solubility in model water, which simulates the composition of the injected water in the target reservoir.

All solutions were prepared with a concentration of 1 wt.%. Solubility was evaluated at a standard temperature visually over a period of 3 days. The results of the study are presented in Table 12.

Table 12. Results of surfactant solubility.

Surfactant	Solubility			
	1 h	1 Day	2 Day	3 Day
Sodium Alkylbenzene Sulfonate	Soluble	Sediment	Sediment	Sediment
Sodium Alkylbenzene Sulfonate	Soluble	Sediment	Sediment	Sediment
Diethanolamide of Lauric Acid	Soluble	Soluble	Opalescence	Opalescence
Sodium Laureth Sulphate	Soluble	Soluble	Soluble	Soluble
Dodecylbenzene Sulfonate	Opalescence	Opalescence	Opalescence	Opalescence
Alpha-Olefin Sulfonate	Soluble	Soluble	Soluble	Soluble
Betaine	Soluble	Soluble	Soluble	Soluble
Oleyl Betaine	Soluble	Soluble	Soluble	Soluble

3.2.2. Evaluation of Phase Behavior

As noted earlier, studies on surfactant phase behaviour at the oil–water interface allow for the assessment of their ability to form microemulsions, which indirectly indicates improved oil displacement efficiency.

It is believed that in optimal surfactant systems, the IFT at the interface approaches 0.001 mN/m. According to the classification presented in the works of M.L. Surguchev, when the interfacial tension at the "oil-surfactant solution" interface decreases, four types of emulsions can form (Figure 17) [14]:

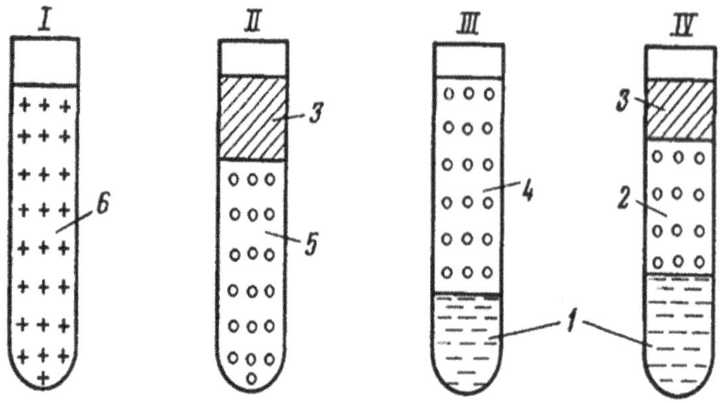

Figure 17. Types of emulsions [14]: 1—water; 2—microemulsion balanced with oil and water; 3—oil; 4—microemulsion balanced with water; 5—microemulsion balanced with oil; 6—unbalanced microemulsion.

Type I: A non-equilibrium solution with a high surfactant concentration, soluble in both water and oil.

Type II: A solution balanced with oil and soluble only in water. Over time, excess oil separates from the solution, forming a stable phase boundary. The interfacial tension at the oil interface is low (0.1–0.001 mN/m), while at the water interface, it is zero. This type is called a solution with an external aqueous phase, sometimes referred to as the "lower phase" or oil-in-water microemulsion.

Type III: A solution balanced with water and soluble only in oil, or a solution with an external hydrocarbon phase, sometimes called the "upper phase" or water-in-oil microemulsion.

Type IV: A solution insoluble in both water and oil, i.e., balanced with oil and water, sometimes referred to as the "middle phase". This solution has very low interfacial tension at both the oil and water interfaces (0.001–0.0001 mN/m), enabling miscible displacement.

To study the phase behaviour of surfactant solutions at the oil–water interface, equal volumes of crude oil from the target reservoir and surfactant solutions at a concentration of 1 wt.% were added to measuring cylinders. The studies were conducted at a temperature of 57 °C and a water-to-oil phase ratio of 1:1.

Figure 18 presents the results of the phase behaviour of surfactants based on sodium laureth sulphate, oleyl betaine, diethanolamide of lauric acid, and betaine.

As can be seen from Figure 18, surfactant samples No. 1 and No. 4 (sodium laureth sulphate and betaine) have the ability to form emulsions. Moreover, sample No. 1 forms a Type IV emulsion, i.e., a solution insoluble in both water and oil. This solution has an IFT of 0.001–0.0001 mN/m. Surfactant sample No. 4 forms a Type II solution, characterised by interfacial tension in the range of 0.1 to 0.001 mN/m. On the other hand, no ability to form an emulsion with oil was observed for samples No. 2 and No. 3 (oleyl betaine and diethanolamide of lauric acid).

Figure 19 presents the results of the phase behaviour of surfactants based on dodecylbenzene sulfonate and alpha-olefin sulfonate.

As can be seen from Figure 18, the surfactant sample based on dodecylbenzene sulfonate (left) forms a Type IV emulsion, i.e., a solution insoluble in both water and oil. This solution has an interfacial tension of 0.001–0.0001 mN/m. The surfactant sample based on alpha-olefin sulfonate (right) did not demonstrate the ability to form an emulsion with the studied oil.

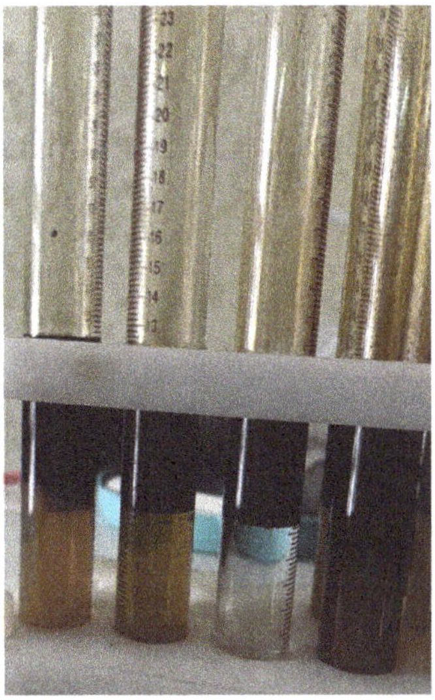

Figure 18. Phase behavior experiment of surfactant samples (based on sodium laureth sulphate, oleyl betaine, diethanolamide of lauric acid, and betaine) at the oil–water interface.

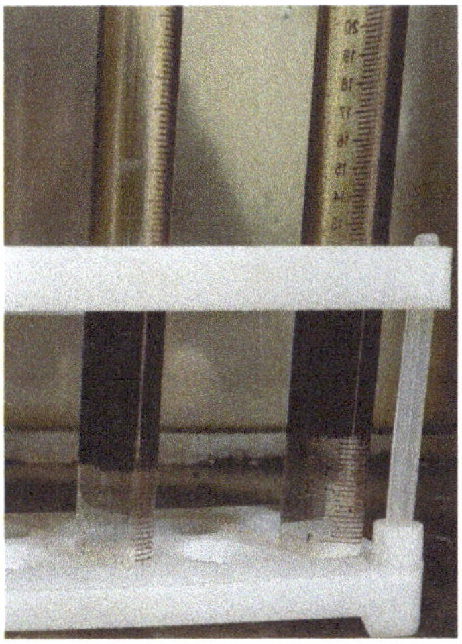

Figure 19. Phase behavior experiment of surfactant samples (based on dodecylbenzene sulfonate and alpha-olefin sulfonate) at the oil–water interface.

Table 13 presents the results of the phase behaviour studies of surfactant solutions and oil.

Table 13. Results of phase behavior studies of surfactant solutions and oil.

Surfactant Type	Emulsion Type	Interfacial Tension, mN/m	Observation
Sodium Laureth Sulphate	Type IV	0.001–0.0001	Forms emulsion
Oleyl Betaine	None	-	No emulsion
Diethanolamide of Lauric Acid	None	-	No emulsion
Betaine	Type II	0.1–0.001	Forms emulsion
Dodecylbenzene Sulfonate	Type IV	0.001–0.0001	Forms emulsion
Alpha-Olefin Sulfonate	None	-	No emulsion

Solutions that demonstrated the ability to form emulsions will be further investigated in subsequent tests.

3.2.3. Study of IFT Dependence on Surfactant Concentration

IFT studies at the "surfactant solution-oil" interface were conducted in the concentration range of 0.1 wt.% to 2 wt.% at the target reservoir temperature. Upon reaching the target temperature, the system automatically records 10 measurements at 10 s intervals. The resulting dataset is subsequently processed to generate a report containing averaged temperature and interfacial tension (IFT) values.

Figure 20 and Table 14 present the results of determining the interfacial tension at the "surfactant solution-oil" interface for various surfactant samples.

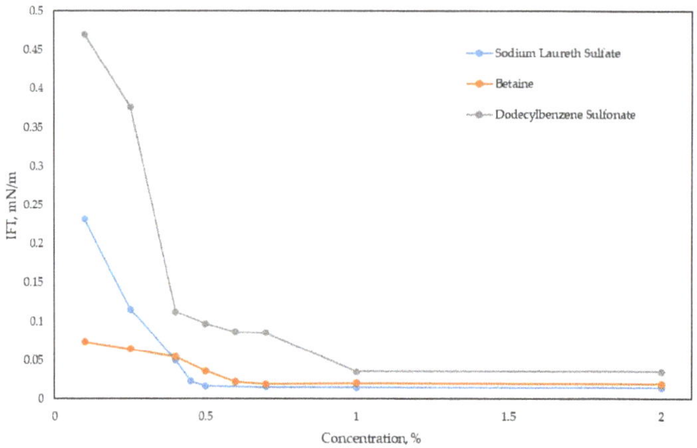

Figure 20. Dependence of interfacial tension at the "surfactant solution-oil" interface on the logarithm of surfactant concentration.

Table 14. Results of interfacial tension determination at the "surfactant solution-oil" interface.

Sodium Laureth Sulfate			Betaine			Dodecylbenzene Sulfonate		
Concentration, wt.%	IFT, mN/m	Temperature, °C	Concentration, wt.%	IFT, mN/m	Temperature, °C	Concentration, wt.%	IFT, mN/m	Temperature, °C
0.1	0.2312	56.9	0.1	0.07361	56.8	0.1	0.47011	56.9
0.25	0.1154	56.8	0.25	0.06422	56.9	0.25	0.37612	56.9
0.4	0.04988	56.9	0.4	0.05423	57.0	0.4	0.11211	56.8
0.45	0.02298	56.9	0.5	0.03599	57.0	0.5	0.09691	56.9
0.5	0.01649	57.0	0.6	0.02211	56.9	0.6	0.08612	56.9
0.7	0.01512	56.9	0.7	0.01923	56.8	0.7	0.08514	56.9
1	0.01462	56.9	1	0.02017	56.9	1	0.03499	57.0
2	0.01407	56.9	2	0.01892	56.9	2	0.03511	57.0

The experimental data demonstrate that all investigated surfactants exhibit significant potential for interfacial tension reduction. A systematic IFT decrease was observed with an increasing surfactant concentration, reaching minimal values within the 0.5–1% concentration range. An optimal IFT reduction to 0.01–0.1 mN/m substantially enhances oil displacement efficiency through two synergistic mechanisms: (1) increased capillary number (Nc) and (2) reduced residual oil saturation in pore spaces. Maximum performance is achieved at concentrations exceeding the CMC, where the combined effect of minimized IFT and enhanced polymer solution viscosity improves both sweep efficiency and ultimate recovery.

Based on the obtained results, the critical micelle concentration was determined for each of the studied surfactants. The CMC for the surfactant based on sodium laureth sulphate was 0.5 wt.%, for the surfactant based on betaine it was 0.6 wt.%, and for the surfactant based on dodecylbenzene sulfonate, the CMC was achieved at a concentration of 1 wt.%. Furthermore, for the surfactants based on sodium laureth sulphate and betaine, the interfacial tension at the "surfactant solution-oil" interface was 2 times and 1.5 times lower, respectively, compared to the surfactant based on dodecylbenzene sulfonate. Therefore, only two surfactant samples—based on sodium laureth sulphate and betaine—will be used for further studies.

3.2.4. Studies of Static Surfactant Adsorption

Table 15 presents the results of static surfactant adsorption studies.

Table 15. Results of static surfactant adsorption studies.

Parameter	Sodium Laureth Sulphate	Betaine
Initial Mass of Solution, g	51.25	48.11
Concentration, wt.%	0.50	0.60
IFT of Initial Solution, mN/m	0.01649	0.02211
IFT of Solution After Contact with Rock, mN/m	0.10912	0.05012
Final Surfactant Solution Concentration, wt.%	0.28	0.41
Static Surfactant Adsorption, µg/g	1127.5	914.1

The obtained values of static adsorption are 1127.5 µg/g for the surfactant based on sodium laureth sulphate and 914.1 µg/g for the surfactant based on betaine. These values can be considered acceptable. Taking into account the obtained values for the CMC and static surfactant adsorption, it is proposed to increase the working and tested surfactant concentration in the solution to 0.7 wt.% for further tests.

Several sources note the influence of temperature and salinity on the behaviour of the "surfactant solution-oil" system. Even a slight shift in temperature within the reservoir can significantly affect the interfacial tension at the "surfactant solution-oil" interface. Figures 21 and 22 show the dependence of interfacial tension at the "surfactant solution-oil" interface on temperature.

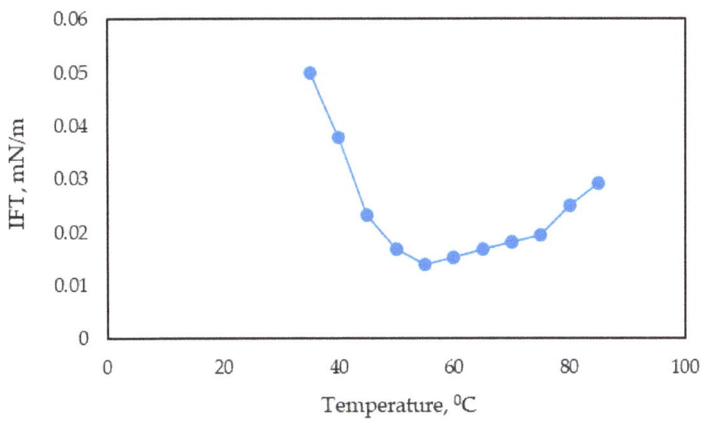

Figure 21. Dependence of interfacial tension at the "surfactant solution-oil" interface on temperature for sodium laureth sulfate.

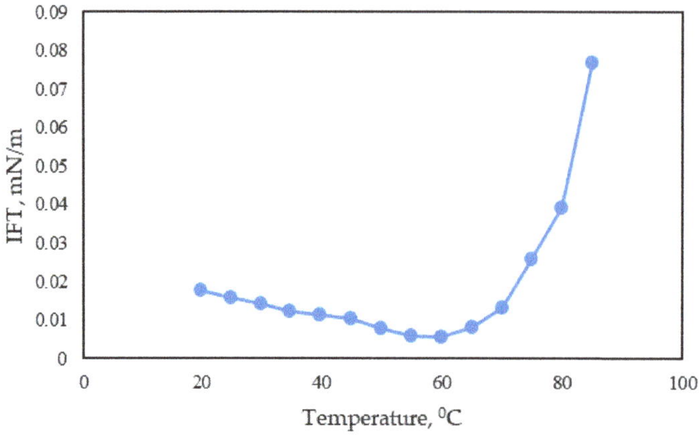

Figure 22. Dependence of interfacial tension at the "surfactant solution-oil" interface on temperature for betaine.

As can be seen from the graphs, for the surfactant based on sodium laureth sulphate, the zone of low interfacial tension is in the temperature range of 50 to 75 °C. On the other hand, for the surfactant based on betaine, an increase in interfacial tension is observed when the temperature reaches 65 °C.

During displacement under reservoir conditions, the salinity of the injected agent will change as a result of contact with the rock and reservoir fluids. To evaluate the influence of salinity on the system, Figures 23 and 24 show the dependence of interfacial tension on the salinity of the water used to prepare the solution.

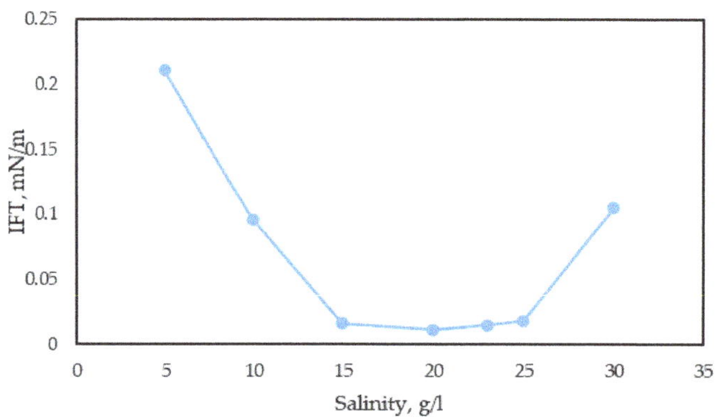

Figure 23. Dependence of interfacial tension at the "surfactant solution-oil" interface on water salinity for sodium laureth sulfate-based surfactant at 57 °C.

As can be seen from the presented graphs, the surfactant based on sodium laureth sulphate maintained low interfacial tension in the salinity range of 15 to 25 g/L for the preparation water, reaching its optimum (lowest value) at 20 g/L. On the other hand, the surfactant based on betaine maintained low interfacial tension across the entire studied range of water salinity, reaching its optimum (lowest value) at 25 g/L.

In practice, this result can be used to modernise the water injection system by utilising alternative water sources or by further purifying the currently used water. Such modernisation would maximise the efficiency of implementing the surfactant-polymer flooding technology.

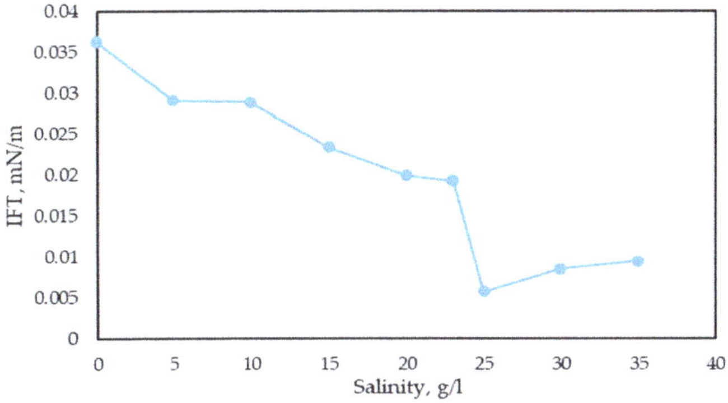

Figure 24. Dependence of interfacial tension at the "surfactant solution-oil" interface on water salinity for betaine-based surfactant at 57 °C.

Considering the geological and physical characteristics of the target reservoir, particularly the reservoir temperature, which varies between 50 and 75 °C in different parts of the formation, only the surfactant based on sodium laureth sulfate will be considered for further studies. This surfactant demonstrated low interfacial tension across the entire temperature range under consideration.

3.3. Studies on the Surfactant–Polymer Formulation

3.3.1. Influence of Surfactant and Polymer Concentration on the Key Properties of the Formulation

When mixing surfactants and polymers, reactions may occur that alter the properties of the injected chemical agent. The key properties affecting the efficiency of flooding are the viscosity of the injected agent and the interfacial tension at the "surfactant solution-oil" interface.

Figure 25 shows the dependence of the effective viscosity of the chemical formulation, including the polymer FP 5205 SH (concentration 0.15 wt.%) and the surfactant based on sodium laureth sulphate, at various surfactant concentrations.

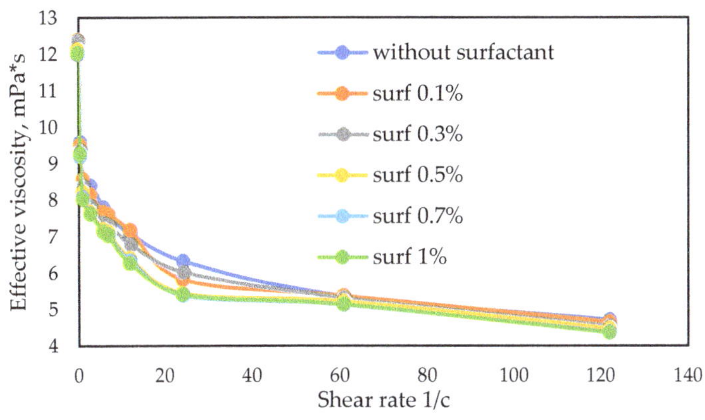

Figure 25. Dependence of effective viscosity of surfactant–polymer solution on shear rate at various surfactant concentrations.

Next, Figure 26 presents the dependence between the effective viscosity of the solution and surfactant concentration at a shear rate of 122.3 s^{-1} and a polymer concentration of 0.15 wt.%. The selected shear rate corresponds to the second Newtonian plateau on the viscosity vs. shear rate graph (under constant conditions, the effective viscosity should remain constant with further increases in shear rate).

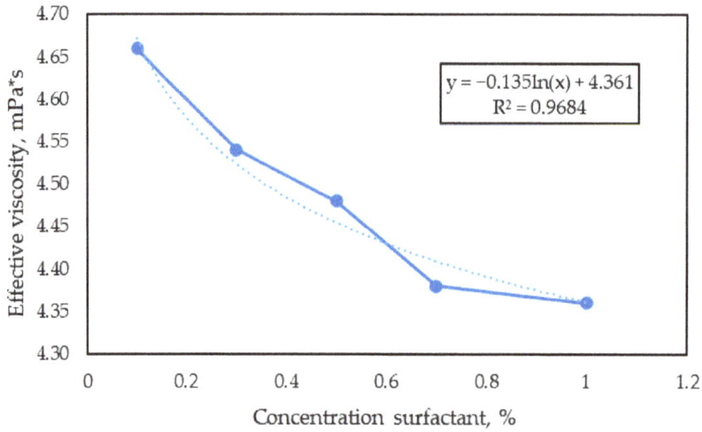

Figure 26. Dependence of effective viscosity of surfactant–polymer solution on surfactant concentration (A logarithmic trend has been added to the dotted line).

As can be seen from the graphs, the influence of the surfactant on the viscosity of the chemical formulation is minimal. Moreover, as the surfactant concentration increases, the effective viscosity stabilises.

Figure 27 shows the dependence of interfacial tension at the "surfactant-polymer solution-oil" interface on the polymer concentration in the formulation. The studies were conducted using the polymer FP 5205 SH and the surfactant based on sodium laureth sulphate (concentration 0.7 wt.%) at the reservoir temperature.

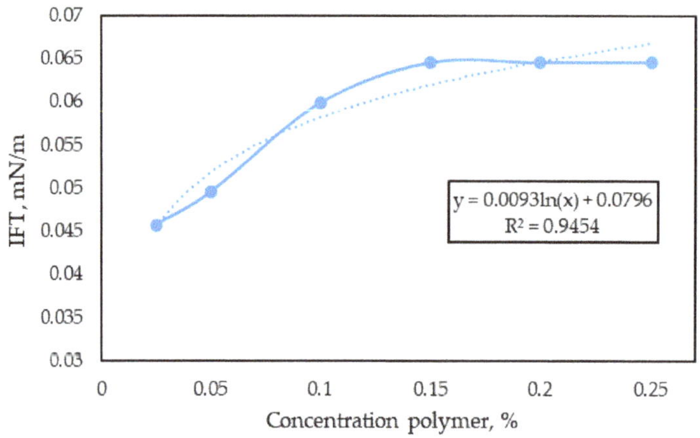

Figure 27. Dependence of interfacial tension at the "surfactant–polymer solution–oil" interface on polymer FP 5205 SH concentration (A logarithmic trend has been added to the dotted line).

As can be seen from the presented graph, increasing the polymer concentration in the chemical composition leads to an increase in interfacial tension. The IFT value changed by more than 35% with the addition of the polymer. However, the order of magnitude of the IFT remained the same, meaning the surfactant's potential did not change significantly. When injecting the surfactant–polymer formulation, the capillary number will increase by 4500 times (reduction in IFT from 32.77 mN/m to 0.065 mN/m; increase in the viscosity of the injected agent from 0.5 mPa·s to 4.36 mPa·s).

3.3.2. Core Flooding Studies of the Surfactant–Polymer Formulation

Figures 28–30 present information on the dynamics of pressure drop changes depending on the pore volume injected for the following cases:

- Water injection before polymer solution injection;
- Polymer solution injection;
- Surfactant–polymer solution injection.

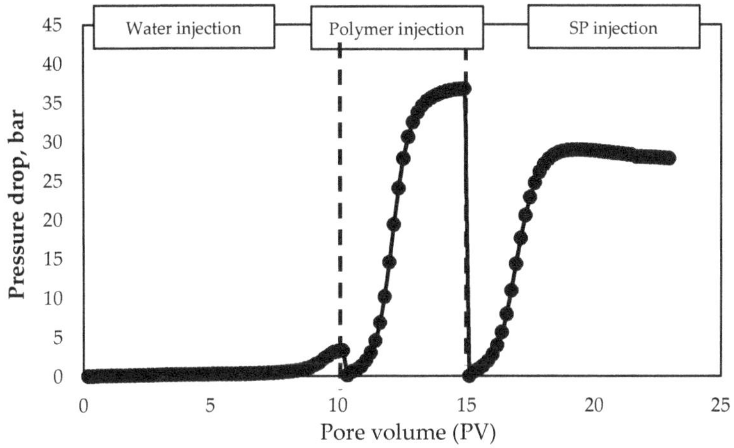

Figure 28. Dynamics of pressure drop change vs. pore volume injected (Experiment No. 1).

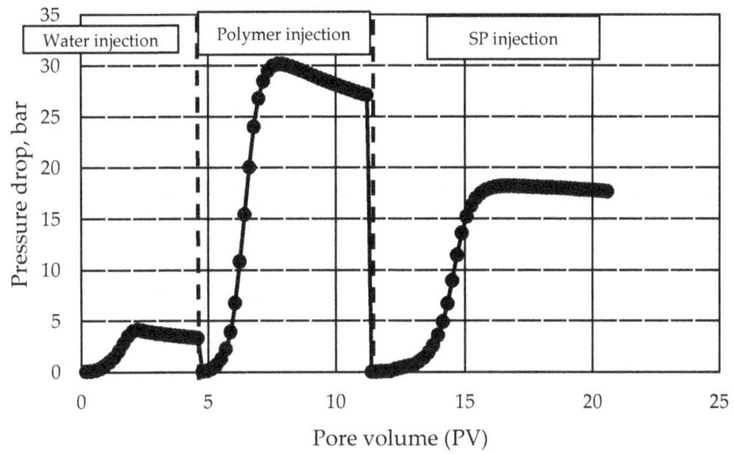

Figure 29. Dynamics of pressure drop change vs. pore volume injected (Experiment No. 2).

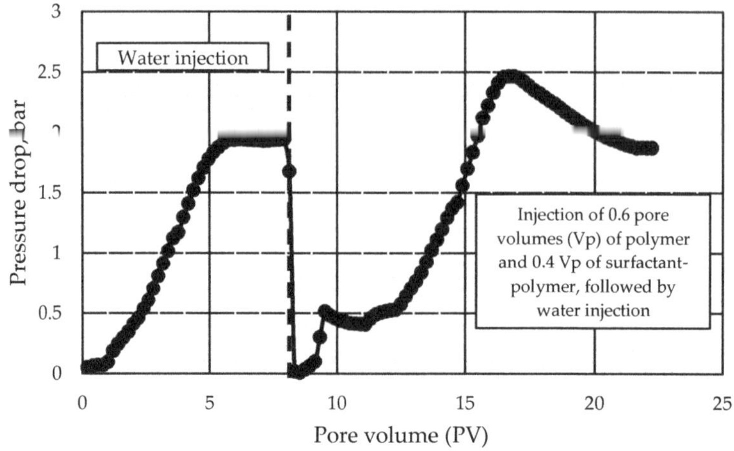

Figure 30. Dynamics of pressure drop change vs. pore volume injected (Experiment No. 3).

Table 16 presents the results of the core flooding studies of the surfactant–polymer formulation

Table 16. Results of the core flooding tests with surfactant-polymer solutions.

Parameter	Exp. 1	Exp. 2	Exp. 3
Core Sample Length and Diameter, cm	3.37/3.02	3.17/3.00	3.07/3.01
Core Sample Porosity, %	22.12	24.5	25.63
Core Sample Permeability, μm^2	0.075	0.116	0.069
Concentration of Polymer, wt.%	0.15	0.15	0.15
Concentration of Surfactant, wt.%	0.7	0.7	0.7
Displaced Oil Volume (by Water), mL	2.9	3.1	2.1
Displacement Coefficient (by Water)	0.54	0.56	0.37
Displaced Oil Volume (with Polymer), mL	0.5	0.6	not evaluated
Displacement Coefficient (by Polymer)	0.63	0.67	not evaluated
Increase in Displacement Coefficient (Polymer)	0.09	0.11	not evaluated
RF (polymer)	11.39	8,05	not evaluated
Displaced Oil Volume (SP), mL	1.2	1.3	0.85
Displacement Coefficient (SP)	0.86	0.91	0.53
Increase in Displacement Coefficient (SP)	0.22	0.24	0.15
RF (SP)	8.62	5.24	1.27
RRF (SP)	not evaluated		0.96

In experiments No. 1 and No. 2, the injection of chemical slugs was carried out continuously until pressure stabilisation and the absence of oil at the outlet of the core holder. As can be seen from the obtained results, the displacement factor after the injection of the polymer and surfactant–polymer slug reached values of 0.86–0.91. At the same time, the increase in the displacement factor after the injection of the polymer slug was 0.09–0.11, while after the injection of the surfactant–polymer slug, the increase was 0.22–0.24.

It is important to note that during the injection of the surfactant–polymer composition, the fluid at the outlet of the core holder is presented as an emulsion (Figure 31). After settling at a temperature of 90 °C, the emulsion separates, and the oil and aqueous phases are separated from each other (Figure 32).

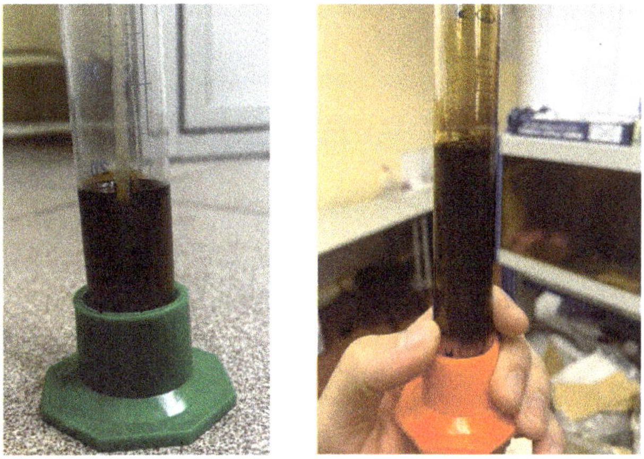

Figure 31. Appearance of the emulsion collected at the outlet of the core holder.

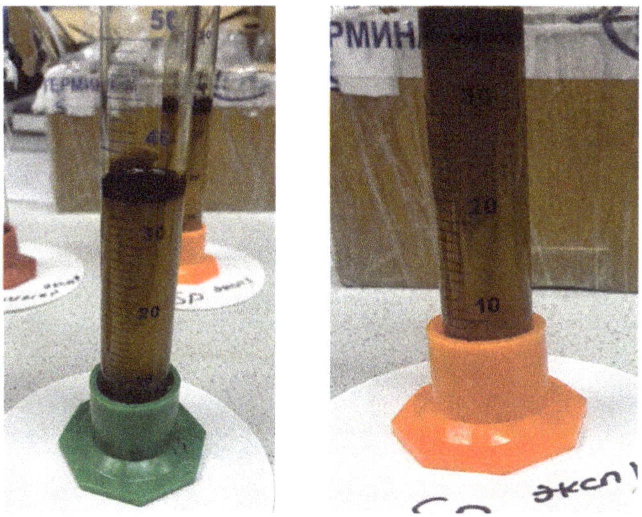

Figure 32. Appearance of the measuring cylinders after emulsion separation at 90 °C.

It is important to note that when switching from the polymer solution injection to surfactant–polymer solution injection, the resistance factor (pressure gradient) decreased by 2.77–2.82. This is explained by both additional oil displacement and the improved characteristics of the formulation.

In experiment No. 3, studies were conducted with a limited injection volume of polymer and surfactant–polymer slugs, equal to 0.5 pore volumes of the core sample.

As can be seen from the results, with a smaller injection volume, the additional displacement coefficient decreases. It is also important to note that no significant increase in the resistance factor was observed at the considered injection volumes.

Thus, summarising the results of the core flooding studies, it can be concluded that the considered surfactant–polymer composition allows for an increase in the oil displacement factor from the core sample by 0.15–0.24, depending on the injection volume. The selected formulation can be recommended for application in water flooding at the target reservoir.

4. Conclusions

1. In the West Siberian Oil and Gas Province, there are currently a lot of mature development oil fields where the implementation of chemical EOR methods, particularly surfactant–polymer flooding, becomes relevant.
2. To develop an effective surfactant–polymer composition that retains its properties under the given geological and physical conditions, and ensures an increase in the oil displacement coefficient, criteria were established, and a laboratory research program was proposed.
3. To improve oil recovery from heterogeneous terrigenous reservoirs during water flooding, a composition was developed consisting of an aqueous solution of an anionic surfactant (sodium laureth sulphate, with ethyl alcohol as a solvent) at a concentration of 0.7 wt.% and partially hydrolysed polyacrylamide (of medium molecular weight, including 5 wt.% sulfonated monomeric additives) at a concentration of 0.15 wt.%.
4. The addition of an anionic surfactant (sodium laureth sulphate) at a concentration of 0.7 wt.% and partially hydrolysed polyacrylamide (of medium molecular weight, including 5% sulfonated monomeric additives) at a concentration of 0.15 wt.% to the injected water reduces the interfacial tension at the "surfactant-polymer solution-oil" interface to 0.065 mN/m and increases the effective viscosity of the injected agent to 4.36 mPa·s (at a shear rate of 122.3 s^{-1} and a temperature of 57 °C).
5. When injecting the surfactant–polymer composition, the capillary number increases by 4500 times (with a reduction in interfacial tension from 32.77 mN/m to 0.065 mN/m and an increase in the viscosity of the injected agent from 0.5 mPa·s to 4.36 mPa·s).
6. The developed surfactant–polymer composition increases the oil displacement coefficient from the core sample by 0.15–0.24. Thus, the proposed composition can be recommended for application in water flooding at the target reservoir.
7. Current research involves hydrodynamic modelling incorporating laboratory-derived parameters. The simulation results will inform the technology implementation design (including well patterns and pore volume injection requirements), with subsequent field deployment planned by the operator at a pilot reservoir.

Author Contributions: Conceptualization, D.P., M.R. and R.B.; Data curation, R.B.; Methodology, D.P. and R.B.; Project administration, D.P.; Supervision, M.R.; Visualization, R.B.; Writing—original draft, R.B.; Writing—review and editing, D.P. and M.R. All authors have read and agreed to the published version of the manuscript.

Funding: This research received no external funding.

Data Availability Statement: The original contributions presented in the study are included in the article, further inquiries can be directed to the corresponding author.

Conflicts of Interest: The authors declare no conflicts of interest.

References

1. Raupov, I.; Rogachev, M.; Sytnik, J. Design of a Polymer Composition for the Conformance Control in Heterogeneous Reservoirs. *Energy* **2023**, *16*, 515. [CrossRef]
2. Mardashov, D.; Bondarenko, A.; Raupov, I. Technique for Calculating Technological Parameters of Non-Newtonian Liquids Injection into Oil Well during Workover. *J. Min. Inst.* **2022**, *258*, 881–894. [CrossRef]
3. Litvinenko, V.; Petrov, E.; Vasilevskaya, D.; Yakovenko, A.; Naumov, I.; Ratnikov, M. Assessment of the Role of the State in the Management of Mineral Resources. *J. Min. Inst.* **2022**, *269*, 95–111. [CrossRef]
4. Eder, L.V.; Provornaya, I.V.; Filimonova, I.V. Problems of Rational Use of Associated Petroleum Gas in Russia. *Geogr. Nat. Resour.* **2019**, *40*, 9–14. [CrossRef]

5. Drozdov, N. Filtration Studies on Cores and Sand Packed Tubes from the Urengoy Field for Determining the Efficiency of Simultaneous Water and Gas Injection on Formation When Extracting Condensate from Low-Pressure Reservoirs and Oil from Oil Rims. *J. Min. Inst.* **2022**, *257*, 783–794. [CrossRef]
6. Pal, S.; Mushtaq, M.; Banat, F.; Al Sumaiti, A.M. Review of Surfactant-Assisted Chemical Enhanced Oil Recovery for Carbonate Reservoirs: Challenges and Future Perspectives. *Pet. Sci.* **2018**, *15*, 77–102. [CrossRef]
7. Sun, C.; Guo, H.; Li, Y.; Song, K. Recent Advances of Surfactant-Polymer (SP) Flooding Enhanced Oil Recovery Field Tests in China. *Geofluids* **2020**, *2020*, 8286706. [CrossRef]
8. Thomas, A. *Essentials of Polymer Flooding*; Wiley: Hoboken, NJ, USA, 2018; Volume 292.
9. Kumar, M.; Hoang, V.T.; Satik, C.; Rojas, D.H. High-Mobility-Ratio Waterflood Performance Prediction: Challenges and New Insights. *SPE Reserv. Eval. Eng.* **2008**, *11*, 186–196. [CrossRef]
10. Wang, Y.; Kovscek, A.R.; Brigham, W.E. Effect of Mobility Ratio on Pattern Behavoir of a Homogenous Porous Medium. *In Situ* **1999**, *23*, 1–20.
11. Chaudhuri, A.; Vishnudas, R. A Systematic Numerical Modeling Study of Various Polymer Injection Conditions on Immiscible and Miscible Viscous Fingering and Oil Recovery in a Five-Spot Setup. *Fuel* **2018**, *232*, 431–443. [CrossRef]
12. Chiappa, L.; Andrei, M.; Lockhart, T.P.; Maddinelli, G.; Burrafato, G.; Pitoni, E. Polymer Design for Relative Permeability Modification Treatments at High Temperature. In Proceedings of the International Symposium on Oilfield Chemistry, Houston, TX, USA, 5–7 February 2003.
13. Dusseault, M.B.; Li, S.; Han, H.; Li, J.; Wu, H. Flooding Thin Low-Permeability Layers With a New Salt-Resistant, Medium-Molecular-Weight Polymer. In Proceedings of the SPE Annual Technical Conference and Exhibition, Anaheim, CA, USA, 11–14 November 2007.
14. Seright, R.S. SPE 129899 Potential for Polymer Flooding Reservoirs with Viscous Oils. *SPE Reserv. Eval. Eng.* **2010**, *13*, 730–740. [CrossRef]
15. Islamov, R.S.; Gazizov, A.A. *Improving Labor Productivity*; JSC "VNIIEG": Moscow, Russia, 2003; Volume 568.
16. Ketova, Y.; Bai, B.; Khizhnyak, G.; Gladkikh, Y.; Galkin, S. Testing of Preformed Particles Polymer Gel Technology on Core Filtration Models to Limit Water Inflows. *J. Min. Inst.* **2020**, *241*, 91. [CrossRef]
17. Zhang, X.; Li, B.; Pan, F.; Su, X.; Feng, Y. Enhancing Oil Recovery from Low-Permeability Reservoirs with a Thermoviscosifying Water-Soluble Polymer. *Molecules* **2021**, *26*, 7468. [CrossRef] [PubMed]
18. Delamaide, E.; Tabary, R.; Rousseau, D. Chemical EOR in Low Permeability Reservoirs. In Proceedings of the SPE EOR Conference at Oil and Gas West Asia, Muscat, Oman, 31 March–2 April 2014.
19. Ghosh, P.; Metidji, M.O.; Dupuis, G.; Wilton, R.; Ravikiran, R.; Bowers, A.; Seright, R. Pushing the Envelope of Polymer Injectivity in Low Permeability Sandstones. In Proceedings of the IOR 2021, Online, 19–22 April 2021; European Association of Geoscientists & Engineers: Bunnik, The Netherlands, 2021; pp. 1–21.
20. Hemmati-Sarapardeh, A.; Schaffie, M.; Ranjbar, M.; Dong, M.; Li, Z. (Eds.) *Chemical Methods*; Elsevier: Amsterdam, The Netherlands, 2022; ISBN 9780128219317.
21. Irfan, M.; Stephen, K.D.; Lenn, C.P. An Experimental Study to Investigate Novel Physical Mechanisms That Enhance Viscoelastic Polymer Flooding and Further Increase Desaturation of Residual Oil Saturation. *Upstream Oil Gas Technol.* **2021**, *6*, 100026. [CrossRef]
22. Popov, V.I. The Origin and Development of Elastic Turbulence in Polymer Flows. *Theor. Found. Chem. Technol.* **2019**, *53*, 189–195. [CrossRef]
23. Hincapie, R.E.; Rock, A.; Wegner, J.; Ganzer, L. Oil Mobilization by Viscoelastic Flow Instabilities Effects during Polymer EOR: A Pore-Scale Visualization Approach. In Proceedings of the SPE Latin America and Caribbean Petroleum Engineering Conference, Buenos Aires, Argentina, 17–19 May 2017.
24. Jin, J.; Qi, P.; Mohanty, K.; Balhoff, M. Experimental Investigation of the Effect of Polymer Viscoelasticity on Residual Saturation of Low Viscosity Oils. In Proceedings of the SPE Improved Oil Recovery Conference, Virtual, 31 August–4 September 2020.
25. Azad, M.S.; Trivedi, J.J. Novel Viscoelastic Model for Predicting the Synthetic Polymer's Viscoelastic Behavior in Porous Media Using Direct Extensional Rheological Measurements. *Fuel* **2019**, *235*, 218–226. [CrossRef]
26. Tolstykh, L.I.; Davletshina, L.F.; Potemkina, K.A. *Polyacrylamide in Oil and Gas Production Processe*; Gubkin Russian State University of Oil and Gas (NRU): Moscow, Russia, 2023; Volume 136.
27. Raupov, I.; Rogachev, M.; Sytnik, J. Overview of Modern Methods and Technologies for the Well Production of High- and Extra-High-Viscous Oil. *Energy* **2025**, *18*, 1498. [CrossRef]
28. *RD-39-0148311-206-85*; Guidelines for the Design and Technical and Economic Analysis of Oil Field Development Using the Polymer Impact Method on the Reservoir. Ministry of Petroleum Industry: Moscow, Russia, 1986.
29. Caulfield, M.J.; Hao, X.; Qiao, G.G.; Solomon, D.H. Degradation on Polyacrylamides. Part I. Linear Polyacrylamide. *Polymer* **2003**, *44*, 1331–1337. [CrossRef]

30. Tesoro, G. Polymer Synthesis, by Paul Rempp and Edward W. Merrill. Huthig and Wepf, Heidelberg and New York, 1991. ISBN 3-85739114-6; Binding: Hard Cover; Price Not Available; 344 pp. *Polym. Adv. Technol.* **1992**, *3*, 47. [CrossRef]
31. Thomas, A.; Gaillard, N.; Favero, C. Some Key Features to Consider When Studying Acrylamide-Based Polymers for Chemical Enhanced Oil Recovery. *Oil Gas Sci. Technol. Rev. D'ifp Energ. Nouv.* **2012**, *67*, 887–902. [CrossRef]
32. Jouenne, S.; Chakibi, H.; Levitt, D. Polymer Stability After Successive Mechanical-Degradation Events. *SPE J.* **2018**, *23*, 18–33. [CrossRef]
33. Al-Shakry, B.; Skauge, T.; Shaker Shiran, B.; Skauge, A. Impact of Mechanical Degradation on Polymer Injectivity in Porous Media. *Polymers* **2018**, *10*, 742. [CrossRef] [PubMed]
34. Doe, P.H.; Moradi-Araghi, A.; Shaw, J.E.; Stahl, G.A. Development and Evaluation of EOR Polymers Suitable for Hostile Environments—Part 1: Copolymers of Vinylpyrrolidone and Acrylamide. *SPE Reserv. Eng.* **1987**, *2*, 461–467. [CrossRef]
35. Surguchev, M.L. *Secondary and Tertiary Methods of Increasing Oil Recovery*; Nedra: Moscow, Russia, 1985; Volume 307.
36. Lake, L.; Johns, R.T.; Rossen, W.R.; Pope, G.A. *Fundamentals of Enhanced Oil Recovery*; Society of Petroleum Engineers: Richardson, TX, USA, 2014; ISBN 978-1-61399-328-6.
37. Mardashov, D. Development of Blocking Compositions with a Bridging Agent for Oil Well Killing in Conditions of Abnormally Low Formation Pressure and Carbonate Reservoir Rocks. *J. Min. Inst.* **2021**, *251*, 617–626. [CrossRef]
38. Mardashov, D.V.; Limanov, M.N. Improving the efficiency of oil well damping in the fields of the Volga-Ural oil and gas province with abnormally low reservoir pressure. *Bull. Tomsk. Polytech. Univ. Asset Geoengin.* **2022**, *333*, 185–194. [CrossRef]
39. Islamov, S.R.; Bondarenko, A.V.; Gabibov, A.F.; Mardashov, D.V. Polymer Compositions for Well Killing Operation in Fractured Reservoirs. In *Advances in Raw Material Industries for Sustainable Development Goals*; CRC Press: London, UK, 2020; pp. 343–351.
40. Korobov, G.Y.; Parfenov, D.V.; Nguyen, V.T. Mechanisms of formation of asphalt resin paraffin deposits and factors of intensity of their formation. *Bull. Tomsk. Polytech. Univ. Asset Geoengin.* **2023**, *334*, 103–116. [CrossRef]
41. Rogachev, M.; Aleksandrov, A. Justification of a Comprehensive Technology for Preventing the Formation of Asphalt-Resin-Paraffin Deposits during the Production of Highlyparaffinic Oil by Electric Submersible Pumps from Multiformation Deposits. *J. Min. Inst.* **2021**, *250*, 596–605. [CrossRef]
42. Korobov, G.Y.; Parfenov, D.V.; Nguyen, V.T. Long-Term Inhibition of Paraffin Deposits Using Porous Ceramic Proppant Containing Solid Ethylene-Vinyl Acetate. *Int. J. Eng.* **2025**, *38*, 1887–1897. [CrossRef]
43. Leusheva, E.; Alikhanov, N.; Brovkina, N. Study on the Rheological Properties of Barite-Free Drilling Mud with High Density. *J. Min. Inst.* **2022**, *258*, 976–985. [CrossRef]
44. Islamov, S.R.; Bondarenko, A.V.; Mardashov, D.V. A Selection of Emulsifiers for Preparation of Invert Emulsion Drilling Fluids. In *Topical Issues of Rational Use of Natural Resources 2019*; CRC Press: Boca Raton, FL, USA, 2019; pp. 487–494.
45. Tananykhin, D.; Grigorev, M.; Simonova, E.; Korolev, M.; Stecyuk, I.; Farrakhov, L. Effect of Wire Design (Profile) on Sand Retention Parameters of Wire-Wrapped Screens for Conventional Production: Prepack Sand Retention Testing Results. *Energy* **2023**, *16*, 2438. [CrossRef]
46. Tananykhin, D.; Grigorev, M.; Korolev, M.; Solovyev, T.; Mikhailov, N.; Nesterov, M. Experimental Evaluation of the Multiphase Flow Effect on Sand Production Process: Prepack Sand Retention Testing Results. *Energy* **2022**, *15*, 4657. [CrossRef]
47. Ghosh, P.; Wilton, R.R.; Bowers, A.; O'Brien, T.; Cao, Y.; Wilson, C.; Metidji, M.O.; Dupuis, G.; Ravikiran, R. Re-Injection of Produced Polymer in EOR Projects to Improve Economics and Reduce Carbon Footprint. In Proceedings of the SPE Improved Oil Recovery Conference, Virtual, 25–29 April 2022.
48. Karnanda, W.; Benzagouta, M.S.; AlQuraishi, A.; Amro, M.M. Effect of Temperature, Pressure, Salinity, and Surfactant Concentration on IFT for Surfactant Flooding Optimization. *Arab. J. Geosci.* **2013**, *6*, 3535–3544. [CrossRef]
49. Podoprigora, D.; Byazrov, R.; Sytnik, J. The Comprehensive Overview of Large-Volume Surfactant Slugs Injection for Enhancing Oil Recovery: Status and the Outlook. *Energy* **2022**, *15*, 8300. [CrossRef]
50. Toro, M.; Khisametdinov, M.; Nuriev, D.; Lutfullin, A.; Daminov, A.; Gaifullin, A.; Puskas, S.; Ordog, T. Initiation of a Surfactant-Polymer Flooding Project at PJSC Tatneft: From Laboratory Studies to Test Injection. In Proceedings of the IOR 2021, Online, 19–22 April 2021; European Association of Geoscientists & Engineers: Bunnik, The Netherlands, 2021; pp. 1–15.
51. Trushin, Y.; Aleshchenko, A.; Arsamakov, M.; Klimenko, A.; Molinier, V.; Jouenne, S.; Kornilov, A.; Sansiev, G. EOR Technology: Surfactant-Polymer Injection to Increase Oil Recovery from Carbonate Reservoir of Kharyaga Oilfield (Russian). In Proceedings of the SPE Russian Petroleum Technology Conference, Virtual, 26–29 October 2020.
52. Yin, S.; Pu, H.; Zhou, S. An Update on Full Field Implementation of Chemical Flooding in Daqing Oilfield, China, and Its Future. In Proceedings of the SPE Improved Oil Recovery Conference, Tulsa, OK, USA, 14–18 April 2018.
53. Delamaide, E. Is Chemical EOR Finally Coming of Age? In Proceedings of the SPE Asia Pacific Oil & Gas Conference and Exhibition, Virtual, 17–19 November 2020.
54. Zhang, Y.; Wei, M.; Bai, B.; Yang, H.; Kang, W. Survey and Data Analysis of the Pilot and Field Polymer Flooding Projects in China. In Proceedings of the SPE Improved Oil Recovery Conference, Tulsa, OK, USA, 11–13 April 2016.

55. Luo, X.; Zhao, C. Practices and Experiences of Seven-Year Polymer Flooding History Matching in China Offshore Oil Field: A Case Study. In Proceedings of the SPE Reservoir Characterisation and Simulation Conference and Exhibition, Abu Dhabi, United Arab Emirates, 9–11 October 2011.
56. Seright, R.S.; Chang, H.L.; Zhang, Z.Q.; Wang, Q.M.; Xu, Z.S.; Guo, Z.D.; Sun, H.Q.; Cao, X.L.; Qi, Q. Discussion and Reply to "Advances in Polymer Flooding and Alkaline/Surfactant/Polymer Processes as Developed and Applied in the People's Republic of China". *J. Pet. Technol.* **2006**, *58*, 80–81. [CrossRef]
57. Gbadamosi, A.O.; Junin, R.; Manan, M.A.; Agi, A.; Yusuff, A.S. An Overview of Chemical Enhanced Oil Recovery: Recent Advances and Prospects. *Int. Nano Lett.* **2019**, *9*, 171–202. [CrossRef]
58. Puskas, S.; Vágó, Á.; Törő, M.; Ördög, T.; Kálmán, G.Y.; Tabajd, R.; Dékány, I.; Dudás, J.; Nagy, R.; Bartha, L.; et al. First Surfactant-Polymer EOR Injectivity Test in the Algyő Field, Hungary. In Proceedings of the 19th European Symposium on Improved Oil Recovery, Stavanger, Norway, 24–27 April 2017.
59. Korobov, G.Y.; Vorontsov, A.A.; Buslaev, G.V. Influence of Inhibitors Based on Surfactants on Processes of Hydrates and Asphalt-Resin-Paraffin Deposits Formation: Experimental Studies. *Int. J. Eng.* **2025**, *38*, 2220–2230.
60. *Башнипинефть РД-39-1-199-79*; Руководство По Проектированию и Применению Метода Заводнения с Водорастворимыми Поверхностно-Активными Веществами (ПАВ). Уфа, Ministry of Petroleum Industry: Moscow, Russia, 1979.
61. *OST 39–195-86*; Oil. Method for Determining the Oil Displacement Coefficient by Water Under Laboratory Conditions. Ministry of Petroleum Industry: Moscow, Russia, 1986.
62. Müller, A.J.; Sáez, A.E. The Rheology of Polymer Solutions in Porous Media. In *Flexible Polymer Chains in Elongational Flow*; Springer: Berlin/Heidelberg, Germany, 1999; pp. 335–393.
63. Puls, C.; Clemens, T.; Sledz, C.; Kadnar, R.; Gumpenberger, T. Mechanical Degradation of Polymers During Injection, Reservoir Propagation and Production-Field Test Results 8 TH Reservoir, Austria. In Proceedings of the SPE Europec Featured at 78th EAGE Conference and Exhibition, Vienna, Austria, 30 May–2 June 2016.
64. Raupov, I.R.; Kone, A.; Podjapolski, E.V.; Milic, J. Justification of Conformance Control in Mature Oil Fields. *Int. J. Eng.* **2025**, *38*, 12–20. [CrossRef]
65. Lapin, K.G.; Toropov, K.V.; Volgin, E.R.; Byazrov, R.R.; Galimov, V.V.; Lagutina, M.A. Study of Polymer Adsorption for the Conditions of the PK Layers of the Russian Deposit. *Exhib. Oil Gas* **2022**, 60–64. [CrossRef]
66. Kenzhekhanov, S.; Yin, X.; Gussenov, I. Microfluidic Study of Polymer Permeability Reduction Mechanisms. In Proceedings of the SPE Annual Technical Conference and Exhibition, SPE Annual Technical Conference and Exhibition, Houston, TX, USA, 3–5 October 2022.
67. Al-Hajri, S.; Mahmood, S.; Abdulelah, H.; Akbari, S. An Overview on Polymer Retention in Porous Media. *Energy* **2018**, *11*, 2751. [CrossRef]

energies

MDPI

Article

Integrated 3D Geological Modeling, Stress Field Modeling, and Production Simulation for CBM Development Optimization in Zhengzhuang Block, Southern Qinshui Basin

Zhong Liu [1], Hui Wang [2,3,*], Xiuqin Lu [1], Qianqian Zhang [1], Yanhui Yang [1], Tao Zhang [1], Chen Zhang [1] and Zihan Wang [1]

[1] Petro China Huabei Oilfield Company, Renqiu 062550, China
[2] School of Emergency Management and Safety Engineering, North China University of Science and Technology, Tangshan 063210, China
[3] School of Energy Resource, China University of Geosciences (Beijing), Beijing 100083, China
* Correspondence: huiwang90@ncst.edu.cn

Abstract: The Zhengzhuang Block in the Qinshui Basin is one of the important coalbed methane (CBM) development areas in China. As high-quality CBM resources become depleted, remaining reserves exhibit complex geological characteristics requiring advanced development strategies. In this study, a multidisciplinary workflow integrating 3D geological modeling (94.85 km^2 seismic data, 973 wells), geomechanical stress analysis, and production simulation was developed to optimize development of the Permian No. 3 coal seam. Structural architecture and reservoir heterogeneity were characterized through Petrel-based modeling, while finite-element analysis identified stress anisotropy with favorable stimulation zones concentrated in southwestern sectors. Computer Modeling Group (CMG) simulations of a 27-well group revealed a rapid initial pressure decline followed by a stabilization phase. A weighted multi-criteria evaluation framework classified resources into three tiers: type I (southwestern sector: 28–33.5 m^3/t residual gas content, 0.8–1.0 mD permeability, 8–12% porosity), type II (northern/central: 20–26 m^3/t residual gas content, 0.5–0.6 mD permeability, 5–8% porosity), and type III (<20 m^3/t residual gas content, <0.4 mD permeability, <4% porosity). The integrated methodology provides a technical foundation for optimizing well patterns, enhancing hydraulic fracturing efficacy, and improving residual gas recovery in heterogeneous CBM reservoirs.

Keywords: coalbed methane; geological modeling; stress field modeling; production simulation; reservoir evaluation

1. Introduction

Coalbed methane (CBM), as a clean and sustainable unconventional resource, plays a crucial role in the global energy transition and carbon reduction [1,2]. In recent years, significant advancements have been made in the development of CBM in China [3,4]. The Zhengzhuang Block in the Qinshui Basin is one of the major CBM development areas in China. As the scale of development expands and technology advances, high-quality resources have been largely depleted, leaving behind reservoirs with increasingly complex geological conditions and presenting considerable challenges for further development [5,6]. Addressing these challenges requires refined three-dimensional (3D) modeling of CBM reservoirs and numerical simulation of dynamic production to comprehensively understand the distribution patterns, productivity characteristics, and controlling factors of CBM reservoirs.

Three-dimensional geological modeling technology, as a vital tool for CBM reservoir characterization [7], integrates multi-source data, such as geological, logging, seismic, well testing, and reservoir engineering information [8–11]. This enables the construction of high-resolution 3D structural and property models, providing an in-depth understanding of the spatial distribution characteristics and accumulation evolution of coal reservoirs [12]. It also provides a solid foundation for numerical simulation and dynamic analysis of reservoir development. With the continuous advancement of stochastic 3D modeling theory, the integration of logging interpretation results and seismic data constraints has become a focal point in CBM modeling research, aiming to establish comprehensive and high-precision geological models for coal reservoirs.

Current research primarily focuses on the optimization of geologically favorable zones through geological characterization, while conventional optimization approaches often neglect the identification of stimulation-favorable zones [13,14]. To improve stimulation effectiveness in these areas, it is essential to develop targeted fracturing strategies based on geostress conditions. Meanwhile, the characterization of residual gas distribution has emerged as a critical research direction for unlocking future reservoir potential [15,16]. Traditional methods estimate residual gas by simply subtracting cumulative production from original gas content at the field scale. Although numerical simulation offers greater accuracy, its application has mostly been restricted to small well clusters due to computational limitations [17,18]. To enhance prediction accuracy for remaining gas reserves, this study proposes an integrated approach that combines localized numerical simulations with broader computational methods. This hybrid methodology utilizes simulation calibration to improve understanding of the spatial distribution of residual gas, providing a technical basis for optimizing subsequent development strategies and enhancing recovery.

Therefore, this study focuses on the No. 3 coal reservoir in the southern Zhengzhuang Block, employing Petrel geological modeling and numerical simulation technologies to conduct systematic characterization of coal reservoir properties. A comprehensive geological model was constructed, integrating sub-models for porosity, permeability, and gas content to identify optimal development zones. Using advanced modeling techniques, the spatial distribution of geostress magnitude and orientation was analyzed to determine stimulation-favorable zones for enhanced reservoir modification. Based on this geological framework, numerical simulations calibrated with historical production data were used to interpret production dynamics, quantify remaining gas, and evaluate both current productivity and untapped potential. An integrated evaluation framework was established by combining development-favorable areas, stimulation-targeted zones, and productivity assessments, ultimately mapping CBM production potential across the study area to support future planning.

The established models provide a scientific foundation for resource exploitation optimization, delivering the following: (1) A technical framework for optimizing well patterns and improving recovery strategies; (2) production performance simulations to forecast gas output and refine extraction plans; and (3) an integrated evaluation system for assessing economic feasibility. Collectively, these contributions support data-driven decision-making in CBM field development, particularly in overcoming the challenges of heterogeneous reservoir stimulation and residual gas recovery [19,20].

2. Geological Settings

The Qinshui Basin, located in North China, is a representative example of a Mesozoic intracratonic rift basin (Figure 1). Its tectonic and sedimentary evolution can be divided into three distinct stages. The first stage, spanning the Pennsylvanian to the Late Triassic, was characterized by prolonged tectonic subsidence and sedimentation. During the second stage in the Jurassic, the basin underwent regional uplift followed by subsidence and subsequent reburial of the strata. The third stage, from the Cretaceous to the present, has been dominated by continuous uplift and extensive erosion [13,21,22].

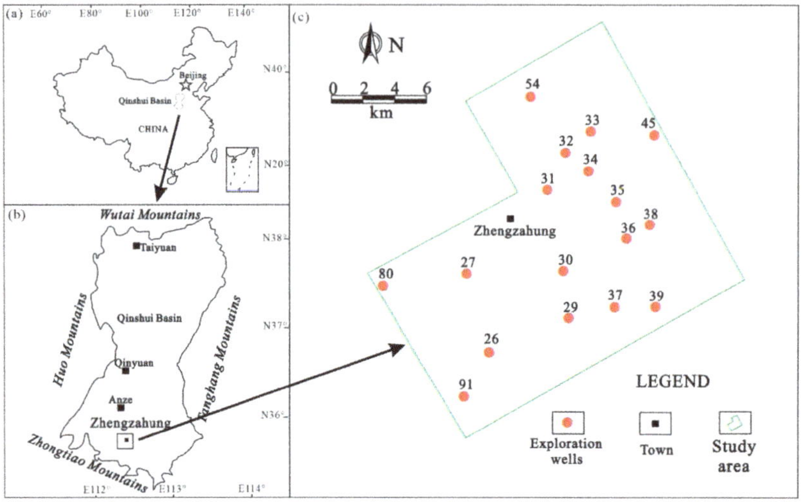

Figure 1. (**a**) Map showing the location of Qinshui Basin in China. (**b**) Location of the Zhengzhuang Field in Qinshui Basin. (**c**) Study area with 3D seismic and locations of the exploration wells in the Zhengzhuang Field (Modified after Wang et al. [23]).

The Pennsylvanian–Permian coal-bearing strata in the basin were subjected to four phases of tectonic deformation, driven by major orogenic events [24]. The first phase, during the Late Permian to Triassic, was associated with the Indosinian orogeny, resulting in compressive deformation under a north–south (N–S) principal stress regime. The second phase, linked to the Yanshanian orogeny in the Jurassic and Early Cretaceous, intensified compressional deformation and reoriented the principal stress to a southeast–northwest (SE–NW) direction. This phase is regarded as crucial for the formation of the basin's structural framework. The third phase, during the Paleogene, was driven by the early Himalayan orogeny, which transitioned the regional stress field to an extensional regime-oriented northwest–west to southeast–east (NWW–SEE). This extensional stress led to the formation of major normal faults, particularly in the eastern Zhengzhuang Field. The final phase, during the late Himalayan orogeny in the Quaternary, saw the stress regime revert to compression, with a principal stress direction-oriented northeast–southwest (NE–SW). This compressional regime is believed to be responsible for the closure of many major faults within the basin [10,24].

As shown in Figure 1b,c, the Zhengzhuang CBM Field, situated in the southern part of the Qinshui Basin, covers an area of approximately 570 km^2 and represents a significant region for coalbed methane development. As shown in Figure 2, the Permo-Carboniferous coal-bearing strata in this field comprise, from bottom to top, the Benxi, Taiyuan, Upper Shihezi, and Shiqianfeng Formations, with the No. 3 coal seam in the Permian Shanxi Formation serving as the primary target for CBM production. The No. 3 coal seam is primarily composed of anthracite, with vitrinite reflectance (Ro, max) values reaching up to 4.3%. The gas generated has reached a total of approximately 300 m^3/t from the Jurassic to Early Cretaceous, with an average gas content of 20.7 m^3/t today, indicating a significant resource potential [23]. The Shanxi Formation, primarily deposited in deltaic environments such as mouth bars, interdistributary bays, and delta plains, consists of interbedded shale, sandstone, sandy shale, and coal. It has an average thickness of approximately 47 m, ranging from 20 to 90 m. The coal-bearing strata are underlain by early Paleozoic marine carbonates and overlain by fluvial deposits of the Triassic [25,26]. Among these, the No. 3 coal seam is the primary focus for CBM exploration and extraction within the Zhengzhuang Field due to its favorable geological and reservoir characteristics.

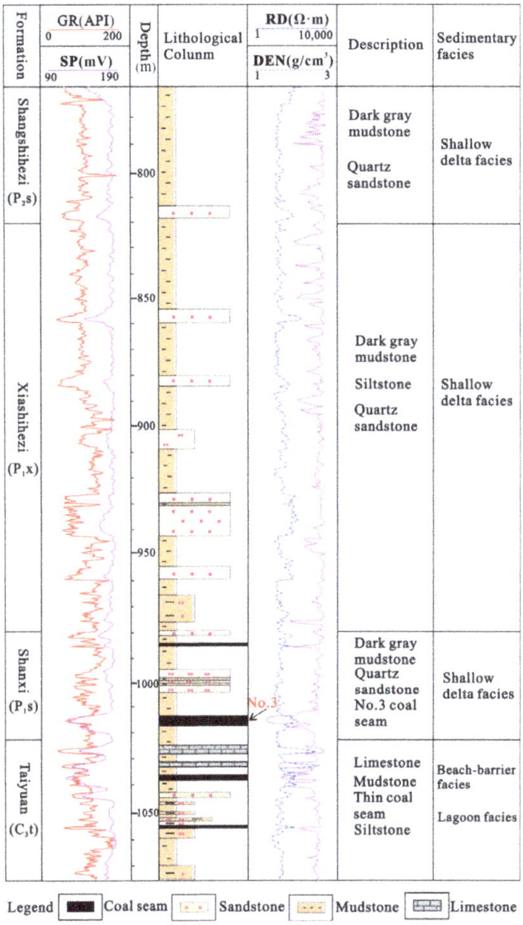

Figure 2. Sedimentary characteristics of the coal-bearing strata in the Zhengzuhang Field (Modified after Wang et al. [23]).

3. Data Sources

The structural interpretation results based on seismic data serve as foundational inputs for geological modeling. This study utilized 3D seismic data covering an area of 94.85 km^2 in the research region to obtain structural morphology information of the target coal seam. The seismic survey area is located in Zhengzhuang, Qinshui County, Shanxi Province. The study area generally exhibits a topographic trend sloping from northwest to southeast, and it is one of the most developed blocks in the Qinshui Basin coalfield.

The study area contains 33 parameter wells and 940 production wells. This research collected drilling and completion data from the 33 parameter wells for geological horizon identification. Core observations from 119 core samples were integrated with selected logging curves—including acoustic interval transit time, density, natural gamma, resistivity, and compensated neutron logs—to determine macroscopic coal lithotypes, coal structure, and rock mechanical properties. Moreover, porosity, permeability, and gas content data were obtained from the 119 core samples distributed across the 33 parameter wells.

For the numerical simulation of production processes, a typical production zone containing 27 active production wells was selected for history matching of gas and water production. These wells commenced production between 2010 and 2012 and had been in continuous operation for four to five years by the data cutoff in December 2017.

4. Results and Discussion

4.1. Three-Dimensional Structural Modeling

The 3D modeling module of the Petrel geological modeling software was used to construct the geological model of the No. 3 coal reservoir in the southern Zhengzhuang Field, based on drilling data and testing results for the No. 3 coal reservoir in the area. The modeling process included the following steps: First, the top-surface model (Figure 3a) and bottom-surface model (Figure 3b) of the No. 3 coal seam were constructed. Subsequently, fault models (Figure 3c) and collapse column models (Figure 3d) were developed. Next, planar gridding was performed using the pillar gridding method, ensuring the grid orientation was generally aligned with the strike of the faults. Considering the practical requirements of the study, including modeling accuracy and computational efficiency, the grid size was set to 25 m × 25 m (Figure 3e). Finally, the three-dimensional structural model of the No. 3 coal reservoir in the study area was completed by integrating the surface models, fault models, and collapse column models (Figure 3f).

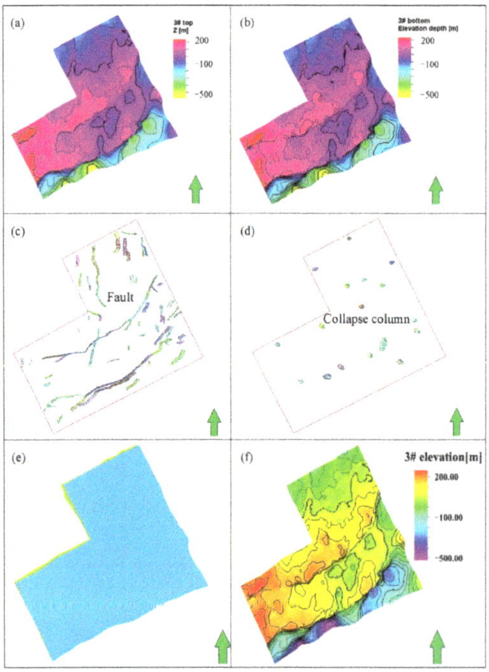

Figure 3. Three-dimensional geological modeling elements and structural characteristics of No. 3 coal seam. (**a**) Top-surface model of the coal seam. (**b**) Bottom-surface model of the coal seam. (**c**) Spatial distribution model of fault systems. (**d**) Collapse column configuration model. (**e**) Discretization grid division (25 m × 25 m resolution). (**f**) Integrated 3D geological structure model incorporating lithological interfaces and tectonic features.

4.2. Modeling of Macroscopic Coal Lithotype

The establishment of a macroscopic coal lithotype model provides critical parameter constraints for subsequent coal seam attribute modeling. The modeling strategy simulates the spatial distribution of macroscopic coal lithotypes using Sequential Gaussian Simulation (SGS), with lithotype identification results from borehole locations serving as foundational data. Previous research has demonstrated quantifiable correlations between coal lithotype characteristics and corresponding well-logging responses, enabling the quantitative identification of lithotypes through logging data interpretation. In this study, the macroscopic lithotypes of 119 core samples from parameter wells within the study area were first determined through systematic core analysis, as illustrated in Figure 4a. Given the established correlation between logging parameters and coal lithotypes, neural network algorithms embedded in Petrel™ were employed to predict lithotypes in non-cored intervals, with two typical examples shown in Figure 4b.

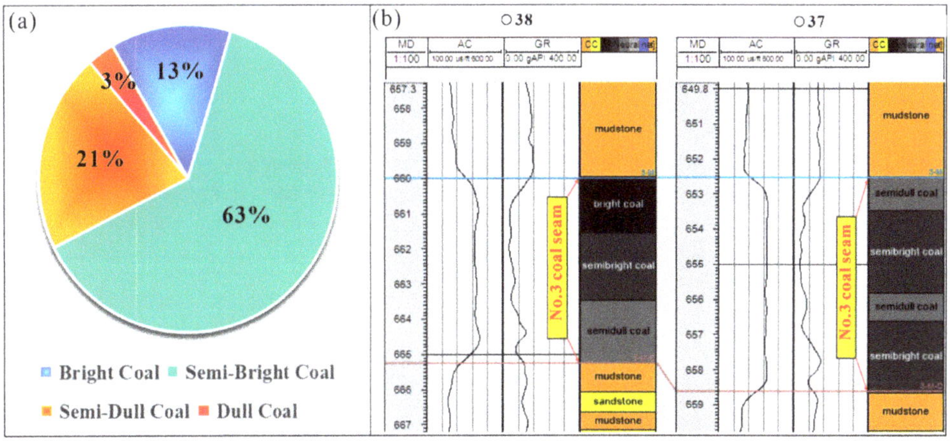

Figure 4. (**a**) Composition of macroscopic coal lithotypes based on core descriptions. (**b**) Representative log-derived lithotype predictions in the study area.

The modeling results of the No. 3 coal seam (Figure 5) reveal distinct spatial heterogeneity in macroscopic lithotype distribution, dominated by semi-bright coal, which covers 38.76% of the area and exhibits laterally continuous depositional patterns in the western and northern sectors. Bright coal constitutes 29.73% of the seam, characterized by clustered distributions in the northern, southeastern, and southwestern regions, with isolated occurrences in the central areas, suggesting localized paleo-environmental controls on woody tissue preservation. Semi-dull coal accounts for 31.51% coverage, displaying well-connected networks in the southern and central zones, indicative of sustained mineral-rich or oxidative swamp conditions. Notably, dull coal is nearly absent (<0.01%), reflecting minimal wildfire events or rapid peat burial during peatification.

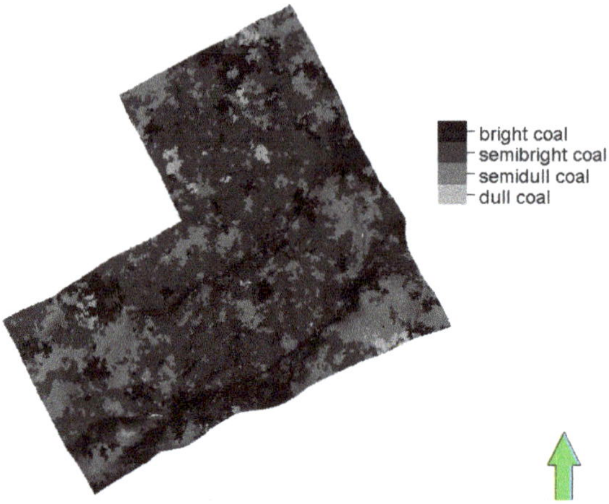

Figure 5. Spatial heterogeneity and distribution of macroscopic coal lithotypes of No. 3 coal seam in the study area.

4.3. Modeling of 3D Geomechanical Stress

The development of a high-fidelity 3D in situ stress model requires seamless integration with a pre-constructed 3D geological framework. The workflow follows a five-stage hierarchical approach (Figure 6): geomechanical grid generation, mechanical property interpolation, natural fracture network characterization, boundary condition calibration, and finite element-based stress field simulation.

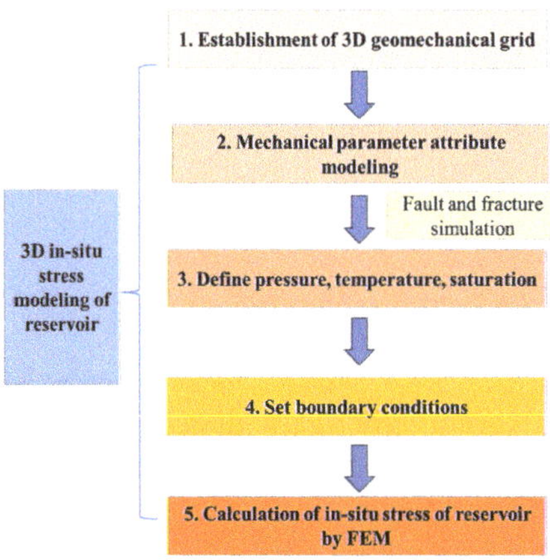

Figure 6. Workflow for 3D geomechanical stress modeling.

4.3.1. Geomechanical Grid Generation

A hexahedral-dominant grid architecture (aspect ratio < 3) was developed with progressive coarsening (scaling factor: 1.5×) from the reservoir core, prioritizing computational efficiency while maintaining ≤5% element distortion threshold. Lateral grid expansion along the I- and J-axes employed symmetric 15-layer scaling (1.5× ratio) with 50 m-thick rigid boundaries to mitigate edge effects, while vertical extension to the surface (Z = −900 m) ensured smooth cell-size transitions through graded coarsening. Underburden optimization constrained global grid dimensions to 197 × 210 × 44 cells (1,820,280 elements), balancing aspect ratio compliance (<3) with minimized computational overhead (Figure 7).

Figure 7. Geomechanical grid of No. 3 coal seam in the study area.

4.3.2. Mechanical Property Interpolation

Well-logging-derived rock mechanical parameterization has emerged as a cost-efficient methodology that provides continuous sampling coverage, enhanced vertical resolution (typically at sub-meter scales), and improved compatibility with reservoir-scale geomechanical modeling workflows [27]. This technique leverages multipole array acoustic logging tools to simultaneously record compressional wave slowness and shear wave slowness measurements across diverse lithological sequences. Through petrophysical inversion of these acoustic datasets, critical elastic mechanical parameters (e.g., Young's modulus, Poisson's ratio, and bulk modulus) can be systematically derived.

(1) Poisson's Ratio (μ)

Poisson's ratio is defined as the absolute ratio of transverse strain to axial strain during uniaxial or triaxial compression within the elastic deformation stage. It is also termed the lateral compression coefficient. A higher Poisson's ratio indicates that the material undergoes greater transverse deformation relative to longitudinal deformation before plastic deformation occurs. The formula is:

$$\mu = \frac{\Delta t_s^2 - 2\Delta t_p^2}{2\left(\Delta t_s^2 - \Delta t_p^2\right)} \quad (1)$$

where Δt_s is shear wave slowness, μs/ft; Δt_p is compressional wave slowness, μs/f.

(2) Shear Modulus (G)

The shear modulus represents the ratio of shear stress to shear strain within the proportional limit of elastic deformation under shear stress. It characterizes the material's ability to resist shear deformation. A higher shear modulus indicates greater rigidity of the material. The formula is:

$$G = \frac{\rho}{\Delta t_s^2} \times 10^6 \quad (2)$$

where G is Shear Modulus, GPa; ρ is density of coal, g/m^3.

(3) Young's Modulus (E)

Young's modulus, also termed the longitudinal elastic modulus, is defined as the ratio of stress to strain within the elastic deformation range. It quantifies a material's stiffness and reflects the rock's ability to resist tensile deformation. The formula is:

$$E = G\left(\frac{3\Delta t_s^2 - 4\Delta t_p^2}{\Delta t_s^2 - \Delta t_p^2}\right) \quad (3)$$

where E is Young's Modulus, GPa.

(4) Bulk Modulus (K)

The bulk modulus is the reciprocal of the compressibility coefficient. The compressibility coefficient measures the relative volume change in a material per 1 MPa increase in confining pressure at constant temperature. The formula for bulk modulus is:

$$K = G\left(\frac{3\Delta t_s^2 - 4\Delta t_p^2}{3\Delta t_p^2}\right) \tag{4}$$

where K is Bulk Modulus, GPa.

(5) Compressive Strength (σ_c)

Compressive strength represents the stress value at which a rock undergoes complete failure under uniaxial pressure. It indirectly reflects the formation fracture strength and is primarily correlated with the rock's elastic modulus and clay content through a statistical relationship. The expression is:

$$\sigma_c = aE(1 - V_{sh}) + bEV_{sh} \tag{5}$$

$$V_{sh} = \frac{4^{\Delta GR} - 1}{3} \tag{6}$$

$$\Delta GR = \frac{GR - GR_{min}}{GR - GR_{max}} \tag{7}$$

where V_{sh} is clay content; a and b are empirical coefficients dependent on lithology, which is estimated at approximately 4.59 and 8.16. GR is gamma logging, API.

(6) Tensile Strength (σ_t)

Tensile strength refers to the ultimate stress at which a rock undergoes complete failure under tension. It influences formation fracture pressure and is closely related to reservoir fracturing parameters. It can be approximately expressed as:

$$\sigma_t = \sigma_c/12 \tag{8}$$

where σ_t is tensile strength, MPa; σ_c is compressive strength, MPa.

(7) Cohesion (τ) and internal friction angle (φ)

Cohesion describes the intrinsic bonding force between adjacent parts within a rock (also termed cohesive strength). A commonly used empirical formula is:

$$\tau = 10^{-12} 0.0054 \frac{1 + 0.78V_{sh}}{\Delta t_p^4}(1 - 2\mu)\rho^2 \frac{1 + \mu}{1 - \mu} \tag{9}$$

The internal friction angle characterizes the frictional properties between rock particles formed by their mutual displacement and bonding. It is defined as the angle between the failure envelope and the horizontal axis in a shear strength diagram.

$$\varphi = 36.545 - 0.4952\tau \tag{10}$$

where τ is cohesion, and φ is the internal friction angle.

4.3.3. Modeling of Reservoir Fracture

Seismic amplitude reflects acoustic impedance contrasts at subsurface geological interfaces. Significant spatial variations in amplitude often correspond to discontinuities such as faults, fractures, or lithological changes. The amplitude contrast method is highly sensitive to abrupt seismic features (e.g., fault zones, high-angle fractures) and effectively captures subtle seismic signal variations. Figure 8a displays ant-tracking attribute results derived from amplitude contrast analysis. The ant-tracking volume reveals clear fracture traces with localized high-density fracture clusters, demonstrating the method's capability to delineate fracture systems.

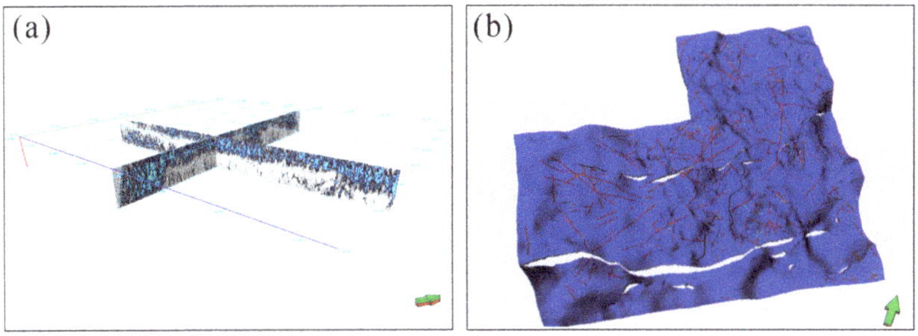

Figure 8. Ant-tracking attribute (**a**) and modeling of reservoir fracture (**b**).

Fracture attributes (including aperture and permeability) in the study area were quantified using empirical relationships:

$$A_f = L \times Normal(0.00001,\ 0.000002) \tag{11}$$

$$P_f = \left(\frac{A_f}{3.2808}\right)^2 \times 10^{15} \tag{12}$$

where A_f is the aperture of fracture, cm; L is the length of fracture, cm; and P_f is fracture permeability, mD.

After upscaling the fracture attribute volume, a deterministic fracture model is generated, as shown in Figure 8b. This model integrates key fracture characteristics (e.g., orientation, density, aperture, and connectivity) into a structured grid format suitable for reservoir simulation and engineering applications.

4.3.4. Boundary Condition Configuration

The geomechanical modeling framework requires simultaneous satisfaction of equilibrium equations, kinematic compatibility, and constitutive relationships, governed by physically constrained boundary conditions. Initial stress field boundary conditions were implemented for efficient regional stress tensor reconstruction, with critical input parameters structured as overburden stress magnitude and stress anisotropy ratios.

Given the abundance of well testing and logging in the Zhengzhuang Field, hydraulic fracturing was used to obtain the vertical principal stress (σ_v, MPa), maximum horizontal principal stress (σ_H, MPa), and minimum horizontal principal stress (σ_h, MPa) [28]. The calculated results of typical well σ_H, σ_h, and σ_v are shown in Table 1, which show that maximum and minimum horizontal stress gradient are 0.0344 MPa/m and 0.0215 MPa/m, respectively.

Table 1. Calculation results of typical well ground stress.

Well	Depth (m)	Top Elevation (m)	σ_H (Mpa)	σ_h (Mpa)	σ_v (Mpa)
14	1122.3	919.11	42.72	26.40	27.50
15	896.71	815.74	26.23	16.87	21.97
19	559.85	746.89	26.86	15.98	13.72

Table 1. Cont.

Well	Depth (m)	Top Elevation (m)	σ_H (Mpa)	σ_h (Mpa)	σ_v (Mpa)
27	747.05	885.91	22.95	14.77	18.30
30	638.5	633.99	14.75	10.51	15.64
31	601.25	674.06	20.78	13.21	14.73
38	659.95	693.19	30.79	18.87	16.17
39	992.9	597.72	27.24	18.92	24.33
64	1242.4	826.31	44.77	27.65	30.44
67	1269.7	924	46.86	29.09	31.11
69	1187.25	1127	29.16	19.19	29.09
76	515.4	904.4	28.45	15.97	12.63
78	702.8	1000	20.17	11.88	17.22
80	752.1	941	21.37	13.45	18.43
82	703.8	937	21.09	13.08	17.24
89	521.6	737	22.75	13.47	12.78
102	1102.4	869.4	42.64	25.99	27.01

4.3.5. Modeling of In-Site Stress

Among numerical simulation methods, the finite element method (FEM) is one of the useful methods to solve geomechanical problems. According to the FEM, the modeling of σ_H, σ_h, and σ_v in the study area are obtained (Figure 9).

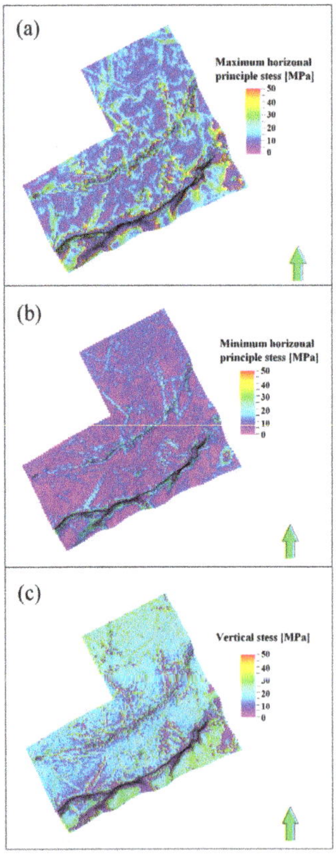

Figure 9. The in situ stress model of the No. 3 coal seam in the study area. (**a**) Maximum horizontal principal stress. (**b**) Minimum horizontal principal stress. (**c**) Vertical principal stress.

4.4. Modeling of Reservoir Properties

This study employs a facies-controlled Sequential Gaussian Simulation (SGS) algorithm for attribute modeling (porosity and permeability) of coal reservoirs. Since sedimentary attributes within the same facies in the study area exhibit similar statistical patterns and value ranges, the modeling process integrates these characteristics to constrain property parameters layer-by-layer and phase-by-phase during stochastic simulation. This approach ensures that the attribute models more accurately reflect the physical property distribution in the study area.

4.4.1. Modeling of Reservoir Porosity

Coal reservoir porosity serves as a fundamental petrophysical parameter governing methane storage capacity and fluid transport mechanisms in CBM systems. This study establishes a novel porosity modeling workflow through systematic integration of density logging measurements. Generally, reservoir porosity can be derived from the relationship between density logs and porosity values [29], as follows:

$$\varphi_f = \frac{\rho_{ma} - \rho_b}{\rho_{ma} - \rho_f} \tag{13}$$

where φ_f is coal porosity (%); ρ_{ma} is density of rock matrix (kg/m^3); ρ_b is bulk density (kg/m^3); and ρ_f is density of pore fluid (kg/m^3).

Figure 10 delineates the reservoir porosity distribution of the No. 3 coal seam. As can be seen, high-porosity zones (8–12%), predominantly distributed across the central-southern and eastern sectors with robust lateral continuity. Discrete high-porosity anomalies (8–10%) in northern structural culminations further reinforce this lithotype-controlled porosity pattern. Moderately porous regions (5–7.5%), characterized by intermittent connectivity in central-north and southwestern domains. Marginally porous zones (1.3–4%) along the northern, southeastern, and western peripheries demonstrate fragmented distributions.

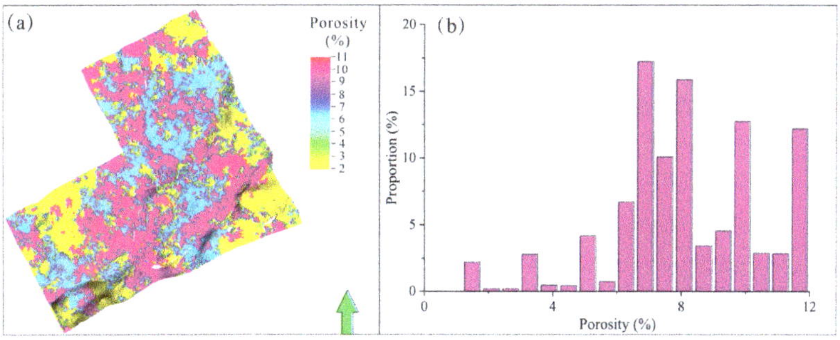

Figure 10. No. 3 coal seam in the study area. (**a**) Reservoir porosity model; (**b**) Permeability distribution histogram.

4.4.2. Modeling of Reservoir Permeability

The permeability calculation of coal seam mainly focuses on fracture permeability. Generally, the fracture aperture and fracture spacing are two key parameters for the calculation of reservoir permeability. The fracture aperture and fracture spacing can be calculated using the following equations, respectively,

$$h_f = \frac{\Delta C}{4C_{mf}} \tag{14}$$

$$h_m = \frac{h_f}{\varphi_f} \quad (15)$$

where h_f is the fracture aperture (cm); ΔC is the difference between deep and shallow conductivity measurements (S/m); C_{mf} is the mud filtrate conductivity (S/m); h_m is fracture spacing (cm); and φ_f is coal porosity (%).

Then, the fracture permeability of No. 3 coal seam was predicted using the Fracture-Stehfest (F-S) method. The fracture permeability of No. 3 coal seam can be calculated as:

$$K_f = \frac{RF \times 8.33 \times 10^6 \times h_f^3}{h_m} = RF \times 8.33 \times 10^6 \times h_f^2 \times \varphi_f \quad (16)$$

where K_f is fracture permeability (μm^2); RF is the correlation coefficient, assigned an empirical value of 0.338×10^4.

Following permeability calculations for individual wells using Equation (16), the permeability data were discretized. The permeability model of the No. 3 coal seam was then constructed (Figure 11) using the SGS method. Model results reveal a tripartite permeability hierarchy: (1) High-permeability sweet spots (0.8–1.0 mD) cluster in the northern and southeastern structural domains; (2) Moderately permeable corridors (0.5–0.6 mD) dominate central-south regions; (3) Low-permeability compartments (0.01–0.04 mD) occupy basin-margin positions (northwest, southeast). The reservoir-scale geometric mean permeability is 0.35 mD.

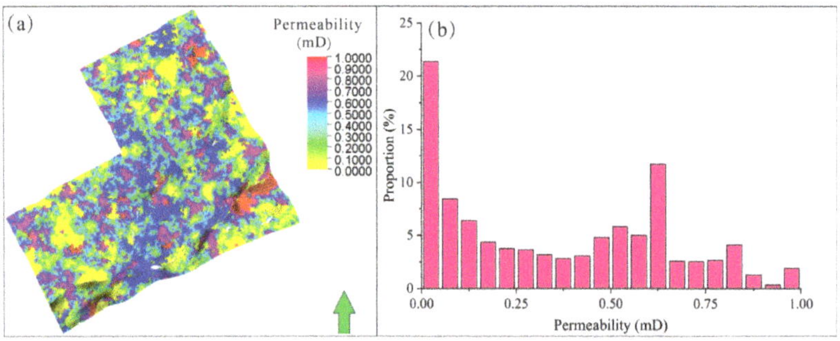

Figure 11. Permeability of No. 3 coal seam in the study area. (**a**) Reservoir permeability model. (**b**) Permeability distribution range histogram.

4.4.3. Modeling of Gas Content

The development of a coal seam gas content model is fundamental for optimizing CBM recovery strategies. A 3D gas content distribution model of the No. 3 coal seam was established by integrating field-measured gas content data from boreholes within and surrounding the study area. As shown in Figure 12, the gas content of the No. 3 coal seam exhibits a range of 14.03–33.54 m^3/t (average 21.7 m^3/t), delineating three distinct domains: high-gas-concentration zones (>28.0 m^3/t) dominate the western and south-eastern structural compartments, intermediate-gas zones (20–28 m^3/t) display spatially sporadic distributions, and low-gas regions (<20 m^3/t) form continuous belts in central and central-northern areas.

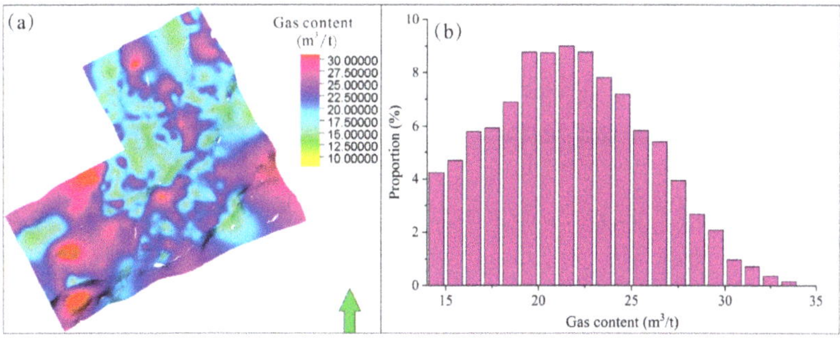

Figure 12. Gas content of the No. 3 coal seam in the study area. (**a**) Reservoir gas content model. (**b**) Gas content distribution range histogram.

4.5. Production Numerical Simulation of Typical Well Group

Due to the large study area, conducting full-field numerical simulations of production processes would involve excessive computational load and coarse accuracy. Numerical simulations of production dynamics were conducted using CMG (v. 2021) software based on the Petrel geological model, focusing on a typical CBM well group in the northern Dongda 3D area (Figure 13). The study area spans 4650 m × 4675 m and includes 27 production wells. These wells were gradually put into operation between 2010 and 2012, accumulating 4–5 years of production history. The peak gas production of individual wells in this area ranged from 500 to 1500 m^3/d, with some wells achieving stable production. The overall drainage–depressurization–desorption–production characteristics of CBM development were evident.

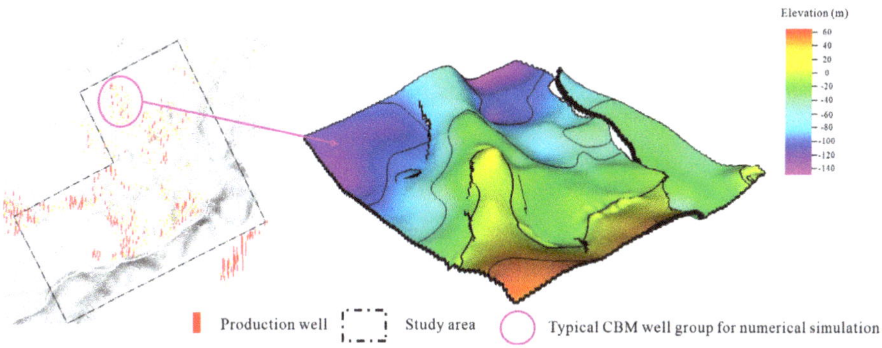

Figure 13. Typical CBM well group for numerical simulation.

4.5.1. Numerical Model Setup

The geological model generated by Petrel was imported into CMG's IMEX simulator, where reservoir and well parameters were configured. Based on actual reservoir data and production dynamics research objectives, grid sizes in the I, J, and K directions were adjusted. The horizontal grid spacing was set to Dx = Dy = 25 m, while the vertical direction was subdivided into 20 grids. Reservoir properties for the area, derived from the geological model, are summarized in Table 2. A dual-porosity dual-permeability (DUALPERM) model was used for initialization, assuming fluid flow in both matrix and fractures, with interporosity transfer driven by diffusion or Darcy flow depending on pressure gradients. The model also captures dynamic gas-water interactions, including CH_4 adsorption/desorption in the coal matrix based on Langmuir isotherms. For history matching, a fixed bottom-hole flowing pressure method was used to match gas and water production across all 27 wells.

Table 2. Reservoir properties for the study area.

Input Parameters	Value
Initial reservoir pressure (Mpa)	4.76–10.89
Permeability (mD)	0.012–1
Porosity (%)	0.0129–0.10
Gas content (m^3/t)	14.1–31.8
Water saturation (%)	100

4.5.2. History Matching

As shown in Figure 14, the simulated cumulative gas and water production of CBM wells in the block were compared with actual field data. The results indicate relatively large errors in early-stage production due to operational instability and frequent adjustments in production strategies, which complicated the matching process. However, the matching accuracy improved as production stabilized. For cumulative gas production after full-scale development, the maximum error was 15.2% (i.e., a matching accuracy > 84.8%). For cumulative water production, the maximum error was 11.6% (i.e., a matching accuracy > 88.4%). Moreover, history matching was performed for gas and water production across all 27 wells. For each single well, the matching accuracy was high during the stable gas decline phase but poorer during early production. This discrepancy arose from rapid changes in reservoir properties and operational adjustments (e.g., drainage rates and gas ramp-up), which affected the simulation of initial gas and water.

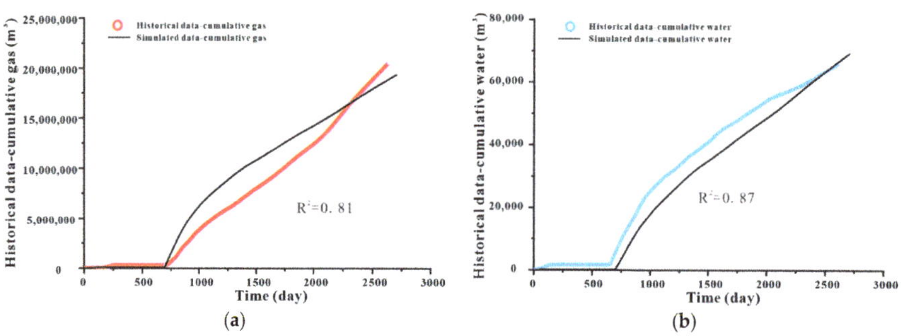

Figure 14. Dynamic History Fitting of Coalbed Methane Well Production. (**a**) Cumulative gas production. (**b**) Cumulative gas production.

4.5.3. Residual Gas Content of Reservoir

Figure 15 illustrates the dynamic changes in reservoir pressure over time within the simulated area. Reservoir pressure declined rapidly during the early years of development but stabilized in later stages. This trend is attributed to two factors: (a) early-stage production required significant depressurization and drainage to initiate gas desorption, but limited desorbed gas provided minimal energy support to the reservoir; (b) In later stages, higher gas desorption rates under lower pressures enhanced energy replenishment, slowing pressure decline. By the simulated cutoff date, the reservoir pressure at the central part of the well group had decreased to approximately 3 MPa, demonstrating effective pressure control by the deployed well pattern. The remaining gas saturation in the study area was obtained (Figure 16), showing relatively minor variations in gas saturation across the overall region. Concurrently, the distribution of residual gas in the study area can be derived from the gas saturation distribution pattern.

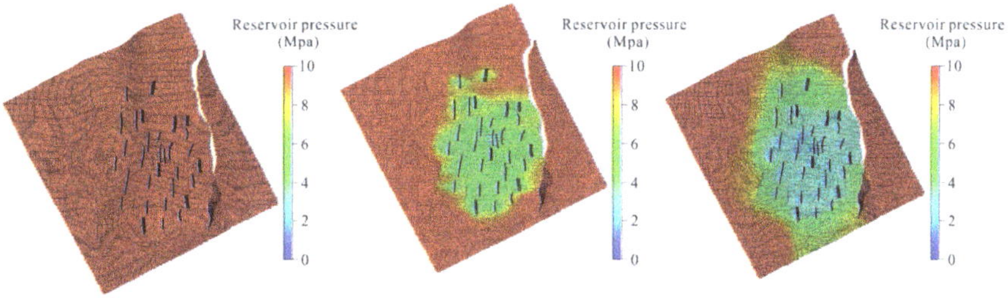

Figure 15. Dynamic variations in reservoir pressure over time.

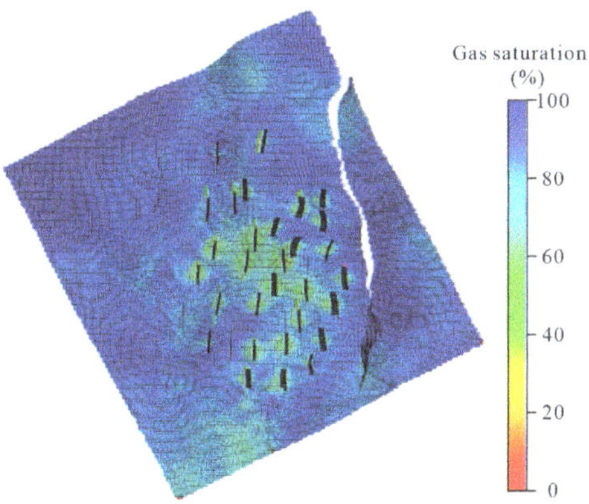

Figure 16. Remaining gas saturation across the simulated area.

4.6. Potential Evaluation of CBM Resources

4.6.1. Residual Resources Assessment

The spatial distribution of residual CBM resources is governed by drainage interference effects from individual wells. The technically recoverable resources available for future development adjustments are determined by subtracting well interference-affected volumes from the total in situ resources within the well pattern. The well-controlled gas volume can be calculated using drainage area and resource concentration:

$$Q_G = d^2 Q \tag{17}$$

where Q_G is gas-in-place per well, m^3; and d is drainage radius, m.

Using actual production data from the study area, the well-controlled recovery factor (η) is derived from cumulative gas production (Q_g):

$$\eta = \frac{Q_g}{Q_G} \tag{18}$$

The equivalent residual gas content (V) is calculated by integrating production performance, resource concentration, and drainage control:

$$V = \frac{Q_G - Q_g}{0.01 \times \mathrm{h} \times \rho} \tag{19}$$

A residual resource distribution map for the No. 3 coal seam (Figure 17) was generated through Kriging interpolation of the residual gas content. As can be seen, the central and western margin zones exhibit lower residual resources, while the western and eastern sectors show higher residual gas concentrations. For large study areas, residual gas content is determined using the aforementioned calculation method, while for small regions such as a typical well-group block, CMG numerical simulation results can be applied to refine residual gas distribution.

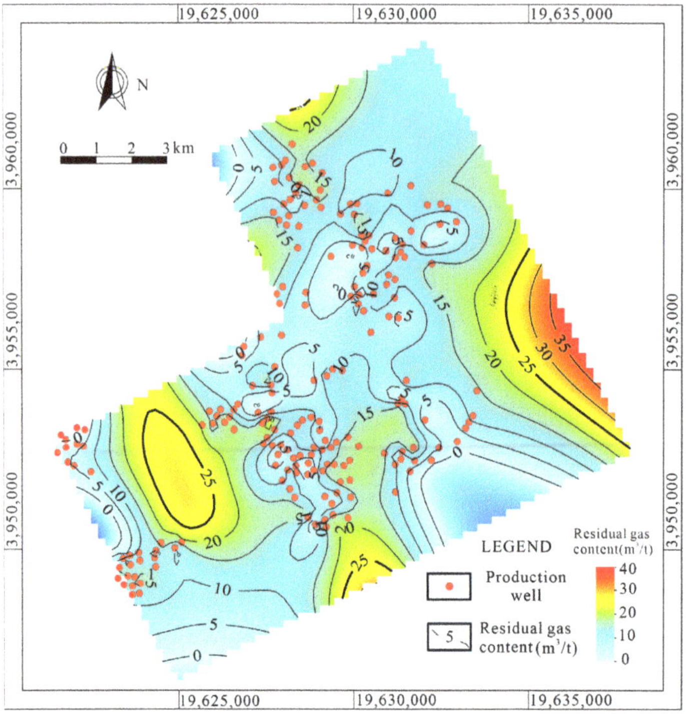

Figure 17. Residual resource distribution of No. 3 coal seam in the study area.

4.6.2. Evaluation of CBM Resource Potential

Based on reservoir modeling results, the CBM resource potential was evaluated through the analytic hierarchy process method, and the details and application example of this method can be found in a previous study [30]. The evaluated workflow of CBM resource potential was displayed in Figure 18. First, three critical assessment projects were established with weighted importance: reservoir properties (0.3), stimulation potential (0.3), and residual gas content (0.4), with weights validated through the CBM exploration and development pilot project of Zhengzhuang Field.

$$U = 0.3U_1 + 0.3U_2 + 0.4U_3 \tag{20}$$

where U is evaluated results of CBM resource potential; and U_1, U_2, and U_3 are evaluated scores of reservoir properties, stimulation potential, and residual gas content, respectively.

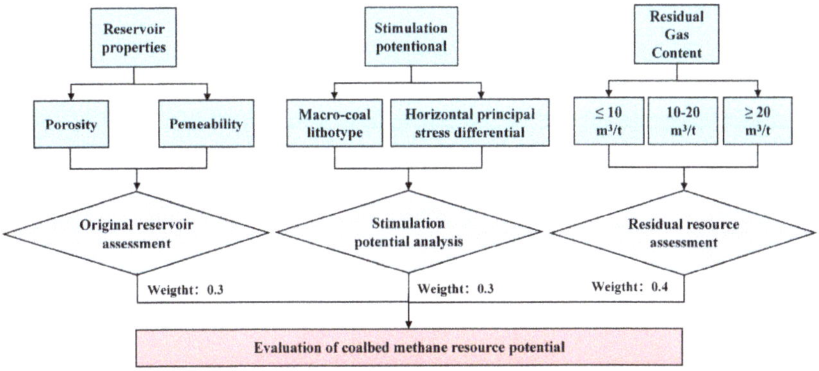

Figure 18. Workflow and weights of key parameters for evaluating CBM resource potential.

And then, threshold values statistically derived from data analysis and expert calibration (Table 3) categorized reservoir parameters into Better, Good, and Bad classes. A tiered scoring system was applied, assigning 1.0, 0.8, and 0.6 points to Better, Good, and Bad categories, respectively.

$$U_1 = u_1 + u_2 \tag{21}$$

$$U_2 = v_1 + v_2 \tag{22}$$

$$U_3 = w_1 \tag{23}$$

where u_1 and u_2 are evaluated scores for porosity, permeability, and gas content, respectively; v_1 and v_2 are evaluated scores for macro-coal lithotype and horizontal principal stress differential, respectively; w_1 is the evaluated score for residual gas content.

Table 3. Weights of key reservoir parameters for evaluating CBM resource potential.

Assessment Projects	Reservoir Parameters	Better	Good	Bad
Reservoir properties	Porosity	≥8%	4–8%	≤4%
	Permeability	≥0.8 mD	0.5–0.8 mD	≤0.5 mD
Stimulation potential	Macro-coal lithotype	Bright coal	Semi-bright coal	Semi-dull coal and dull coal
	Horizontal principal stress differential	≥10 MPa	5–10 MPa	≤5 MPa
Residual resource assessment	Residual gas content	≥20 m^3/t	10–20 m^3/t	≤10 m^3/t

Integrated reservoir assessment delineated three CBM prospectively tiers within the study area (Figure 19). The type I reservoirs, primarily concentrated in the southwestern part of the study, showcase high residual resources (≥20 m^3/t) coupled with optimal reservoir conditions, including relative high porosity (8–12%) and permeability (≥0.8 mD). Type II transitional reservoirs prospects, across the northern and southern transitional part of the study area, exhibit moderate porosity (5–8%), permeability (0.5–0.8 mD), and gas content (10–20 m^3/t). Type III marginal reservoirs, constrained by low gas content (<10 m^3/t), demonstrate limited commercial viability without substantial technological intervention.

Integrated reservoir assessment delineated three CBM prospectively tiers within the study area (Figure 19). The type I reservoirs, primarily concentrated in the southwestern part of the study, showcase high residual resources (≥20 m^3/t) coupled with optimal reservoir conditions, including relatively high porosity (8–12%) and permeability (≥0.8 mD). Type II transitional reservoir prospects, across the northern and southern transitional part of the study area, exhibit moderate porosity (5–8%), permeability (0.5–0.8 mD), and gas content (10–20 m^3/t). Type III marginal reservoirs, constrained by low residual gas content (<20 m^3/t), low permeability (<0.4 mD), low porosity (<4%), demonstrate limited commercial viability without substantial technological intervention.

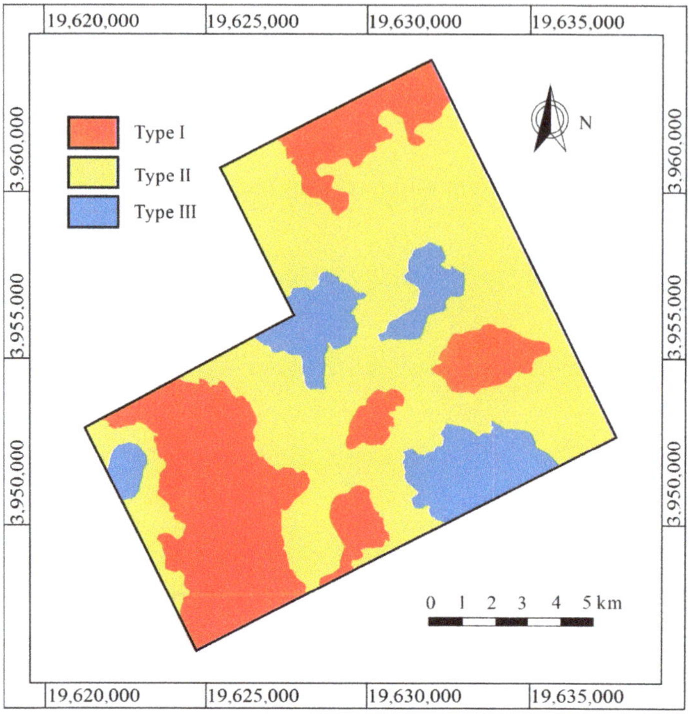

Figure 19. Evaluation results of CBM resource potential.

5. Conclusions

This study establishes a comprehensive workflow for CBM reservoir evaluation and development optimization in the Zhengzhuang Block. Key findings include:

(1) The 3D geological model effectively characterized spatial heterogeneity of the No. 3 coal seam, identifying semi-bright coal as dominant, with porosity (1.3–12%) and permeability (0.01–1 mD) variations controlled by fracture networks.

(2) Stress field modeling revealed significant stress anisotropy (σ_H/σ_h gradients: 0.0344/0.0215 MPa/m), delineating stimulation-favorable zones in southwestern structural compartments.

(3) Numerical simulations revealed that 84.8% of cumulative gas production and 88.4% of water production were matched historically, with residual gas content (14–33.5 m^3/t) concentrated in structurally elevated western and southeastern regions.

(4) A multi-criteria evaluation framework categorized resources into three tiers: Type I, Type II, and Type III, among which Type I areas are prioritized for future development. This integrated approach establishes a technical basis for optimizing well patterns, improving hydraulic fracturing efficiency, and enhancing residual gas recovery in heterogeneous CBM reservoirs.

Author Contributions: Methodology, X.L., Q.Z. and Z.W.; Software, C.Z.; Validation, Z.L., X.L., Y.Y., T.Z. and Z.W.; Formal analysis, Z.L.; Investigation, Q.Z., T.Z. and C.Z.; Resources, Z.L.; Writing—original draft, H.W.; Funding acquisition, H.W. All authors have read and agreed to the published version of the manuscript.

Funding: This research was funded by the National Key R&D Program of China (2024YFC2909400), the National Natural Science Foundation of China [42402180, 42202195], and the Hebei Natural Science Foundation [D2024209008].

Data Availability Statement: The original contributions presented in this study are included in the article. Further inquiries can be directed to the corresponding author.

Conflicts of Interest: Authors Zhong Liu, Xiuqin Lu, Qianqian Zhang, Yanhui Yang, Tao Zhang, Chen Zhang and Zihan Wang were employed by the Petro China Huabei Oilfield Company. The remaining authors declare that the research was conducted in the absence of any commercial or financial relationships that could be construed as a potential conflict of interest.

References

1. Pashin, J.C. Geology of North American coalbed methane reservoirs. In *Coal Bed Methane (Second Edition): Theory and Applications*; Elsevier Inc.: Amsterdam, The Netherlands, 2020. [CrossRef]
2. Cai, Y.; Liu, D.; Yao, Y.; Li, J.; Qiu, Y. Geological Controls on Prediction of Coalbed Methane of No.3 Coal Seam in Southern Qinshui Basin, North China. *Int. J. Coal Geol.* **2011**, *88*, 101–112. [CrossRef]
3. Ma, X.; Zhou, A.; Cheng, X.; Cheng, C.; Zhao, W. Multi-Field Coupling Models of Coal and Gas and Their Engineering Applications to CBM in Deep Seams: A Review. *Energies* **2024**, *17*, 6221. [CrossRef]
4. Wei, C.; Qin, Y.; Wang, G.; Jiang, B.; Zhang, Z. Simulation Study on Rvolution of Coalbed Methane Reservoir in Qinshui Basin, China. *Int. J. Coal Geol.* **2007**, *72*, 53–69. [CrossRef]
5. Li, C.; Sun, L.; Zhao, Z.; Zhang, J.; Li, Y.; Meng, Y.; Wang, L. A Method for Evaluating Coalbed Methane Reservoir Productivity Considering Drilling Fluid Damage. *Energies* **2024**, *17*, 1686. [CrossRef]
6. Leng, K.; Guan, B.; Liu, W.; Jiang, C.; Cong, S.; Xin, Y. Research Progress of Coalbed Methane Extraction. *Energy Rep.* **2024**, *12*, 5728–5746. [CrossRef]
7. Mondalt, D.; Sang, S.; Han, S.; Zhou, X.; Zhao, F.; Zhang, J.; Gao, W. Coalbed Methane Reservoir Properties Assessment and 3D Static Modeling for Sweet-spot Prediction in Dahebian Block, Liupanshui Coal field, Guizhou Province, Southwestern China. *Heliyon* **2024**, *10*, e35481. [CrossRef]
8. Cao, L.; Yao, Y.; Liu, D.; Yang, Y.; Wang, Y.; Cai, Y. Application of Seismic Curvature Attributes in the Delineation of Coal Texture and Deformation in Zhengzhuang Field, Southern Qinshui Basin. *AAPG Bull.* **2020**, *104*, 1143–1166. [CrossRef]
9. Chen, S.; Chen, P.; Tang, D.; Tao, S.; Zhang, T. Identification of Thin-layer Coal Texture Using Geophysical Logging Data: Investigation by Wavelet Transform and Linear Discrimination Analysis. *Int. J. Coal Geol.* **2021**, *239*, 103727. [CrossRef]
10. Liu, D.; Yao, Y.; Wang, H. Structural Compartmentalization and Its Relationships with Gas Accumulation and Gas Production in the Zhengzhuang Field, Southern Qinshui Basin. *Int. J. Coal Geol.* **2022**, *259*, 104055. [CrossRef]
11. Di, C.; Wei, Z.; Wang, K.; Deng, P.; Chen, B.; Shen, L.; Chen, Z. The impact of pressurization-induced decrease of capillary pressure and residual saturation on geological carbon dioxide storage. *J. Clean. Prod.* **2025**, *486*, 144573. [CrossRef]
12. Chen, Y.; Tang, D.; Xu, H.; Li, Y.; Meng, Y. Structural Controls on Coalbed Methane Accumulation and High Production Models in the Eastern Margin of Ordos Basin, China. *J. Nat. Gas Sci. Eng.* **2015**, *23*, 524–537. [CrossRef]
13. Wang, B.; Zhang, Q.; Qu, Z.; Zhang, Y. South Anze Structure and Its Control on Coalbed Methane Aggregation in the Qinshui Basin and the Mechanism of Syncline Gas Enrichment in the Qinshui Basin. *Energies* **2023**, *16*, 4521. [CrossRef]
14. Huang, S.; Liu, D.; Gomez-Rivas, E.; Griera, A.; Gan, Q.; Wang, M.; Xing, Y.; Zhao, Y. Experimental Insights into the Nucleation and Propagation of Hydraulic Fractures in Anthracite Coalbed Methane Reservoirs. *Earth Energy Sci.* 2025; *in press*. [CrossRef]
15. Zhou, F.D.; Yao, G.Q.; Tyson, S. Impact of Geological Modeling Processes on Spatial Coalbed Methane Resource Estimation. *Int. J. Coal Geol.* **2015**, *146*, 14–27. [CrossRef]
16. Zhang, B.; Deng, Z.; Fu, X.; Yin, K. A Study on Three-Phase Gas Content in Coal Reservoirs and Coalbed Methane-Water Differential Distribution in the Western Fukang Mining Area, Xinjiang, China. *ACS Omega* **2021**, *6*, 3999–4012. [CrossRef]
17. Yue, P.; Qu, S.; Zhang, Y.; Tu, J.; Ning, W.; Duan, J. Numerical Simulation Study of the Upper Aquifer Impact on the Coalbed Methane Production. *Front. Energy Res.* **2022**, *9*, 789021. [CrossRef]
18. Quan, F.; Wei, C.; Feng, S.; Hu, Y. Physical Experiment and Numerical Simulation of the Depressurization Rate for Coalbed Methane Production. *ACS Omega* **2020**, *5*, 22661–22673. [CrossRef] [PubMed]

19. Rahman, M.K.; Suarez, Y.; Chen, Z.; Rahman, S.S. Unsuccessful Hydraulic Fracturing Cases in Australia: Investigation into Causes of Failures and Their Remedies. *J. Pet. Sci. Eng.* **2007**, *55*, 213–227. [CrossRef]
20. Zhang, Y.P.; Yang, Y.H.; Shao, G.L.; Chen, L.W.; Wei, N.; Zhang, L.W. Problems and Countermeasures in the Exploitation of High Coal Rank CBM Horizontal Wells in Fanzhuang–Zhengzhuang Area of Qinshui Basin. *Nat. Gas Ind.* **2017**, *37*, 46–54. [CrossRef]
21. Su, X.; Lin, X.; Liu, S.; Zhao, M.; Song, Y. Geology of Coalbed Methane Reservoirs in the Southeast Qinshui Basin of China. *Int. J. Coal Geol.* **2005**, *62*, 197–210. [CrossRef]
22. Yang, X.; Tang, S.; Xi, Z.; Zhang, S.; Zhang, Q.; Zhang, K. Dynamic Evolution and Differential Enrichment of Deep Coalbed Methane: A Case Study in Qinshui Basin. *Int. J. Coal Geol.* **2025**, *299*, 104696. [CrossRef]
23. Wang, H.; Yao, Y.; Huang, C.; Liu, D.; Cai, Y. Fault Development Characteristics and Their Effects on Current Gas Content and Productivity of No. 3 Coal Seam in the Zhengzhuang Field, Southern Qinshui Basin, North China. *Energy Fuel* **2021**, *35*, 2268–2281. [CrossRef]
24. Qin, Y.; Jiang, B.; Wang, J.; Wu, C.; Ju, Y. Coupling Control of Tectonic Dynamical Conditions to Coalbed Methane Reservoir Formation in the Qinshui Basin, Shanxi, China. *Acta Geolog. Sin.* **2008**, *82*, 1355–1362.
25. Shao, L.; Wang, X.; Wang, D.; Li, M.; Wang, S.; Li, Y.; Shao, K.; Zhang, C.; Gao, C.; Dong, D.; et al. Sequence Stratigraphy, Paleogeography, and Coal Accumulation Regularity of Major Coal-Accumulating Periods in China. *Int. J. Coal Sci. Technol.* **2020**, *7*, 240–262. [CrossRef]
26. Wang, H.; Yao, Y.; Liu, D.; Pan, Z.J.; Yang, Y.; Cai, Y. Fault-sealing Capability and Its Impact on Coalbed Methane Distribution in the Zhengzhuang Field, Southern Qinshui Basin, North China. *J. Nat. Gas Sci. Eng.* **2016**, *28*, 613–625. [CrossRef]
27. Allawi, R.H. Geomechanical Model of Ratawi Shale Reservoir. *Int. J. Coal Sci. Technol.* **2024**, *35*, 1–16. [CrossRef]
28. Huang, S.; Liu, D.; Yao, Y.; Gan, Q.; Cai, Y.; Xu, L. Natural Fractures Initiation and Fracture Type Prediction in Coal Reservoir under Different In-situ Stresses during Hydraulic Fracturing. *J. Nat. Gas Sci. Eng.* **2017**, *43*, 69–80. [CrossRef]
29. Sun, Y.; Zhang, J.; Yu, Z.; Liu, Z.; Yin, P. WOA (Whale Optimization Algorithm) Optimizes Elman Neural Network Model to Predict Porosity Value in Well Logging Curve. *Energies* **2022**, *15*, 4456. [CrossRef]
30. Wang, Y.; Liu, D.; Cai, Y.; Yao, Y.; Pan, Z. Constraining coalbed methane reservoir petrophysical and mechanical properties through a new coal structure index in the southern Qinshui Basin, northern China: Implications for hydraulic fracturing. *AAPG Bull.* **2020**, *104*, 1817–1842. [CrossRef]

energies

MDPI

Review

Research on the Application of Biochar in Carbon Sequestration: A Bibliometric Analysis

Shizhao Zhang [1,2], Shuzhi Wang [1,2,*], Jiayong Zhang [1], Bao Wang [3], Hui Wang [1,2], Liwei Liu [1], Chong Cao [1,2], Muyang Shi [4] and Yuhan Liu [1]

1 School of Emergency Management and Safety Engineering (SEMS), North China University of Science and Technology (NCUST), Tangshan 063210, China; zsz17330403282@163.com (S.Z.); zyj815@163.com (J.Z.); huiwang90@ncst.edu.cn (H.W.); llw2190477378@163.com (L.L.); caochong@ncst.edu.cn (C.C.); 15530573853@163.com (Y.L.)

2 Tangshan Key Laboratory of Geological Resource Development and Disaster Prevention, School of Emergency Management and Safety Engineering, North China University of Science and Technology, Tangshan 063210, China

3 PetroChina Jidong Oilfield Company, Tangshan 063000, China; wangbao@petrochina.com.cn

4 School of Civil and Architectural Engineering (SCAE), North China University of Science and Technology (NCUST), Tangshan 063210, China; smy18633046423@163.com

* Correspondence: wangszh@ncst.edu.cn; Tel.: +86-180-0129-6100

Abstract: Driven by global carbon neutrality goals, biochar has gained significant attention due to its stable carbon sequestration capabilities and environmental benefits. This research employs bibliometric tools such as VOSviewer 1.6.16, Citespace 6.2 R6, and Scimago Graphica to systematically analyze 2076 publications from the Web of Science Core Collection between 2007 and 2024, aiming to clarify the evolutionary trajectory, research hotspots, and international collaboration patterns of biochar carbon sequestration research while identifying future knowledge gaps for innovation. Research results reveal a three-stage developmental characteristic: 2007–2014 was a slow accumulation period for fundamental mechanism exploration, 2015–2020 was an accelerated expansion period driven by policies like the Paris Agreement, and 2021 to the present marks an exponential growth phase of interdisciplinary integration due to global carbon market consolidation. China and the United States are core producing countries, though inter-institutional deep collaboration remains insufficient. Research hotspots have progressively shifted from early biochar preparation and carbon stability to multiple waste materials (such as rice straw and urban carbon sequestration waste) and co-pyrolysis technologies (significantly emerging since 2022), with machine learning applications in process optimization becoming a nascent direction. The study recommends increasing cross-disciplinary research funding, establishing biochar raw material pollution standards, and promoting coordinated policies that combine biochar carbon sequestration with agricultural efficiency to support global carbon reduction objectives. Notably, the research's reliance on the Web of Science Core Collection may limit coverage of non-English literature and regional studies. By quantitatively analyzing technological evolution and collaboration networks, this study provides a data-driven framework for optimizing biochar carbon sequestration strategies, helping bridge the gap between laboratory potential and actual climate emission reduction, and offering focused action pathways for policymakers and researchers.

Keywords: biochar; carbon sequestration; bibliometrics; multi-waste conversion; co-pyrolysis; soil carbon sequestration

1. Introduction

In recent years, under the combined influence of human activities and natural factors, global climate change has been intensifying, with frequent extreme climate events posing a severe challenge to the sustainable development of ecosystems and human society, prompting countries worldwide to actively explore countermeasures [1–3]. The 2015 Paris Agreement proposed controlling the global average temperature rise to within 2 °C compared to pre-industrial levels, and striving to limit it to 1.5 °C [4], which requires a significant reduction in greenhouse gas emissions [5,6].

Carbon sequestration technology, as a key negative emission strategy to mitigate greenhouse gas emissions, shows significant differences among various technological paths in terms of the sequestration cycle, cost-effectiveness, and environmental synergy. Geological sequestration, while having a large-capacity advantage, faces challenges in leak monitoring [7,8]. Mineral carbonation can permanently fix Carbon dioxide (CO_2) without leakage risks, but its slow reaction speed and high energy consumption limit widespread application [9,10]. Ocean sequestration has enormous capacity but may cause ecological risks [11,12]. In comparison, biochar is reshaping the carbon sequestration technology landscape with its unique closed-loop carbon cycle characteristics. Biochar is formed through pyrolysis of biomass under oxygen-limited conditions, creating a stable aromatic carbon structure [13], with high carbon content and thermal stability [14]. It can store CO_2 for hundreds of years [15] and achieve high cost-effectiveness with relatively low energy requirements [16]. Research indicates that converting 373 million tons of annual agricultural waste into biochar could capture approximately 550 million tons of CO_2 [17], highlighting its undeniable potential in carbon cycle regulation. Moreover, biochar can improve soil structure, enhance nutrient retention capabilities, and regulate microbial activity, facilitating a win–win scenario for carbon sequestration and agricultural sustainable development [18,19].

Over the past two decades, biochar has garnered widespread attention in the field of environmental remediation as an eco-friendly and multifunctional material [20–23]. However, existing research has primarily focused on its role in improving soil fertility and enhancing soil carbon sequestration, with a notable lack of systematic and comprehensive analysis of biochar in carbon storage [15]. Biochar is often merely described as a supplement to soil improvement characteristics, with insufficient in-depth exploration of its potential as a critical carbon fixation method. As global demand for reducing carbon dioxide emissions becomes increasingly urgent, and the potential of biochar in reducing greenhouse gas emissions gains recognition, a review of systematic applications, international research status, hotspots, and development trends in biochar carbon fixation has become particularly necessary. Against this background, this study employs bibliometric analysis methods to provide a unique perspective and comprehensive understanding of the biochar carbon fixation field, charting a course for future research.

As a crucial tool for quantitatively analyzing the scientific literature structure and evolutionary patterns [24], bibliometrics can precisely identify research hotspot evolution, core knowledge producers, and interdisciplinary interaction modes through high-frequency keyword co-occurrence, collaborative network analysis, and temporal mapping [25]. Compared to methods like meta-analysis focusing on parameter quantitative integration, bibliometrics' core advantage lies in its macro-scale analysis capability to systematically reveal knowledge network characteristics of technological development trajectories (such as knowledge flow direction and interdisciplinary nodes) and social network attributes (like institutional collaboration patterns and national competitiveness landscapes), rather than being limited to statistical parameter inference.

This panoramic analysis method is particularly suitable for revealing technological evolution paths and innovation ecosystem requirements in the biochar carbon sequestration field. This study integrates global literature from the Web of Science core database, utilizing VOSviewer, Citespace, and Scimago Graphica to construct co-occurrence networks and timeline maps for institutions, countries, authors, and keywords. The method of selection is primarily based on two considerations: first, while quantitative analyses of parameters like biochar application range and pH effects have been established through existing meta-research, unstructured issues such as technological adoption barriers and research hotspot migration still require bibliometric revelation; second, technical innovation for agricultural, waste-source, and nano-structured biochar necessitates establishing a technological evolution analysis framework from three dimensions: knowledge production subjects, knowledge transmission carriers, and knowledge dissemination networks.

The research results will provide a theoretical basis for optimizing biochar carbon sequestration strategies; offer decision-making support for policymakers in balancing technological feasibility, economic costs, and ecological risks; and identify key interdisciplinary knowledge integration nodes to guide technological innovation.

2. Materials and Methods

2.1. Data Sources and Retrieval Strategy

This research used the Web of Science Core Collection (WoSCC) as the data source, as it indexes over 9000 high-impact journals globally [26], covering core fields such as environmental science and soil science, and is one of the most widely used databases in bibliometric research. Compared to other databases like Scopus and PubMed, it is widely considered the most comprehensive and reliable bibliometric analysis database [27,28]. Although databases like Scopus can provide Supplementary Information, Web of Science has advantages in cross-disciplinary literature coverage and data standardization, making it suitable for constructing a systematic bibliometric analysis framework. The search strategy was as follows: TS = (("biochar" OR "bio-char") AND ("carbon sequestration" OR "carbon storage" OR "carbon sink" OR "carbon removal" OR "CO_2 sequestration" OR "CO_2 storage")) NOT TS = ("geological sequestration" OR "enhanced oil recovery"). The search was conducted on 22 April 2025.

2.2. Screening Criteria

Based on a systematic retrieval strategy, the earliest retrievable literature dates back to 2007. Up to now, 2245 related papers have been obtained. Since 2025 data have not yet been fully included, we excluded 2025 data. We limited the results to "Article" and "Review Article" published between 2007 and 2024. Because the number of non-English articles in the literature is very small and lacks representativeness, the language was set to English. This selection process ultimately formed a core dataset containing 2076 literature references, laying the foundation for final analysis. The search and analysis process is shown in Figure 1.

2.3. Data Analysis

This research adopts a multi-tool collaborative analysis method, achieving data visualization through three bibliometric tools. VOSviewer 1.6.16 was selected for co-occurrence network visualization, leveraging its proven stability in processing large-scale bibliometric datasets and interpretability advantages in traditional keyword clustering [29]. Although advanced artificial intelligence support tools excel in processing unstructured data and predicting research trends, VOSviewer remains irreplaceable in transparent, rule-based co-occurrence analysis, which is crucial for reproducible bibliometric research. Complementing VOSviewer is CiteSpace, a Java application developed by Professor Chen Chaomei's team, which demonstrates outstanding value in revealing discipline evolution paths and research hotspot transitions through its longitudinal analysis advantages [30]. Therefore, the article specifically employs CiteSpace 6.2 R6's journal overlay mapping, keyword emergence intensity detection, and cited literature emergence intensity detection functions to systematically reveal the knowledge diffusion trajectory of the research field. To optimize international collaboration networks, this research introduced Scimago Graphica for spatial processing, which exhibits significant professional advantages in visualization layout and interactive analysis of complex networks. Through the collaborative application of these three tools, co-occurrence networks of institutions, countries, authors, and keywords, as well as timeline maps, were constructed.

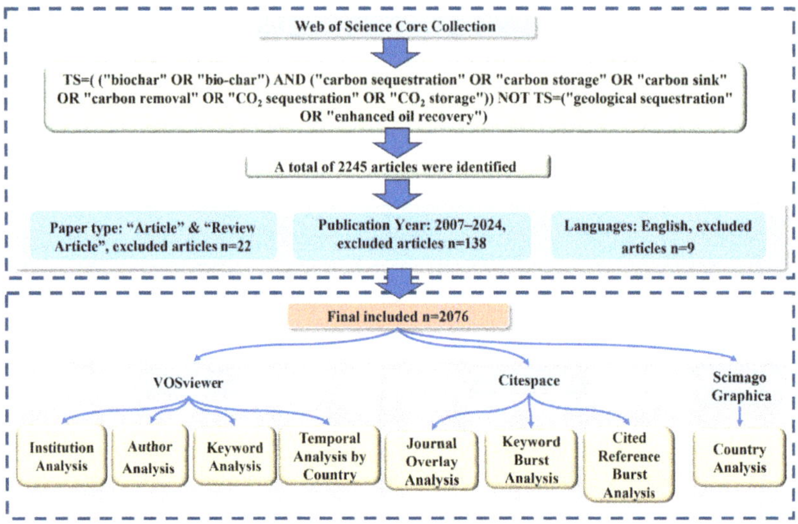

Figure 1. Screening Flow Chart.

3. Results

3.1. Annual Publication Volume Analysis

The development status, knowledge accumulation, and maturity of a research field can be measured by changes in its literature output [31]. This study systematically retrieved and counted the annual publication volume of the literature related to biochar carbon sequestration in the Web of Science Core Collection (WoSCC) from 2007 to 2024, selecting a total of 2075 papers. Figure 2A shows that the number of papers in the literature exhibits significant staged characteristics. During the initial accumulation period from 2007 to 2014, academic interest in this field showed a moderate growth trend, with an average annual publication of 34.89 papers. After entering 2015, the annual publication volume entered an acceleration period, with the average annual growth rate increasing by approximately 256% compared to the previous stage, while maintaining a steady upward trend. Since 2021, the data curve has demonstrated a notable exponential growth characteristic, with the annual publication volume breaking through 200 papers, and it is expected to continue growing. The citation frequency of the literature also shows a steady growth trend. To better display the increment trend, a fitting method was used to calculate the relationship between the number of publications and years ($Y = 1.275X^2 - 5122.35X + 5144823.975$, $R^2 = 0.9856$). According to this equation, the number of articles will continue to increase (Figure 2B).

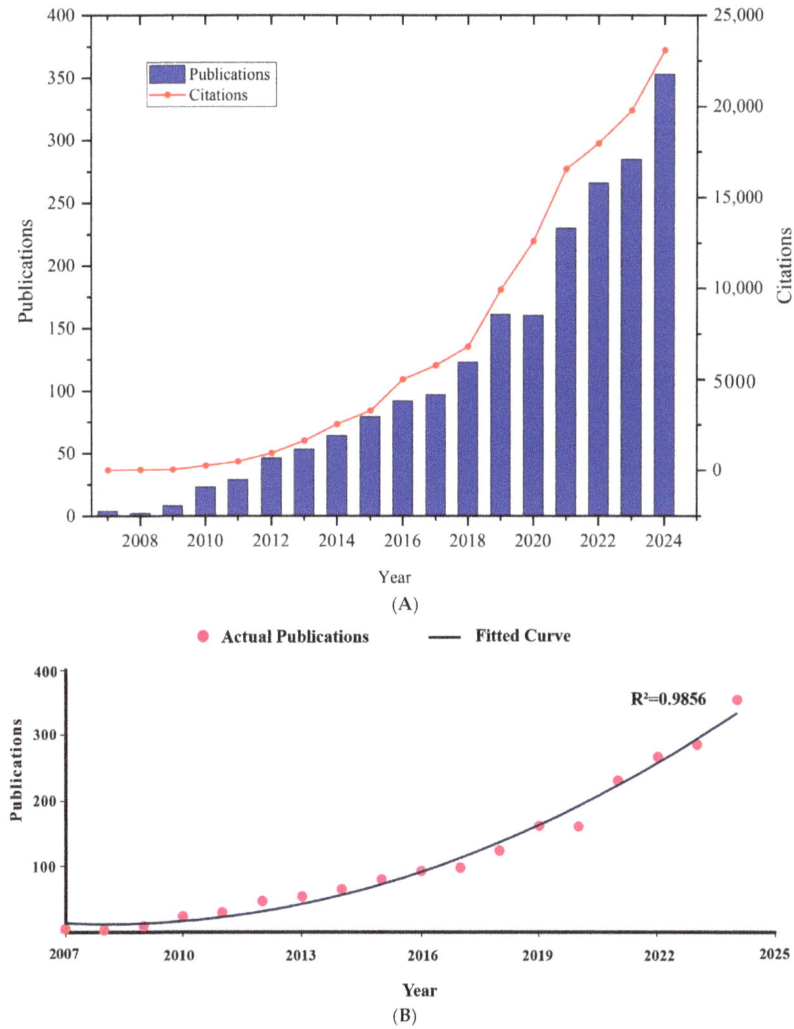

Figure 2. (**A**) Global publication number and citation count in the field of biochar carbon sequestration from 2007 to 2024. (**B**) The red scatter points in the figure represent the actual annual number of publications from 2007 to 2024, and the solid line is the prediction curve based on data fitting. The coefficient of determination for the fitted model (R^2 = 0.9856) indicates that the model can describe the growth trend of publications with high accuracy, reflecting the continuous and rapid development of the research field.

3.2. *Journal Analysis*

The 2075 literature references are distributed across 376 journals, with significant variations in publication volume. Based on Bradford's Law, a cluster analysis of the journal distribution in the biochar carbon sequestration technology field identified 12 core journals (Table 1). Science of the Total Environment, with 167 published articles, emerged as the most productive journal, far surpassing others. It primarily publishes research on comprehensive environmental studies, environmental pollution and remediation, and environmental technologies and applications, and is considered a top-tier environmental science journal. Among these journals, 58% have an impact factor (IF) $\geq$ 5, indicating a high quality level. The publishers are predominantly from Elsevier, accounting for 58.3% of the publications. From a disciplinary perspective, the included journals cover environmental science, soil science, energy and fuels, agronomy, environmental engineering, and green sustainable science and technology (Figure 3). Environmental science journals published the highest number of articles, serving as the primary platform for research findings and innovative ideas in this field.

Table 1. Top 10 Publication Sources.

Rank	Journal	Number of Documents	IF (2024)	Publishers	Core Research Directions
1st	Science of the Total Environment	164	8.2	Elsevier	Environmental Science; Pollution Control
2nd	Journal of Environmental Management	67	8.0	Academic	Environmental Management; Sustainability Policies
3rd	Journal of Cleaner Production	64	9.7	Elsevier	Cleaner Production Technologies; Circular Economy
4th	Agronomy-Basel	59	3.3	MDPI	Agronomy; Sustainable Agriculture
5th	Sustainability	58	3.3	MDPI	Comprehensive Sustainability Studies
6th	Global Change Biology Bioenergy	48	5.9	Wiley	Climate Change and Bioenergy
7th	Chemosphere	41	8.8	Elsevier	Environmental Chemistry; Pollutant Studies
8th	Bioresource Technology	40	9.7	Elsevier	Bioresource Technology; Bioenergy
9th	Biochar	39	13.1	Springer Nature	Biochar Applications; Soil Amendment
10th	Geoderma	39	5.6	Elsevier	Soil Science; Land Management
11	Bioresource Technology	39	9.7	Elsevier	Biological Resource Conversion and Environmental Biotechnology
12	Agriculture Ecosystems & Environment	38	3.9	Elsevier	Agricultural Ecosystem Sustainability

Dual mapping overlay is a visualization analysis technique aimed at analyzing, comparing, and contrasting publication portfolio characteristics [32]. This method enables cross-disciplinary attributes between research sources and targets and has been widely applied in bibliometric research [30]. The journal dual mapping overlay represents the thematic distribution of journals (Figure 4), with citing journals on the left, cited journals on the right, and colored paths indicating citation relationships [33,34]. The journal dual mapping overlay analysis reveals the interdisciplinary knowledge dissemination pattern of biochar in carbon sequestration research. Three main citation pathways were identified, reflecting interactions across different disciplines. The citing journals' domains are (Veterinary, Animal, Science), while the cited journals are (Plant, Ecology, Zoology), (Environmental, Toxicology, Nutrition), and (Chemistry, Materials, Physics), with cited journals converging towards citing journals. Beyond these mainstream citation pathways, other citation paths exist, albeit with weaker associations. This suggests that biochar research is developing in a multidisciplinary direction.

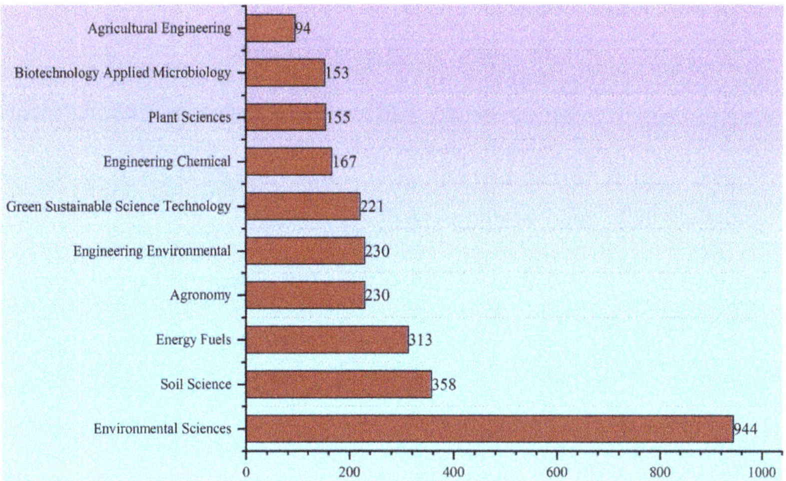

Figure 3. Bar Chart of Subject Distribution. The bar chart illustrates the frequency of subject classifications assigned to 2076 selected publications. Each publication may correspond to one or more subject categories; counts represent the total number of occurrences of each subject across all documents, reflecting the interdisciplinary nature of the dataset.

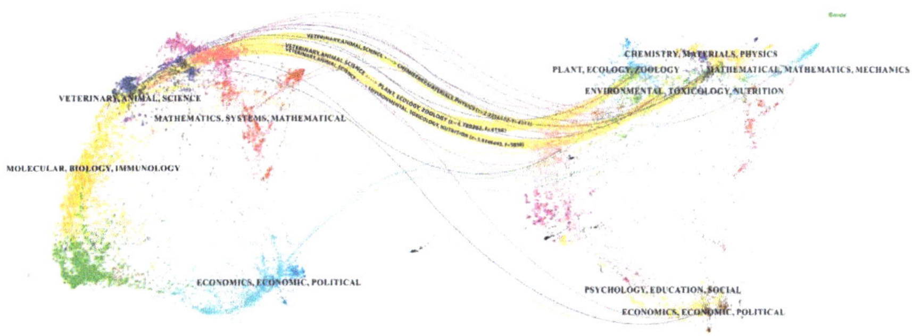

Figure 4. Journal Overlay Double Map. In this map, cited journals are displayed on the right side, while citing journals are located on the left side. Wider lines represent the primary citation pathways, and different colors of the lines denote connections from different regions.

3.3. *Institutional Analysis*

Network analysis of the collaboration among 2464 global research institutions reveals that in the 2076 literature entries analyzed, 15 institutions contributing over 15 papers accounted for 29.3% of the total literature volume (Table 2). Furthermore, 9 of the top 15 institutions are from China. Taking the Chinese Academy of Sciences (CAS) as an example, it comprehensively leads with 116 published papers, 5492 citations, and a Total Link Strength (TLS) of 104. Total Link Strength represents the sum of link intensities between a node (such as keywords, authors, countries) and all other nodes, and in institutional collaboration network analysis, it indicates the total number of joint publications between institutions, serving as a core metric for measuring network node association intensity.

Notably, while 3/5 of the top 15 institutions by publication volume are Chinese universities, Chinese institutions generally have lower citation rates. For instance, Cornell University from the United States, despite having only 25 papers, has 5853 citations, with the highest citation per paper rate of 234.12 among all institutions.

Table 2. Top 15 Publishing Institutions.

Rank	Institution	Countries/ Regions	Publication	Citation	Citation per Publication	TLS
1st	Chinese Academy of Sciences	China	116	5492	47.34	104
2nd	Zhejiang University	China	60	4872	81.20	45
3rd	Northwest A&F University	China	52	1713	32.94	27
4th	Nanjing Agricultural University	China	48	2525	52.60	34
5th	University of Chinese Academy of Sciences	China	44	2115	48.07	60
6th	University of Edinburgh	England	38	3248	85.47	30
7th	China Agricultural University	China	30	1235	41.17	29
8th	Zhejiang A&F University	China	30	1933	64.34	22
9th	King Saud University	Saudi Arabia	29	2392	82.48	15
10th	Shanghai Jiao Tong University	China	29	2546	87.79	21
11th	University of Florida	U.S.	28	4433	158.32	14
12th	University of Western Australia	Australia	28	2444	87.29	35
13th	Chinese Academy of Agricultural Sciences	China	27	768	28.44	24
14th	Aarhus University	Denmark	27	1249	46.25	19
15th	Cornell University	U.S.	25	5853	234.12	22

From the institutional overlay visualization map (Figure 5), among the 37 institutions with ≥15 papers, the Chinese Academy of Sciences has the largest node with dense connection lines, demonstrating high productivity and strong collaborative attributes. Institutions like Zhejiang University, Nanjing Agricultural University, and Northwest A&F University form a regional research cluster around the Chinese Academy of Sciences, with most of these high-intensity collaborations being domestic institutions. Chinese institutions have relatively less collaboration with high-productivity U.S. institutions such as University of Florida, University of Massachusetts, Iowa State University, and USDA ARS. In terms of spatio-temporal evolution, U.S. institutions generally emerged earlier in the research, while Chinese institutions typically emerged later.

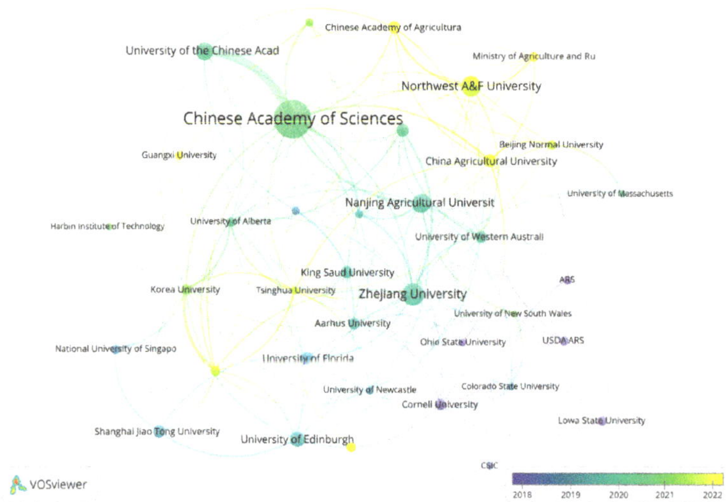

Figure 5. Institutional Timeline Overlay. Each node represents an institution, with larger nodes indicating more publications, and the node color ranging from blue–purple for institutions that entered the field earlier to yellow–green for those that joined later.

3.4. National Analysis

A total of 100 countries have published articles in the field of carbon sequestration, with 48 institutions screening out ≥10 articles. China is the country with the most publications, with a total of 811 papers, followed by the United States (395 papers), Australia (174 papers), Germany (156 papers), India (154 papers), Canada (100 papers), Italy (94 papers), United Kingdom (87 papers), Pakistan (84 papers), and Republic of Korea (82 papers), with developed countries accounting for 70% (Figure 6). In terms of H-index, the United States (102) ranks first, followed by China (97), Australia (64), Germany (54), Canada (46), India (45), United Kingdom (41), Italy (38), Republic of Korea (36), and Pakistan (35). In terms of average citations per paper, the top three countries are Republic of Korea (128.54 citations), Canada (119.3 citations), and Australia (109.67 citations), while the United States (105.31 citations), United Kingdom (95.46 citations), and Germany (82.37 citations) also demonstrate high research impact (Figure 6). However, China has the lowest number of citations among the top 10 countries in terms of publication volume (only 51.72).

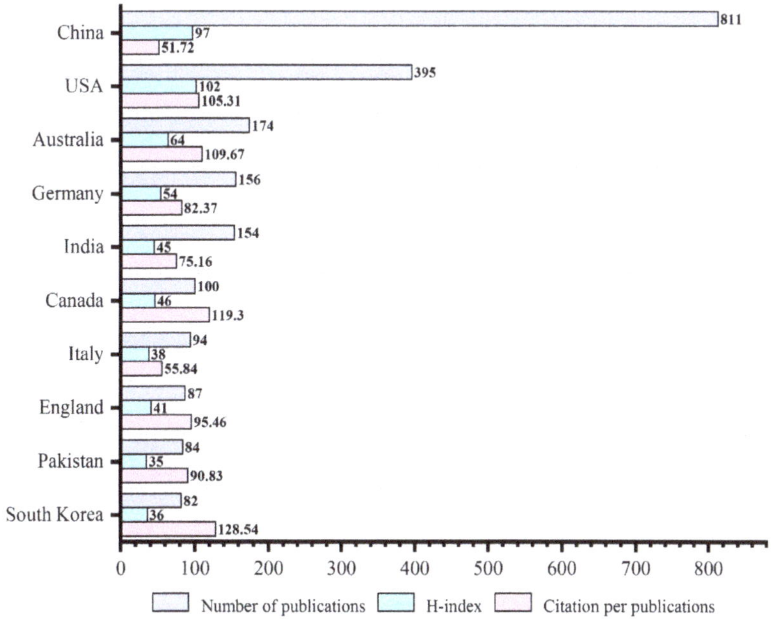

Figure 6. Total Publications, H-index, and Average Citations per Publication for Top 10 Countries from 2007 to 2024.

As shown in Figure 7, China had a low publication volume in the early stage, with a relatively slow article growth rate from 2010 to 2015. After 2015, it entered a growth phase, especially after 2020, showing exponential growth. The United States demonstrates a steady growth trend, reaching its peak in 2023, with a slight decrease in 2024. Australia, Germany, and India had relatively low publication volumes but experienced small explosive growth after 2021.

Figure 8 displays the overall landscape of global research collaboration, visualizing the intensity and nature of cooperation between countries. Countries are represented by circles, with size indicating the number of published papers. Connection line colors range from yellow to red, with redder colors indicating higher cooperation intensity. China and the United States are the two core players, with their interconnecting line being extremely red, demonstrating the highest cooperation intensity in the global research network. China's total cooperation intensity is 634, while the United States' total connection intensity is 439, reflecting a leading international research collaboration. China collaborates closely with countries like Republic of Korea, Saudi Arabia, and Australia, mostly geographically proximate nations, with a tendency towards regional research collaboration, as indicated by the reddish connection lines. Countries collaborating with the United States span across Asia, Europe, and South America, with a more extensive and international collaboration nature, as shown by connection lines varying from yellow to red. Both China and the United States maintain close cooperation with Republic of Korea, with reddish lines highlighting the tight trilateral collaboration. Europe is the most densely connected collaboration network, with connection lines among many countries rich in color and predominantly reddish, reflecting high-intensity multilateral research cooperation, closely related to geographical proximity and deep academic exchange traditions.

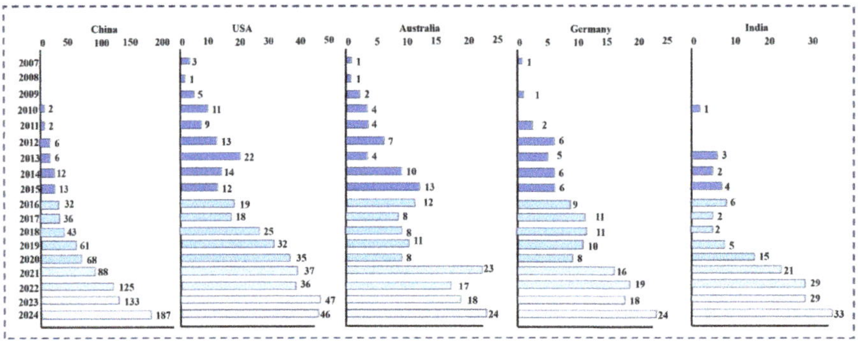

Figure 7. Annual Growth Trend of Top 5 Countries.

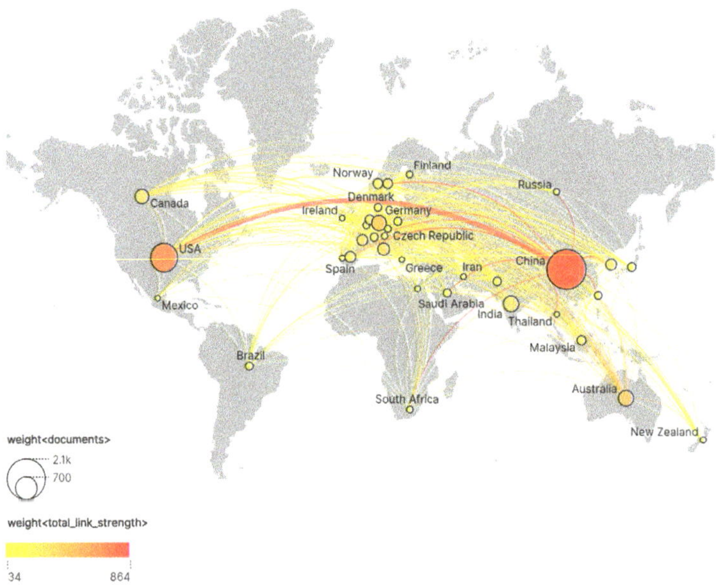

Figure 8. National Collaboration Network. Each country is represented by a circle, with the circle's size indicating the number of papers published. The lines connecting countries range in color from yellow to red, with more intense red signifying closer collaboration.

Figure 9's country time overlay network reveals that the United States and Australia were the earliest countries to research this field, followed by developed countries like the United Kingdom, Denmark, and Germany. Developing countries such as China, India, and Pakistan entered the field more recently.

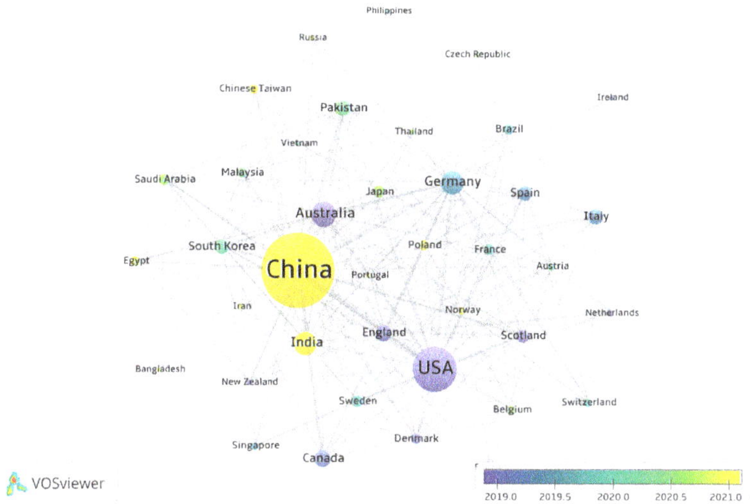

Figure 9. National Time Overlay Network. Each node represents a country, with larger nodes indicating more publications from that country. Countries that entered the field earlier are represented by nodes with more bluish-purple colors, while those joining later are depicted with more yellowish-green colors.

3.5. Author Analysis

The study included a total of 8943 authors. According to the screening criteria ($\geq$10 publications and $\geq$100 citations), 29 core authors were ultimately identified. After removing isolated nodes lacking cooperative relationships, the cooperation network contained 27 authors, forming four research clusters with distinctive characteristics (Figure 10). Based on Price's law for selecting core authors, if an author's number of papers $M = 0.749\sqrt{P_{max}}$, they belong to the core author group, where Pmax represents the number of papers by the most productive author. Therefore, Table 3 lists the top 4 authors by publication volume. Among these four authors, Yong Sik Ok from a Korean university has the highest H-index and the most citations, indicating his greatest influence.

The interaction and collaboration between different clusters are not tight-knit, with the green cluster showing the most cohesive collaborative relationships, primarily focusing on research about biochar applications in rice fields and agricultural production [35–38]. The red cluster exhibits relatively low internal collaboration, with its representative scholar Lehmann Johannes dedicating long-term research to biochar in soil improvement [39,40]. He and Professor Yong Sik Ok are the only authors with collaborations spanning all four clusters. The yellow cluster centers around Professor Yong Sik Ok, characterized by notably interdisciplinary research [41,42]. This scholar has not only promoted the technological transformation of biochar across multiple industries but has also recently optimized the biochar functional design through machine learning [43–45]. His academic impact indicators, including publication volume, citation frequency, and H-index, are at the top level (Figure 11, Table 3). Several authors in the blue cluster have relatively frequent collaborations, with their research primarily oriented towards biochar applications in water purification and heavy metal adsorption [46,47].

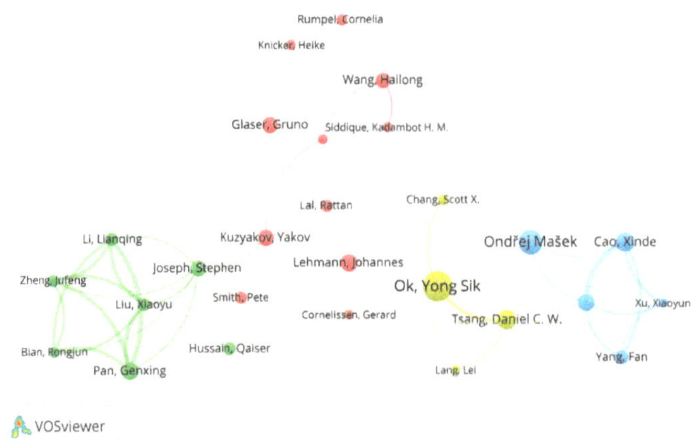

Figure 10. Author Collaboration Network. Each node represents an author, with larger nodes indicating a higher number of publications, and the thickness of the connections is proportional to the collaborative relationship between authors.

Table 3. Top 4 Authors by Publication Volume.

Ranking	Author	Publications	Citations	H-Index	Affiliated Institution
1st	Yong Sik Ok	30	7795	145	Korea University
2nd	Ondřej Mašek	24	1544	49	University of Edinburgh
3rd	Tsang, Daniel C. W.	19	1537	129	Hong Kong University of Science and Technology
4th	Cao, Xinde	19	2429	78	Shanghai Jiao Tong University

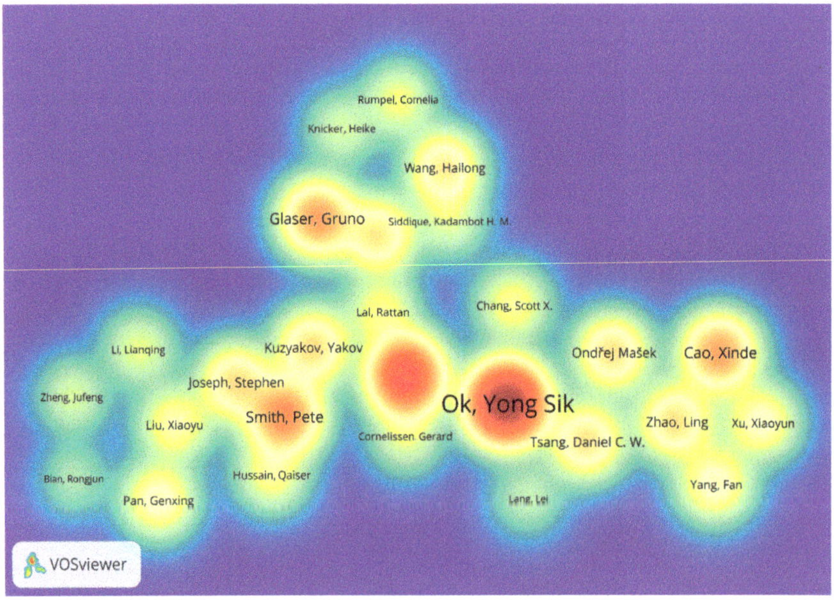

Figure 11. Author Citation Density Map. Nodes closer to red indicate authors with more citations, while nodes closer to light green represent authors with fewer citations.

3.6. Keyword Co-Occurrence

To ensure the accuracy and consistency of research result analysis and avoid potential biases caused by subtle differences in terminology usage, this study standardized the keywords found in the retrieved literature. "CO_2 storage" and "CO_2 sequestration" were jointly merged into "carbon sequestration"; "CO_2" was merged into "carbon dioxide"; "climate" and "climate change" were integrated into "climate-change"; and "life cycle assessment" was merged into "life-cycle assessment".

High-frequency and high-ranking keywords directly reflect the core research content and hot issues in the field. By tracking the changes in keyword frequency over time, research hotspots can be captured and traced. Table 4 shows the top 15 keywords, with biochar being the most frequent, appearing 1208 times, with a total connection intensity of 10,736, followed by carbon sequestration, appearing 878 times, with a total connection intensity of 8004.

Table 4. Top 15 Keywords.

Rank	Keywords	Occurrences	Total Link Strength
1st	biochar	1208	10,736
2nd	carbon sequestration	878	8004
3rd	pyrolysis	388	3560
4th	black carbon	341	3345
5th	biomass	319	3048
6th	soil	301	2705
7th	carbon	237	2050
8th	charcoal	214	2040
9th	nitrogen	213	2018
10th	pyrolysis temperature	209	2072
11th	stability	202	1948
12th	organic—matter	202	1880
13th	greenhouse—gas emissions	184	1856
14th	temperature	184	1768
15th	adsorption	172	1564

There were 7238 keywords in total. When the keyword threshold was below 15, there were three clusters, but the large number of keywords was not conducive to in-depth analysis. Setting the keyword appearance threshold to over 25 times still resulted in three clusters, but with significantly different keyword distributions. When the threshold was set between 30 and 60 times, the clusters reduced to two. Upon further increasing the keyword appearance threshold to 70 times, the keyword clusters returned to three, with significantly reduced interference keywords. At this point, only core keywords in the carbon sequestration field related to biochar emerged, with a total of 54 high-frequency terms identified. Carbon sequestration and biochar were the most focused keywords. Based on keyword co-occurrence clustering, current biochar research directions can be divided into three categories: the red area's main keywords include biochar, carbon sequestration, pyrolysis, adsorption, carbon, soil, biomass, waste, etc., indicating biochar technology development and application; the blue area focuses on black carbon stability mechanisms, with high-frequency keywords like black carbon, stability, oxidation, mineralization, and mechanisms; the green area's keywords include organic carbon, microbial community, greenhouse gas emissions, and climate change, primarily reflecting organic carbon dynamics and climate feedback mechanisms (Figure 12A).

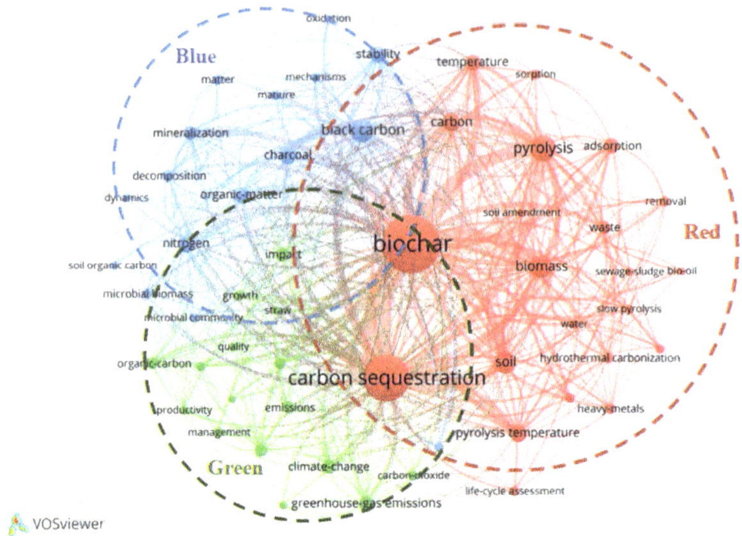

(A)

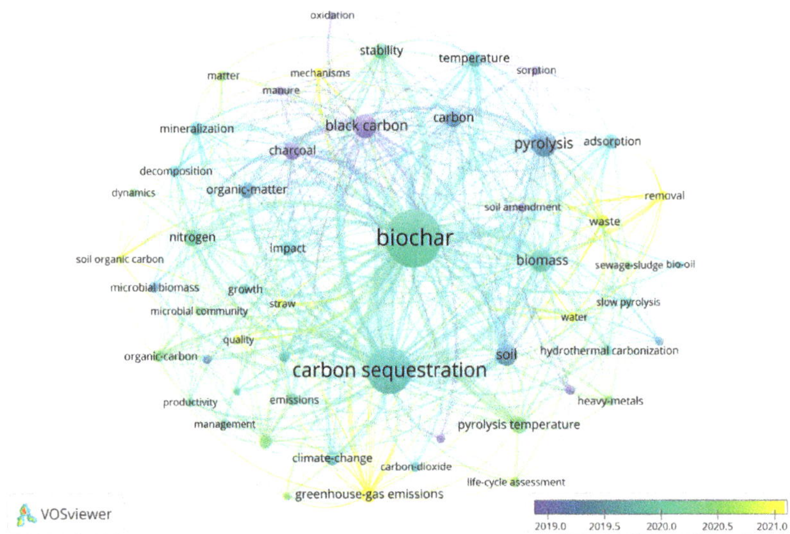

(B)

Figure 12. (**A**) Keyword Co-occurrence Network. Each node represents a keyword, node colors are related to clustering, node size increases with occurrence frequency, and line thickness is proportional to collaboration relationships between authors. (**B**) Keyword Time-Overlaid Network. Each node represents a keyword, node size increases with keyword frequency, keywords appearing earlier are closer to blue–purple colors, while keywords participating later shift towards yellow–green colors.

To reveal the temporal evolution trend of research themes, an overlay visualization map was constructed (Figure 12B). Purple indicates keywords that appeared relatively early, while yellow represents recently emerging keywords. The overlay visualization map shows that keywords like charcoal, black carbon, sorption, oxidation, and chemical properties appear in purple, indicating they emerged earlier; keywords such as removal, waste, water, straw, quality, and soil organic carbon appear in yellow–green; greenhouse gas emissions and mechanisms appear in bright yellow, indicating these themes have recently attracted increasing attention.

Keyword citation burst refers to keywords with rapidly increasing citation volumes. Burst detection is a useful analytical method for discovering keywords of special focus from relevant scientific communities during a specific period [48]. The red portion represents the time period of citation bursts [49], and keywords with strong burst intensity can represent research hotspots and frontiers [50]. Using CiteSpace to explore keywords with strong burst intensity from 2007 to 2024, Figure 13 lists the top 25 keywords in the biochar carbon sequestration field. These represent rapidly growing topics in biochar research. By analyzing the 25 high-burst-intensity keywords between 2007 and 2024, a clear evolutionary path in carbon sequestration research emerges. Early research centered on black carbon and charcoal, with burst intensities of 51.98 and 47.05, respectively, reflecting the academic community's strong interest in carbon stability mechanisms. Keywords like oxidation (intensity 12.92) and bioavailability further revealed black carbon's resistance to decomposition in the environment. These foundational studies provided theoretical support for subsequent applications.

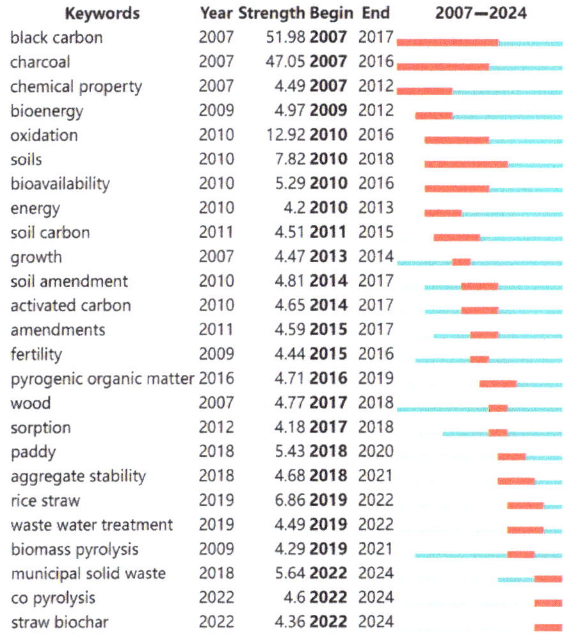

Figure 13. Top 25 Keywords with Highest Citation Burst Rates. Red segments indicate the start and end years of sudden increase duration.

From the early 2010s, the research focus gradually shifted to agricultural practices. Keywords like soil carbon and soil amendment significantly increased in frequency, indicating biochar research transitioned from laboratory stages to field trials. Scientists conducted application research on straw biochar in rice field systems, verifying carbon sequestration effects and collecting scalable application data. Between 2022 and 2024, keywords like rice straw, municipal solid waste, and co-pyrolysis remained active, showing research moving towards multi-waste collaborative conversion. Specific examples include co-pyrolysis technologies applying rice straw biochar with municipal waste. Simultaneously, keywords like sorption and aggregate stability continued to appear, involving optimization of biochar's physicochemical properties. Early high-frequency terms like bioenergy and chemical property had relatively shorter active periods. Current research expanded to exploring microbial community influences on organic carbon decomposition and observing relationships between biochar application and greenhouse gas emissions, encompassing comprehensive ecosystem-level impact assessments.

3.7. Co-Citation Analysis

When two papers appear simultaneously in the citations of a third paper, they are considered to have a co-citation relationship [51]. Important publications typically attract significant attention and interest, which can be reflected through citations [52], and co-citation analysis serves as a crucial method for detecting specific domain structures and evolutionary pathways [53]. This article generated a literature co-citation visualization map using VOSviewer software (Figure 14) and listed the top 10 co-cited papers using Cite Space (Table 5), with only the first authors cited. In the biochar research field, a series of foundational literature collectively constructs the knowledge landscape. While all these papers are related to biochar, they focus on different perspectives.

Table 5. Top 10 Co-Cited References.

First Author	Article	Journal	Year	Co-Citation	Total Citation
Tomczyk, A. [54]	Biochar physicochemical properties: pyrolysis temperature and feedstock kind effects	Environmental Science and Bio/Technology	2020	115	1606
Wang, J.Y. [55]	Biochar stability in soil: meta-analysis of decomposition and priming effects	Global Change Biology Bioenergy	2016	105	897
Lehmann, J. [40]	Biochar effects on soil biota—A review	Soil Biology and Biochemistry	2011	101	4017
Lehmann, J. [56]	Biochar in climate change mitigation	Nature Geoscience	2021	93	433
Leng, L.J. [57]	Biochar stability assessment methods: A review	Science of The Total Environment	2019	76	469
El-Naggar, A. [41]	Biochar application to low fertility soils: A review of current status, and future prospects	Geoderma	2019	75	727
Smith, P. [58]	Soil carbon sequestration and biochar as negative emission technologies	Global Change Biology	2016	70	658
Borchard, N. [59]	Biochar, soil and land-use interactions that reduce nitrate leaching and N_2O emissions: A meta-analysis	Science of the Total Environment,	2021	69	361
Jeffery, S. [60]	A quantitative review of the effects of biochar application to soils on crop productivity using meta-analysis	Agriculture, Ecosystems & Environment	2011	65	1917
Woolf, D. [61]	Sustainable biochar to mitigate global climate change	Nature Communications	2010	65	1761

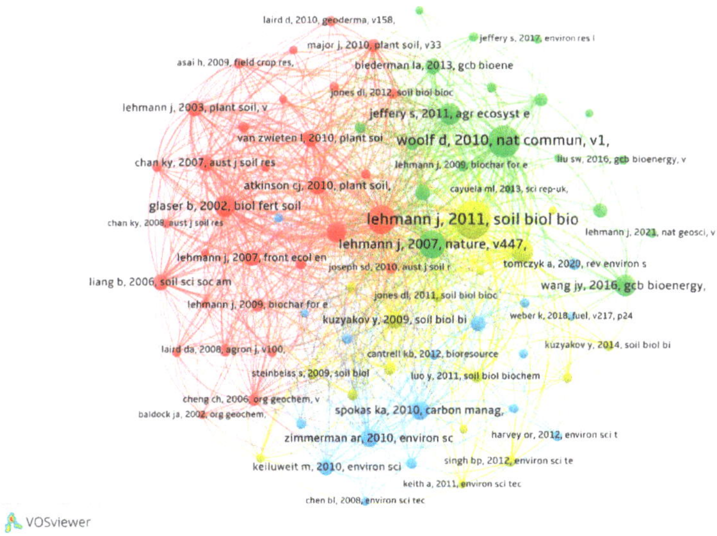

Figure 14. Co-Citation Network of Literature. Each node represents a literature source, node size is positively correlated with citation frequency, each color represents a cluster, and the thickness of connections between nodes is proportional to the co-citation frequency.

Tomczyk et al. (2020) [54] systematically revealed the influence of pyrolysis temperature (300–700 °C) and raw material type on biochar's physicochemical properties, finding that high temperatures significantly enhance specific surface area and ash content while reducing volatile matter and CEC. Woody material-based biochar demonstrates higher carbon sequestration and pollutant adsorption potential due to high carbon content and porous structure, while manure-based biochar is more suitable for soil improvement due to high ash content and CEC. Wang et al. (2016) [55] conducted a meta-analysis of 128 isotope tracing studies, revealing biochar's dual-pool decomposition mode in soil, showing an overall negative priming effect on soil organic carbon mineralization but significantly promoting mineralization in sandy soils, quantifying biochar's long-term carbon sequestration potential and environmental factors' impact on its stability.

Lehmann's team, from a microbial ecology perspective, first unveiled biochar's porous structure's regulatory mechanism on rhizosphere microbiota, discovering its promotion of beneficial microbial colonization through allelopathic substance adsorption [56]. Leng et al. (2019) [57] systematically reviewed biochar stability assessment methods, proposing a multi-dimensional evaluation system combining elemental analysis, thermal oxidation stability, and molecular markers, finding high correlation between H/C ratio and mineralization model results, and recommending thermal oxidation stability indicators to improve prediction accuracy.

El-Naggar, through field trials and meta-analysis, systematically demonstrated biochar's potential to enhance nutrient utilization efficiency in low-fertility soils, innovatively proposing the "designed biochar" concept to optimize material performance [38]. Smith et al. (2016) [58] examined biochar from a negative emissions technology perspective, proving its synergistic benefits in carbon sequestration and soil improvement through life cycle assessment. Borchard et al. (2019) [59] meta-analyzed 88 studies, finding biochar can reduce N_2O emissions by 38% and NO_3^- leaching by 13%, with particularly significant effects in paddy and sandy soils, though diminishing over time.

Jeffery et al. (2011) [60] meta-analyzed 16 studies, discovering biochar generally increases crop yield by 10% on average, with effects significantly influenced by soil pH and texture. Poultry manure-based biochar showed the best performance (28%), while solid biochar demonstrated negative effects. Woolf et al. (2010) [61] proposed that biochar, through sustainable biomass pyrolysis and soil sequestration, could achieve an annual 12% greenhouse gas emission reduction potential, with century-long benefits significantly higher than direct biomass power generation.

All research focuses on biochar's carbon sequestration potential and its environmental/agricultural synergistic benefits, serving climate change mitigation and sustainable development goals.

3.8. Analysis of Highly Cited References

Highly cited literature is a direct reflection of academic recognition of research achievements. According to Price's law and ESI database standards (top 1% in citation frequency), such literature often possesses high theoretical innovation or practical guidance value. By analyzing the citation networks of these publications, one can reveal knowledge diffusion paths and the formation mechanism of academic influence. Therefore, this article selected the top 20 publications (top 1% in citation frequency) for analysis (Table 6). Among these, 10 are review articles and 10 are research articles, with review articles ranking relatively high.

From the temporal distribution perspective, research on biochar began in 2007, with research enthusiasm significantly increasing around 2010 (five papers in 2010). That year had a high number of related publications. Research remained active in 2014–2015 (three papers in 2014, three papers in 2015) and continued to follow up until 2020, indicating sustained attention to biochar research over the past decade. In terms of research content, numerous publications extensively discussed biochar applications in environmental and agricultural fields. On one hand, they focused on its function as an adsorbent in treating soil and water pollutants, with some of the literature specifically exploring biochar's application in soil and water pollution management [42,62,63]. On the other hand, they examined how preparation conditions (such as pyrolysis temperature) influence its physicochemical properties [54,64]. Simultaneously, many publications analyzed biochar's effects as a soil amendment on soil properties, plant growth, and microbial communities, including the agronomic value of green waste biochar as a soil improver [65–71]. Additionally, several review articles systematically summarized biochar's preparation, modification, properties, and environmental applications, reviewing research progress and future directions [72–76].

The overall research formed a coherent narrative from fundamental property exploration to practical application potential, consistently focusing on biochar's applications in pollution management and soil improvement, providing solid theoretical support and directional guidance for subsequent in-depth research.

Reference literature burst detection is another method to track and capture research hotspots. References with the highest citation counts indicate special attention during a specific period and are typically considered the foundational research frontier of a field [77]. Figure 15 shows the top 25 references with the strongest citation bursts. The article "Biochar Sequestration in Terrestrial Ecosystems—A Review" has the earliest burst time, while "Biochar effects on soil biota—A review" has the highest burst intensity. Both are from the same author; Lehmann J. Tomczyk's work "Biochar physicochemical properties: pyrolysis temperature and feedstock kind effects" began bursting in 2020 and continues until 2024.

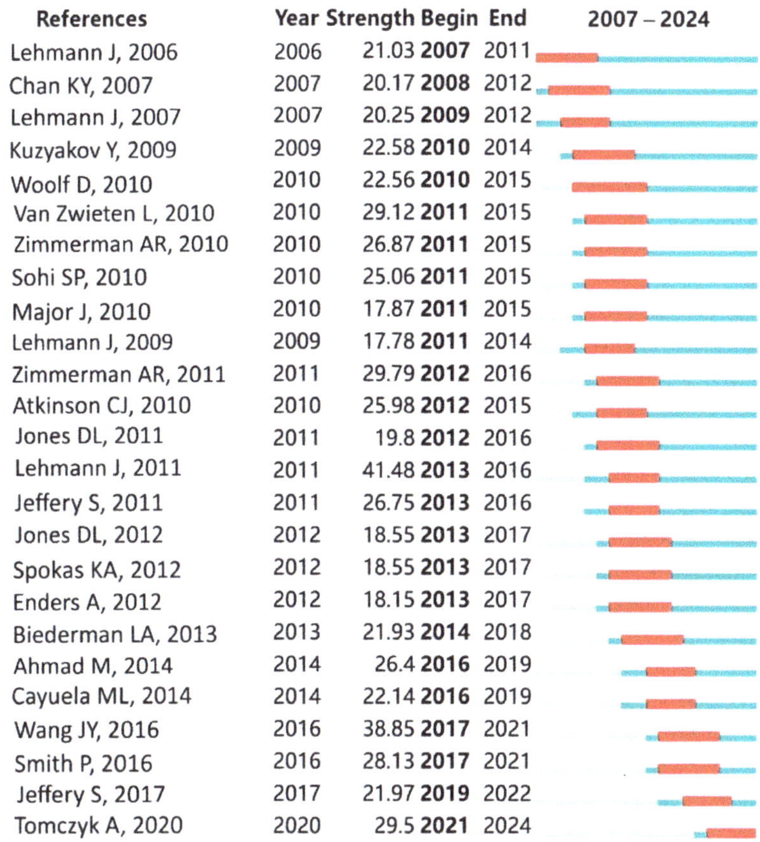

Figure 15. The 25 Most Highly Cited References. Red segments indicate the start and end years of citation surge periods.

Table 6. Top 20 Most Cited References.

First Author	Article	Journal	Year	Cited Times	Document Type
Ahmad, M. [42]	Biochar as a sorbent for contaminant management in soil and water: A review	Chemosphere	2014	3762	Review
Keiluweit, M. [78]	Dynamic Molecular Structure of Plant Biomass-Derived Black Carbon (Biochar)	Environmental Science & Technology	2010	2453	Article
Mohan, D. [62]	Organic and inorganic contaminants removal from water with biochar, a renewable, low cost and sustainable adsorbent—A critical review	Bioresource Technology	2014	1985	Review
Wang, J.L. [72]	Preparation, modification and environmental application of biochar: A review	Journal of Cleaner Production	2019	1684	Review
Tomczyk, A. [54]	Biochar physicochemical properties: pyrolysis temperature and feedstock kind effects	Environmental Science and Bio/Technology	2020	1671	Review

Table 6. *Cont.*

First Author	Article	Journal	Year	Cited Times	Document Type
Tan, X.F. [63]	Application of biochar for the removal of pollutants from aqueous solutions	Chemosphere	2015	1618	Review
Atkinson, C.J. [66]	Potential mechanisms for achieving agricultural benefits from biochar application to temperate soils: a review	Plant and Soil	2010	1582	Review
Chan, K.Y. [65]	Agronomic values of greenwaste biochar as a soil amendment greenwaste	Australian Journal of Soil Research	2007	1579	Article
Lehmann, J. [79]	Bio-energy in the black	Frontiers in Ecology and the Environment	2007	1563	Article
Biederman, L.A. [76]	Biochar and its effects on plant productivity and nutrient cycling: a meta-analysis	Global Change Biology Bioenergy	2013	1361	Review
Kambo, H.S. [73]	A comparative review of biochar and hydrochar in terms of production, physico-chemical properties and applications	Renewable and Sustainable Energy Reviews	2015	1283	Review
Warnock, D.D. [80]	Mycorrhizal responses to biochar in soil—concepts and mechanisms	Plant and Soil	2007	1133	Article
Gul, S. [74]	Physico-chemical properties and microbial responses in biochar-amended soils: Mechanisms and future directions	Agriculture, Ecosystems & Environment	2015	1113	Review
Mukherjee, A. [81]	Surface chemistry variations among a series of laboratory-produced biochars	Geoderma	2011	1013	Article
Fuss, S. [75]	Negative emissions-Part 2: Costs, potentials and side effects	Environmental Research Letters	2018	970	Review
Joseph, S.D. [69]	An investigation into the reactions of biochar in soil	Australian Journal of Soil Research	2010	952	Article
Gaskin, J.W. [67]	EFFECT OF LOW-TEMPERATURE PYROLYSIS CONDITIONS ON BIOCHAR FOR AGRICULTURAL USE	Transactions of the ASABE	2008	944	Article
Singh, B. [68]	Characterisation and evaluation of biochars for their application as a soil amendment	Australian Journal of Soil Research	2010	896	Article
Al-Wabel, M.I. [64]	Pyrolysis temperature induced changes in characteristics and chemical composition of biochar produced from conocarpus wastes	Bioresource Technology	2013	868	Article
Chan, K.Y. [70]	Using poultry litter biochars as soil amendments	Australian Journal of Soil Research	2008	864	Article

4. Discussion

4.1. Evolutionary Stages of Biochar Technology

In this era of information explosion, keeping up with industry frontiers and understanding the latest research findings is increasingly challenging [82]. To present a global scientific output study of biochar from 2007 to 2024, this paper employs bibliometric analysis to provide a novel method for organizing the knowledge structure of specific research.

The literature growth in this field exhibits a clear three-stage characteristic (Figure 2): The initial stage (2007–2014) had an average annual publication volume of less than 29 papers, reflecting that technological R&D was still in the exploratory stage. The research focus during this period was on basic scientific verification, such as the launch of the CSIRO farmland carbon sequestration experiment in Australia in 2007 [83]. The release of the European Union's White Paper "Biochar Climate Mitigation Potential" in 2009 [84] led to a year-on-year surge of 225% in the number of publications in 2010, reflecting the academic community's systematic exploration of biochar carbon sequestration mechanisms driven by policy.

The acceleration stage (2015–2020) saw an average annual output increase to 117 papers, a threefold increase compared to the previous stage. This shift is significantly correlated with the implementation of the Paris Agreement. In 2015, the Paris Agreement listed biochar as a negative emission technology option [4], and the literature growth entered an accelerated channel. Key industrialization events such as the breakthrough in pyrolysis technology by Germany's Pyreg company in 2016 [85] and the development of Cool Terra nano-char in the United States in 2017 [86] promoted the extension of research topics to engineering optimization and carbon metrology methodologies. The establishment of the AgriProve carbon trading platform in 2018 further triggered explosive growth in carbon credit accounting model research. The significant increase in citation frequency during this stage further confirms the strengthening effect of technological maturity on academic influence.

During the outbreak stage (2020–2024), the accelerated integration of the global carbon market after 2021 directly triggered high-speed growth in the number of publications. Combining the annual growth trend chart of the top 5 countries (Figure 7), it is not difficult to find that China contributed the main growth momentum. This is directly related to the top-level design of the "dual carbon" strategy—special central government financial support, green financial policy innovation, and the directed allocation of scientific research resources [87–91] have formed systematic support for technological breakthroughs. This institutional advantage may provide sustained momentum for the development of the field.

4.2. Heterogeneity and Institutional Logic of Sino–U.S. Scientific Research Output

China and the United States far surpass other countries in scientific paper output, being the most productive nations in the field. While their overall collaboration frequency ranks high, deep cooperation between core research institutions remains limited (Figure 9). The Chinese Academy of Sciences, despite establishing a broad international collaboration network, primarily concentrates its cooperation intensity with domestic top-tier universities. Collaboration with U.S. institutions like University of Florida, University of Massachusetts, Iowa State University, and USDA ARS is comparatively minimal (Figure 5), a "shallow collaboration" phenomenon potentially linked to differences in research evaluation systems and institutional technology transfer mechanisms.

The two countries contribute approximately 60% of global research outcomes but demonstrate differentiated development paths. Chinese research institutions have achieved scale superiority in total paper volume, yet the United States maintains leadership in research impact (such as the H-index) (Figure 7). Examining average citation frequencies,

China ranks relatively low internationally. While China leads in total paper count, each paper receives fewer citations compared to the U.S., primarily due to four factors: Extensive Chinese research faces barriers from language limitations and insufficient international promotion, struggling to enter global academic systems. Chinese institutions have low representation in top-tier international journals, and domestic scholars tend to cite domestic literature, constraining international citation volume. International collaborations remain concentrated among few elite institutions, and Sino–U.S. cooperation has diminished due to policy influences, narrowing knowledge exchange channels [92]. High retraction rates in international journals, data fabrication issues, and domestic evaluation systems emphasizing paper quantity further impact international credibility [93].

Improvement requires multiple approaches: expanding international collaboration networks by deepening engagement with European and American institutions, exploring partnerships with emerging research countries like India and Brazil, and encouraging scholars with international backgrounds. Increasing English-language paper proportions, supporting internationalized journals like Research (co-established by the Chinese Academy of Sciences and U.S. partners), and utilizing open-access platforms to enhance paper visibility are further approaches. Additional approaches include reforming research evaluation standards by reducing paper quantity assessments, increasing support for originality and long-term research, and strengthening academic integrity oversight.

4.3. Insufficient Author Collaboration Seriously Impedes Knowledge Dissemination

The current author collaboration network in the biochar carbon sequestration field exhibits significant clustering characteristics, accompanied by insufficient cross-cluster cooperation. Through author co-occurrence analysis (Figure 10), it can be seen that 27 core authors form four relatively independent research clusters. However, except for a few core scholars like Professor Yong Sik Ok and Lehmann Johannes, inter-cluster collaboration lines are sparse. Research teams have relatively close internal cooperation, but there is little interaction and collaboration between different teams, with knowledge mostly circulating within their own small circles and cross-team interaction and knowledge sharing being relatively limited.

This collaborative mode's limitations have created a dual impediment to field development. First, knowledge dissemination faces barriers, making it difficult to achieve cross-fusion of frontier achievements in specific sub-domains, such as biochar stability mechanisms, pyrolysis process optimization, and machine learning-driven material design, through inter-cluster collaboration. Second, the risk of redundant research increases, as each cluster independently explores common methodological issues like carbon sequestration efficiency assessment and environmental risk modeling, failing to establish a standardized research framework.

To address the problem of insufficient collaboration between author clusters, a routine collaborative network can be established, leveraging regional advantageous research institutions. For example, using institutions like the Chinese Academy of Sciences and Cornell University as pivots, regional alliances can be formed by combining universities and research institutes in Asia, North America, and other regions. By regularly hosting "Interdisciplinary Biochar Forums" and other activities, geographical and disciplinary boundaries can be broken down, promoting face-to-face communication among authors from agriculture, materials, environmental, and other clusters. This approach can help new discoveries in soil microbiology and frontier achievements in machine learning material design collide and generate collaborative innovation, effectively resolving knowledge dissemination barriers and redundant research challenges.

4.4. Technical Hotspots and Trends in Biochar Carbon Sequestration

In keyword burst detection (Figure 13), black carbon and charcoal show particularly high burst intensities, with values quite close to each other. Their colors are also the darkest in the keyword time overlay graph (Figure 12B), indicating an especially early appearance. By consulting the literature, this may be because in early biochar research (2007–2016), the "biochar" concept was not yet unified and was often mixed with terms like "charcoal" and "black carbon". For example, historically remaining combustion residues in natural soil (such as black carbon) were once classified as precursors of biochar [94]. Their outbreaks ended in 2016 and 2017, respectively, which may be related to the publication of the "Standardized Product Definition and Product Testing Guidelines for Biochar for Soil" (Version 2.1) by the IBI in 2015 (International Biochar Initiative. Standardized Product Definition and Product Testing Guidelines for Biochar That Is Used in Soil) [95]. This guideline specified the organic carbon content, preparation conditions, and functional uses of biochar from a technical standard level, marking the definition of biochar as authoritative and globally applicable.

The basic principle of biochar carbon sequestration is to convert unstable organic carbon in biomass into highly stable aromatic structures through thermochemical conversion, and reduce its decomposition and release in the environment through physical and chemical interactions, thereby achieving long-term carbon storage [96]. Laboratory cultivation experiments show that biochar's mean residence time (MRT) is 617–2829 years, with a half-life of approximately 1400 years [97], but the actual carbon sequestration age is significantly influenced by field environmental aging, microbial activity, and soil heterogeneity [98]. For instance, in a long-term field trial in Germany, high-dose biochar (40 Mg ha^{-1}) was applied to sandy soil, combined with digestate or compost. Results showed that initial soil organic carbon (SOC) storage increased by 61 Mg ha^{-1}; but after four years, due to insufficient soil physical protection, 38 Mg ha^{-1} of SOC disappeared. Nine years later, SOC storage in biochar-improved soil was only 7 Mg ha^{-1} higher than the control group, and black carbon storage almost returned to its original level [99]. This indicates that in sandy soil, biochar's physical stability is insufficient for long-term carbon sequestration, and the actual carbon sequestration effect may be far lower than theoretical expectations.

In the keyword overlay diagram, the highlight yellow of "climate change" and "greenhouse gas emissions" demonstrates that biochar production and improvement have been increasingly applied to mitigate global climate change, a trend driven by the international policy of the 2015 Paris Agreement [3], as well as China's growing influence in global climate governance. China's proposal of the "dual carbon" target (carbon peaking and carbon neutrality) in 2020 further promoted the practical application of biochar in climate governance. Biochar's function in achieving carbon neutrality is primarily through carbon fixation and emission reduction [100,101].

Combining keyword burst detection (Figure 13) and keyword network co-occurrence (Figure 12A), biochar stability has long been a focus of attention. The preparation raw materials are a crucial factor affecting biochar stability [102]. Keyword burst detection shows that keywords like "wood", "rice straw", "waste water treatment", and "municipal solid waste" burst after 2017 (Figure 13), with "straw biochar" specifically bursting after 2022. This indicates that biochar raw materials are currently transitioning from single sources to diverse waste materials, including crop residues [103], forestry residues [104], municipal solid waste [105], sewage sludge, manure [106], and biomass oil distillation residues [107]. This transformation is driven by rapid industrialization, widespread environmental pollution concerns, and the need for proper waste disposal. In 2015, the United Nations Environment Programme (UNEP) proposed integrating lifecycle methods with circular economy [108], emphasizing policy tools that cover the entire chain. This compelled cities to seek alternative treatment solutions, with biochar production being preferred due to its carbon sequestration and resource recycling functions. Through comparison with other recent bibliometric analyses on biochar, we found that Karki (2024) [109] and Hao et al. (2025) [110] both proposed applications of biochar in waste management, although they only studied biochar's application in specific types of waste. Particularly, Hao et al. (2025) focused on tea leaf waste, which was not detected in our keyword co-occurrence network (Figure 12A) or keyword burst detection (Figure 13). However, their findings are consistent with our current conclusions, both discovering that biochar has received widespread attention and significant enhancement in applications across various waste types.

However, this transformation also has drawbacks. The diversification of raw materials means varying pyrolysis temperatures and carbon fixation capabilities [111], leading to wide-ranging pH, specific surface area, and ash content indicators [112,113], making it challenging to develop universal industry standards. More seriously, it may risk pollutant migration. For instance, biochar produced from urban wastewater and municipal solid waste may contain excessive heavy metals [114], which could cause soil pollution when applied as fertilizer [102,115]. While the International Biochar Initiative (IBI) emphasizes raw material sustainability, it has not clearly defined pollutant thresholds [116].

In the highly cited literature, several articles investigate the impact of pyrolysis on physicochemical property stability (Table 6). The latest burst in literature detection is Professor Tomczyk A's article "Biochar physicochemical properties: pyrolysis temperature and feedstock kind effects", which is still ongoing (Figure 15). This indicates that the academic community has recently been particularly focused on pyrolysis technology. Co-pyrolysis, with the latest keyword burst (Figure 13), suggests that co-pyrolysis technology for biochar has received significant attention in recent years. This is likely closely related to the diversification of biochar raw materials. Co-pyrolysis is a technique of heating multiple biomass or waste materials under oxygen-deficient conditions, which can significantly improve biochar quality, including enhanced carbon fixation capacity [117,118]. Márquez et al. (2024) [119] conducted a bibliometric analysis of pyrolysis and concluded that co-pyrolysis has become a hot topic of comprehensive research since 2018, a conclusion consistent with ours.

However, we did not find co-pyrolysis literature among highly cited references. This may be because co-pyrolysis is a recently emerging technology, and the literature citations take time to accumulate. Despite this, it remains an emerging technology currently constrained by raw material complexity, pollution risks, yield and energy efficiency trade-offs, and technological promotion challenges [120], requiring further process optimization and long-term benefit verification.

Soil is the largest terrestrial carbon reservoir [121,122], with surface soil storing about two-thirds (approximately 1500 Pg) of global carbon, which is three times the atmospheric carbon pool [55,123]. Therefore, any slight changes in the soil carbon reservoir can have tremendous impacts on global greenhouse gas emissions [124]. The co-occurrence keyword graph showing connections between biochar and soil, soil organic carbon (Figure 12A) demonstrates that biochar is a core technology for soil carbon sequestration, making significant contributions. Numerous studies indicate that biochar application can reduce soil greenhouse gas emissions [87]. When applied as a fertilizer, biochar can also suppress methane (CH_4) and nitrous oxide (N_2O) release.

In addition to preserving substantial carbon sinks, the keyword burst detection (Figure 13) reveals that keywords like Soil, Soil carbon, amendments, and fertility emerged around 2010, with high burst duration and intensity. This suggests that biochar applications in soil were among the earliest areas of focus. The burst time of bioavailability coincides with soil, indicating scholars' particular interest in biochar's impact on soil microbial communities. Wang et al.

(2024) [125] pointed out that new eco-fertilizers prepared from organic raw materials have attracted widespread attention and application in recent years. As a soil amendment, biochar can enhance soil carbon sequestration capacity [126], making it a continuing research hotspot. The dark-blue keywords related to soil in the keyword time overlay diagram (Figure 12A) further confirmed this perspective.

Gross' meta-analysis synthesizing 64 studies and 736 treatments demonstrated that biochar application increases soil organic carbon storage by an average of 29% (field trials), with even higher carbon storage in experiments lasting over 5 years [127]. However, Han et al. (2013) [128] found that while biochar application can provide significant soil carbon sequestration, it also involves considerable uncertainty. The mechanisms underlying biochar's initial effects on soil carbon sequestration remain unclear [129].

Biochar technology represents an emerging interdisciplinary frontier and trend. Its disciplinary distribution spans environmental science, soil science, energy and fuel, agronomy, environmental engineering, and green sustainable science and technology, with environmental science being the most prominent (Figure 3). The biochar analysis in Figure 4 reveals deep interactions across environmental science, chemistry, and earth sciences, and even includes limited pathways involving psychology, education, socioeconomics, economics, and political science. This indicates a shift from single technological research towards comprehensive eco-economic system regulation.

Professor Yong Sik Ok from a Korean university has become a core leader in biochar research through exceptional publication volume, citation frequency, and H-index (Table 3). With the highest collaboration intensity, his emerging research directions potentially represent future development trends. His team has recently focused on machine learning (ML)-driven biochar technology optimization [43–45]. In the current era of internet big data and artificial intelligence (AI), machine learning demonstrates significant potential in biochar preparation process optimization, performance prediction, and application scenario expansion, marking an important future development direction.

5. Conclusions

This study employs bibliometric analysis to assess the current status and research prospects of international cooperation in biochar carbon sequestration. The analysis includes a comprehensive evaluation of the distribution of publication years and the identification of prolific authors and their institutions, international collaboration patterns, publication sources, high-frequency keywords, and related citation rates. This method aids in identifying and exploring research hotspots and emerging trends in the field. Future research directions should closely address the key gaps identified through bibliometric analysis: (1) Focus on the growth potential of research related to "biochar-microorganism interactions" and form interdisciplinary teams, leveraging the high-producing author groups identified through quantitative analysis, to elucidate the underlying microscopic mechanisms, thereby filling the current knowledge gap. (2) Biochar carbon sequestration research has progressively advanced from early mechanistic exploration to engineering applications, currently focusing on waste resource utilization and technological innovation. However, constrained by factors such as raw material diversity and the imperfect long-term carbon sequestration monitoring system, the field faces practical challenges such as a lack of standards. In the future, researchers can utilize improved machine learning models to assist in biochar preparation and conduct tests in complex soil systems to screen for biochar suited to specific application scenarios. This will help to thoroughly clarify the mechanisms and influencing factors of biochar carbon sequestration, and then accurately predict the environmental impact of biochar's life cycle. (3) At the policy level, there is a need to expedite the establishment of thresholds for heavy metals and other pollutants in raw materials, support "biochar + agriculture" demonstration projects, and promote it as an important technical pathway to achieve carbon neutrality.

Researchers should leverage these insights to prioritize high-impact collaborations, advocate for standardized assessment frameworks, and involve policymakers in scaling biochar from pilot studies to climate mitigation strategies. By bridging these gaps, the field can transition from fragmented findings to systematic, globally coordinated carbon sequestration solutions.

6. Limitations

This research has certain limitations in terms of data sources and analytical methods. On the data source side, the study is based solely on the Web of Science database, without incorporating Scopus, patent databases, or regional academic resources. This approach may overlook research outcomes from non-English or non-mainstream journals, and the data are limited to 2024, with non-English literature excluded, which affects the comprehensiveness, timeliness, and cultural diversity of the conclusions. Additionally, the international collaboration network analysis relies exclusively on bibliometric data, without integrating non-literature factors such as funding support or policy guidance, thus providing limited insight into the driving mechanisms of collaboration modes. Furthermore, while widely used analysis software like Vosviewer and Citespace were employed, these methods cannot fully exploit the data potential as digital technologies advance. Future research could consider developing or utilizing more advanced analysis software and Supplementing Data from alternative databases or custom data sources to optimize the study.

Supplementary Materials: The following supporting information can be downloaded at: https://www.mdpi.com/article/10.3390/en18112745/s1.

Author Contributions: Conceptualization, S.Z. and S.W.; methodology, S.Z.; software, J.Z.; validation, L.L., C.C., M.S. and H.W.; formal analysis, S.Z.; investigation, Y.L.; resources, H.W. and B.W.; data curation, J.Z.; writing—original draft preparation, S.Z.; writing—review and editing, S.W.; visualization, J.Z.; supervision, S.W.; project administration, S.W.; funding acquisition, S.W. All authors have read and agreed to the published version of the manuscript.

Funding: This research was funded by the Natural Science Foundation of Hebei province of China (grant nos. D2023209016; D2024209008), the Research Project of the Education Department of Hebei Province (grant no. BJ2020037), the Innovation Capacity Enhancement Project of Hebei Province (grant no. 23564201D), and the National College Student Innovation and Entrepreneurship Training Program (grant no. 202410081021).

Acknowledgments: The authors are profoundly grateful to Yajie Ma from North China University of Science and Technology for his exceptional scholarly guidance, insightful academic advice, and unwavering support throughout the preparation of this research manuscript. Heartfelt appreciation is also extended to Haotian Xun for his critical technical assistance and invaluable contributions in software implementation and operational support.

Conflicts of Interest: Author Bao Wang was employed by the PetroChina Jidong Oilfield Company. The remaining authors declare that the research was conducted in the absence of any commercial or financial relationships that could be construed as a potential conflict of interest.

References

1. Prince Adamu, T.S.; Zhang, Z.; Amena, A. Multi-Feature Data Fusion-Based Load Forecasting of Electric Vehicle Charging Stations Using a Deep Learning Model. *Energies* **2023**, *16*, 1306.
2. Nandhini, K.; Krishnamurthy, K.; Nandhini, K.; Jadoun, V.K.; Shrivastava, A.; Vidya, S.R.; Kudva, G. Optimal Placement and Sizing of Electric Vehicle Charging Infrastructure in a Grid-Tied DC Microgrid Using Modified TLBO Method. *Energies* **2023**, *16*, 1781. [CrossRef]

3. Zhang, S.; Liu, J.; Li, L.; Kassabi, N.; Hamdi, E. Petrophysical and Geochemical Investigation-Based Methodology for Analysis of the Multilithology of the Permian Longtan Formation in Southeastern Sichuan Basin, SW China. *Energies* **2024**, *17*, 766. [CrossRef]
4. United Nations Framework Convention on Climate Change. *The Paris Agreement: Article 2*; United Nations: New York, NY, USA, 2015.
5. Knutti, R.; Rogelj, J.; Sedláček, J.; Fischer, E.M. A scientific critique of the two-degree climate change target. *Nat. Geosci.* **2016**, *9*, 13–18. [CrossRef]
6. Rogelj, J.; den Elzen, M.; Höhne, N.; Fransen, T.; Fekete, H.; Winkler, H.; Shaeffer, R.; Riahi, K.; Meinshausen, M. Paris Agreement climate proposals need a boost to keep warming well below 2 °C. *Nature* **2016**, *534*, 631–639. [CrossRef]
7. Cui, Y.; Bai, J.; Liao, S.; Cao, S.; Liu, F. Suggestions on the Development of Environmental Monitoring Technology of CO_2 Geological Storage and Leakage under the Background of China's "Double-Carbon" Strategy. *Atmosphere* **2023**, *14*, 51. [CrossRef]
8. Gao, S.; Jia, S.; Zhu, Y.; Zhao, L.; Cao, Y.; Qi, X.; Guan, F. Study on Quantitative Assessment of CO_2 Leakage Risk Along the Wellbore Under the Geological Storage of the Salt Water Layer. *Energies* **2024**, *17*, 5302. [CrossRef]
9. Yin, Y.; Zhang, L.; Cao, X.; Li, Q.; Gan, M.; Wang, Y. Progress, challenges, and prospects of CO_2 mineral sequestration in basalt: A critical review. *Appl. Energy* **2025**, *381*, 125127. [CrossRef]
10. Zhang, S.; Ben Abdallah, K.; Li, L.; Hamdi, E.; Liu, J. Multiphase Flow Controlled by Synergistic Injection-Production Pressure: Enabling CO_2 Geo-Sequestration with Additional Gas Recovery from Vertically Heterogeneous Depleted Shale Reservoirs. *Fuel* **2025**, *398*, 135591. [CrossRef]
11. Mahmoudnia, A. The role of PFAS in unsettling ocean carbon sequestration. *Environ. Monit. Assess.* **2023**, *195*, 310. [CrossRef]
12. Du, J.L.; Tian, J. Distinct carbon sequestration in the glacial Pacific despite vigorous deep ocean circulation. *Earth Planet. Sci. Lett.* **2025**, *660*, 119344. [CrossRef]
13. Zhang, T.; Tang, Y.; Li, H.; Hu, W.; Cheng, J.; Lee, X. A bibliometric review of biochar for soil carbon sequestration and mitigation from 2001 to 2020. *Ecotoxicol. Environ. Saf.* **2023**, *264*, 115438. [CrossRef] [PubMed]
14. Feng, D.; Wang, S.; Zhang, Y.; Zhao, Y.; Sun, S.; Chang, G.; Lai, X.; Tan, H.; Qin, Y. Review of carbon fixation evaluation and emission reduction effectiveness for biochar in China. *Energy Fuels* **2020**, *34*, 10583–10606. [CrossRef]
15. Manyà, J.J. Pyrolysis for Biochar Purposes: A Review to Establish Current Knowledge Gaps and Research Needs. *Environ. Sci. Technol.* **2012**, *46*, 7939–7954. [CrossRef] [PubMed]
16. Chicaiza-Ortiz, C.; Zhang, P.; Zhang, J.; Zhang, T.; Yang, Q.; He, Y. CO_2-enhanced methane production by integration of bamboo biochar during anaerobic co-digestion. *J. Environ. Manag.* **2025**, *373*, 123603. [CrossRef]
17. Salma, A.; Fryda, L.; Djelal, H. Biochar: A Key Player in Carbon Credits and Climate Mitigation. *Resources* **2024**, *13*, 31. [CrossRef]
18. Morales, M.M.; Comferford, N.; Guerrini, I.A.; Falcão, N.P.S.; Reeves, J. Sorption and desorption of phosphate on biochar and biochar–soil mixtures. *Soil Use Manag.* **2013**, *29*, 306–314. [CrossRef]
19. Lin, M.; Li, F.; Li, X.; Rong, X.; Oh, K. Biochar-clay, biochar-microorganism and biochar-enzyme composites for environmental remediation: A review. *Environ. Chem. Lett.* **2023**, *21*, 1837–1862. [CrossRef]
20. Arias, C.M.; Silva, L.F.S.; Soares, M.R.; Forti, V.A. A bibliometric analysis on the agricultural use of biochar in Brazil from 2003 to 2021: Research status and promising raw materials. *Renew. Agric. Food Syst.* **2023**, *38*, e124. [CrossRef]
21. Nan, H.; Wang, L.; Luo, D.; Zhang, Y.; Liu, G.; Wang, C. A bibliometric analysis of biochar application in wastewater treatment from 2000 to 2021. *Int. J. Environ. Sci. Technol.* **2023**, *20*, 13957–13974. [CrossRef]
22. Phiri, Z.; Moja, N.T.; Nkambule, T.T.L.; de Kock, L.A. Utilization of biochar for remediation of heavy metals in aqueous environments: A review and bibliometric analysis. *Heliyon* **2024**, *10*, e25785. [CrossRef] [PubMed]
23. Jin, F.; Lu, J.; Sun, F.; Yang, F.; Li, Z. Application and development of sludge-based materials for environmental pollution remediation: A bibliometric review from 2004 to 2024. *RSC Adv.* **2025**, *15*, 8072–8087. [CrossRef] [PubMed]
24. Pritchard, A. Statistical bibliography or bibliometrics. *J. Doc.* **1969**, *25*, 348–349.
25. Chen, C.; Leydesdorff, L. Patterns of connections and movements in dual-map overlays: A new method of publication portfolio analysis. *J. Assoc. Inf. Sci. Technol.* **2013**, *65*, 334–351. [CrossRef]
26. Shen, Z.; Wu, H.; Chen, Z.; Hu, J.; Pan, J.; Kong, J.; Lin, T. The global research of artificial intelligence on prostate cancer: A 22-year bibliometric analysis. *Front. Oncol.* **2022**, *12*, 843735. [CrossRef]
27. Marzi, G.; Caputo, A.; Garces, E.; Dabić, M. A Three Decade Mixed-Method Bibliometric Investigation of the IEEE Transactions on Engineering Management. *IEEE Trans. Eng. Manag.* **2018**, *67*, 4–14. [CrossRef]
28. Wu, H.; Wang, Y.; Tong, L.; Yan, H.; Sun, Z. Global research trends of ferroptosis: A rapidly evolving field with enormous potential. *Front. Cell Dev. Biol.* **2021**, *9*, 646311. [CrossRef]
29. Xie, L.; Chen, Z.; Wang, H.; Zheng, C.; Jiang, J. Bibliometric and visualized analysis of scientific publications on atlantoaxial spine surgery based on Web of Science and VOSviewer. *World Neurosurg.* **2020**, *137*, 435–442. [CrossRef]
30. Xu, D.; Wang, Y.L.; Wang, K.T.; Wang, Y.; Dong, X.R.; Tang, J. A scientometrics analysis and visualization of depressive disorder. *Curr. Neuropharmacol.* **2021**, *19*, 766–786. [CrossRef]

31. Liu, H.; Hong, R.; Xiang, C.; Lv, C.; Li, H. Visualization and analysis of mapping knowledge domains for spontaneous combustion studies. *Fuel* **2020**, *262*, 116598. [CrossRef]
32. Ebrahimi, P.; Abbasi, S.; Pashaei, R.; Bogusz, A.; Oleszczuk, P. Investigating impact of physicochemical properties of microplastics on human health: A short bibliometric analysis and review. *J. Hazard. Mater.* **2022**, *424*, 127787. [CrossRef] [PubMed]
33. Chen, C. Science Mapping: A Systematic Review of the Literature. *J. Data Inf. Sci.* **2017**, *2*, 1–40. [CrossRef]
34. Miao, Y.; Zhang, Y.; Yin, L. Trends in hepatocellular carcinoma research from 2008 to 2017: A bibliometric analysis. *PeerJ* **2018**, *6*, e5477. [CrossRef]
35. Liu, Z.; Zhang, W.; Ma, R.; Li, S.; Song, K.; Zheng, J.; Wang, Y.; Bian, R.; Zhang, X.; Pan, G. Biochar-plant interactions enhance nonbiochar carbon sequestration in a rice paddy soil. *Commun. Earth Environ.* **2023**, *4*, 494. [CrossRef]
36. Zhang, A.; Cui, L.; Pan, G.; Li, L.; Hussain, Q.; Zhang, X.; Zheng, J.; Crowley, D. Effect of biochar amendment on yield and methane and nitrous oxide emissions from a rice paddy from Tai Lake plain, China. *Agric. Ecosyst. Environ.* **2010**, *139*, 469–475. [CrossRef]
37. Zhang, A.; Bian, R.; Pan, G.; Cui, L.; Hussain, Q.; Li, L.; Zheng, J.; Zhang, X.; Han, X.; Yu, X. Effects of biochar amendment on soil quality, crop yield and greenhouse gas emission in a Chinese rice paddy: A field study of 2 consecutive rice growing cycles. *Field Crops Res.* **2012**, *127*, 153–160. [CrossRef]
38. Bian, R.; Joseph, S.; Cui, L.; Pan, G.; Shi, X.; Li, L.; Liu, X.; Zhang, A.; Rutlidge, H.; Wong, S.; et al. A three-year experiment confirms continuous immobilization of cadmium and lead in contaminated paddy field with biochar amendment. *J. Hazard. Mater.* **2014**, *272*, 121–128. [CrossRef]
39. Schmidt, M.W.; Torn, M.S.; Abiven, S.; Dittmar, T.; Guggenberger, G.; Janssens, I.A.; Kleber, M.; Kögel-Knabner, I.; Lehmann, J.; Manning, D.A.; et al. Persistence of soil organic matter as an ecosystem property. *Nature* **2011**, *478*, 49–56. [CrossRef]
40. Lehmann, J.; Rillig, M.C.; Thies, J.; Masiello, C.A.; Hockaday, W.C.; Crowley, D. Biochar effects on soil biota—A review. *Soil Biol. Biochem.* **2011**, *43*, 1812–1836. [CrossRef]
41. El-Naggar, A.; Lee, S.S.; Rinklebe, J.; Farooq, M.; Song, H.; Sargam, A.K.; Zimmerman, A.R.; Ahmad, M.; Shaheen, S.; Ok, Y.S. Biochar application to low fertility soils: A review of current status, and future prospects. *Geoderma* **2019**, *337*, 536–554. [CrossRef]
42. Ahmad, M.; Rajapaksha, A.U.; Lim, J.E.; Zhang, M.; Bolan, N.; Mohan, D.; Vithanage, M.; Lee, S.S.; Ok, Y.S. Biochar as a sorbent for contaminant management in soil and water: A review. *Chemosphere* **2014**, *99*, 19–33. [CrossRef] [PubMed]
43. Wang, S.J.; Li, J.; Ok, Y.S. Machine Learning Predicts Biochar Aging Effects on Nitrous Oxide Emissions from Agricultural Soils. *ACS Agric. Sci. Technol.* **2024**, *4*, 888–898. [CrossRef]
44. Zhang, C.; Hu, T.; Deng, M.; Ok, Y.S.; Wang, H.; Liu, Y.; Zhang, L. Spectral-Based Machine Learning Enables Rapid and Large-Scale Adsorption Capacity Prediction of Heavy Metals in Soil. *ACS ES&T Eng.* **2024**, *4*, 2657–2667.
45. Yuan, X.; Suvarna, M.; Liu, J.Y.; Lim, J.Y.; Pérez-Ramírez, J.; Wang, X.; Ok, Y.S. Active Learning-Based Guided Synthesis of Engineered Biochar for CO_2 Capture. *Environ. Sci. Technol.* **2024**, *58*, 6628–6636. [CrossRef] [PubMed]
46. Inyang, M.; Gao, B.; Yao, Y.; Xue, Y.; Zimmerman, A.R.; Pullammanappallil, P.; Cao, X. Removal of heavy metals from aqueous solution by biochars derived from anaerobically digested biomass. *Bioresour. Technol.* **2012**, *110*, 50–56. [CrossRef]
47. Inyang, M.I.; Gao, B.; Yao, Y.; Xue, Y.; Zimmerman, A.; Mosa, A.; Pullammanappallil, P.; Cao, X. A review of biochar as a low-cost adsorbent for aqueous heavy metal removal. *Crit. Rev. Environ. Sci. Technol.* **2016**, *46*, 406–433. [CrossRef]
48. Guo, Y.-M.; Huang, Z.-L.; Guo, J.; Li, H.; Guo, X.-R.; Nkeli, M.J. Bibliometric Analysis on Smart Cities Research. *Sustainability* **2019**, *11*, 3606. [CrossRef]
49. Huang, L.; Kelly, S.; Lv, K.J.; Giurco, D. A systematic review of empirical methods for modelling sectoral carbon emissions in China. *J. Clean. Prod.* **2019**, *215*, 1382–1401. [CrossRef]
50. Chen, X.; Lin, B.Q. Towards carbon neutrality by implementing carbon emissions trading scheme: Policy evaluation in China. *Energy Policy* **2021**, *157*, 112510. [CrossRef]
51. Boyack, K.W.; Klavans, R. Co-citation analysis, bibliographic coupling, and direct citation: Which citation approach represents the research front most accurately? *J. Assoc. Inf. Sci. Technol.* **2010**, *61*, 2389–2404. [CrossRef]
52. Ho, Y.-S. Top-cited articles in chemical engineering in Science Citation Index Expanded: A bibliometric analysis. *Chin. J. Chem. Eng.* **2012**, *20*, 478–488. [CrossRef]
53. Liao, H.; Tang, M.; Luo, L.; Li, C.; Chiclana, F.; Zeng, X.-J. A Bibliometric Analysis and Visualization of Medical Big Data Research. *Sustainability* **2018**, *10*, 166. [CrossRef]
54. Tomczyk, A.; Sokołowska, Z.; Boguta, P. Biochar physicochemical properties: Pyrolysis temperature and feedstock kind effects. *Rev. Environ. Sci. Biotechnol.* **2020**, *19*, 191–215. [CrossRef]
55. Wang, J.; Xiong, Z.; Kuzyakov, Y. Biochar stability in soil: Meta-analysis of decomposition and priming effects. *Glob. Change Biol. Bioenergy* **2016**, *8*, 512–523. [CrossRef]
56. Lehmann, J.; Cowie, A.; Masiello, C.A.; Kammann, C.; Woolf, D.; Amonette, J.E.; Cayuela, M.L.; Camps-Arbestain, M.; Whitman, T. Biochar in climate change mitigation. *Nat. Geosci.* **2021**, *14*, 883–892. [CrossRef]

57. Leng, L.; Huang, H.; Li, H.; Zhou, J.; Zhou, W. Biochar stability assessment methods: A review. *Sci. Total Environ.* **2019**, *647*, 210–222. [CrossRef]
58. Smith, P. Soil carbon sequestration and biochar as negative emission technologies. *Glob. Change Biol.* **2016**, *22*, 1315–1324. [CrossRef]
59. Borchard, N.; Schirrmann, M.; Cayuela, M.L.; Kammann, C.; Wrage-Mönnig, N.; Estavillo, J.M.; Fuertes-Mendizábal, T.; Sigua, G.; Spokas, K.; Ippolito, J.A.; et al. Biochar, soil and land-use interactions that reduce nitrate leaching and N2O emissions: A meta-analysis. *Sci. Total Environ.* **2019**, *651*, 2354–2364. [CrossRef]
60. Jeffery, S.; Verheijen, F.G.A.; van der Velde, M.; Bastos, A.C. A quantitative review of the effects of biochar application to soils on crop productivity using meta-analysis. *Agric. Ecosyst. Environ.* **2011**, *144*, 175–187. [CrossRef]
61. Woolf, D.; Amonette, J.; Street-Perrott, F.; Lehmann, J.; Joseph, S. Sustainable biochar to mitigate global climate change. *Nat. Commun.* **2010**, *1*, 56. [CrossRef]
62. Mohan, D.; Sarswat, A.; Ok, Y.S.; Pittman, C.U., Jr. Organic and inorganic contaminants removal from water with biochar, a renewable, low cost and sustainable adsorbent—A critical review. *Bioresour. Technol.* **2014**, *160*, 191–202. [CrossRef] [PubMed]
63. Tan, X.; Liu, Y.; Zeng, G.; Wang, X.; Hu, X.; Gu, Y.; Yang, Z. Application of biochar for the removal of pollutants from aqueous solutions. *Chemosphere* **2015**, *125*, 70–85. [CrossRef] [PubMed]
64. Al-Wabel, M.I.; Al-Omran, A.; El-Naggar, A.H.; Nadeem, M.; Usman, A.R.A. Pyrolysis temperature induced changes in characteristics and chemical composition of biochar produced from conocarpus wastes. *Bioresour. Technol.* **2013**, *131*, 374–379. [CrossRef] [PubMed]
65. Chan, K.Y.; Van Zwieten, L.; Meszaros, I.; Downie, A.; Joseph, S. Agronomic values of greenwaste biochar as a soil amendment. *Aust. J. Soil Res.* **2007**, *45*, 629–634. [CrossRef]
66. Atkinson, C.J.; Fitzgerald, J.D.; Hipps, N.A. Potential mechanisms for achieving agricultural benefits from biochar application to temperate soils: A review. *Plant Soil* **2010**, *337*, 1–18. [CrossRef]
67. Gaskin, J.W.; Steiner, C.; Harris, K.; Das, K.C.; Bibens, B. Effect of low-temperature pyrolysis conditions on biochar for agricultural use. *Trans. ASABE* **2008**, *51*, 2061–2069. [CrossRef]
68. Singh, B.; Singh, B.P.; Cowie, A.L. Characterisation and evaluation of biochars for their application as a soil amendment. *Aust. J. Soil Res.* **2010**, *48*, 516–525. [CrossRef]
69. Joseph, S.D.; Camps-Arbestain, M.; Lin, Y.; Munroe, P.; Chia, C.H.; Hook, J.; van Zwieten, L.; Kimber, S.; Cowie, A.; Singh, B.P. An investigation into the reactions of biochar in soil. *Aust. J. Soil Res.* **2010**, *48*, 501–515. [CrossRef]
70. Chan, K.Y.; Van Zwieten, L.; Meszaros, I.; Downie, A.; Joseph, S. Using poultry litter biochars as soil amendments. *Aust. J. Soil Res.* **2008**, *46*, 437–444. [CrossRef]
71. Huang, S.; Liu, D.; Gomez-Rivas, E.; Griera, A.; Gan, Q.; Wang, M.; Xing, Y.; Zhao, Y. Experimental Insights into the Nucleation and Propagation of Hydraulic Fractures in Anthracite Coalbed Methane Reservoirs. *Earth Energy Sci.* 2025; *in press*. [CrossRef]
72. Wang, J.; Wang, S. Preparation, modification and environmental application of biochar: A review. *J. Clean. Prod.* **2019**, *227*, 1002–1022. [CrossRef]
73. Kambo, H.S.; Dutta, A. A comparative review of biochar and hydrochar in terms of production, physico-chemical properties and applications. *Renew. Sustain. Energy Rev.* **2015**, *45*, 359–378. [CrossRef]
74. Gul, S.; Whalen, J.K.; Thomas, B.W.; Sachdeva, V.; Deng, H.Y. Physico-chemical properties and microbial responses in biochar-amended soils: Mechanisms and future directions. *Agric. Ecosyst. Environ.* **2015**, *206*, 46–59. [CrossRef]
75. Fuss, S.; Lamb, W.F.; Callaghan, M.W.; Hilaire, J.; Creutzig, F.; Amann, T.; Beringer, T.; Garcia, W.D.; Hartmann, J.; Khanna, T. Negative emissions-Part 2: Costs, potentials and side effects. *Environ. Res. Lett.* **2018**, *13*, 063002. [CrossRef]
76. Biederman, L.A.; Harpole, W.S. Biochar and its effects on plant productivity and nutrient cycling: A meta-analysis. *Glob. Change Biol. Bioenergy* **2013**, *5*, 202–214. [CrossRef]
77. Gao, Y.; Ge, L.; Shi, S.; Shi, Z.; Sun, Y.; Liu, M.; Wang, B.; Shang, Y.; Wu, J.; Tian, J. Global trends and future prospects of e-waste research: A bibliometric analysis. *Environ. Sci. Pollut. Res.* **2019**, *26*, 17809–17820. [CrossRef]
78. Keiluweit, M.; Nicöl, P.; Johnson, M.G.; Kleber, M. Dynamic Molecular Structure of Plant Biomass-Derived Black Carbon (Biochar). *Environ. Sci. Technol.* **2010**, *44*, 1253–1259. [CrossRef]
79. Lehmann, J. Bio-Energy in the Black. *Front. Ecol. Environ.* **2007**, *5*, 381–387. [CrossRef]
80. Warnock, D.D.; Lehmann, J.; Kuyper, T.W.; Rillig, M.C. Mycorrhizal Responses to Biochar in Soil – Concepts and Mechanisms. *Plant Soil* **2007**, *300*, 9–20. [CrossRef]
81. Mukherjee, A.; Zimmermann, A.R.; Harris, W. Surface Chemistry Variations among a Series of Laboratory-Produced Biochars. *Geoderma* **2011**, *163*, 247–255. [CrossRef]
82. Pei, Z.; Chen, S.; Ding, L.; Liu, J.; Cui, X.; Qiu, F. Current perspectives and trend of nanomedicine in cancer: A review and bibliometric analysis. *J. Control. Release* **2022**, *352*, 211–241. [CrossRef] [PubMed]

83. Meier, E.; Thorburn, P.; Biggs, J.; Palmer, J.; Dumbrell, N.; Kragt, M. Using machine learning with case studies to identify practices that reduce greenhouse gas emissions across Australian grain production regions. *Agron. Sustain. Dev.* **2023**, *43*, 29. [CrossRef]
84. Meyer, S.; Genesio, L.; Vogel, I.; Schmidt, H.-P.; Soja, G.; Shackley, S.; Verheijen, F.G.A.; Glasen, B. Biochar standardization and legislation harmonization. *J. Environ. Eng. Landsc. Manag.* **2017**, *25*, 175–191. [CrossRef]
85. BSR WATER. *Palette of Solutions for Nutrient Recycling in the Baltic Sea Region*; Helsinki Commission: Helsinki, Finland, 2021; ISBN 978-951-0-06818-7.
86. Cool Planet Energy Systems. *Cool Planet Announces Launch of Cool Terra™ Biochar Soil Amendment: Reduces Atmospheric CO_2 and Increases Crop Yields*; Business Wire: San Francisco, CA, USA, 2013.
87. Chen, C.M. CiteSpace II: Detecting and visualizing emerging trends and transient patterns in scientific literature. *J. Am. Soc. Inf. Sci. Technol.* **2006**, *57*, 359–377. [CrossRef]
88. Zhao, X.; Ma, X.W.; Chen, B.Y.; Shang, Y.; Song, M.L. Challenges toward carbon neutrality in China: Strategies and countermeasures. *Resour. Conserv. Recycl.* **2022**, *176*, 105959. [CrossRef]
89. Liu, Z.; Deng, Z.; He, G.; Wang, H.L.; Zhang, X.; Lin, J.; Qi, Y.; Liang, X. Challenges and opportunities for carbon neutrality in China. *Nat. Rev. Earth Environ.* **2022**, *3*, 141–155. [CrossRef]
90. Li, L.; Ma, S.; Liu, X.; Liu, J.; Lu, Y.; Zhao, P.; Kassabi, N.; Hamdi, E.; Elsworth, D. Coal measure gas resources matter in China: Review, challenges, and perspective. *Phys. Fluids* **2024**, *36*, 071301. [CrossRef]
91. Li, L.; Liu, J.; Wang, W.; Zhao, P.; Lu, Y. Variational Reservoir Characteristics of the Wufeng–Longmaxi Formation from Different Sedimentary Region and Its Implications in Southeastern Sichuan Basin, China. *ACS Omega* **2023**, *8*, 20684–20696. [CrossRef]
92. Wang, X.; Xu, S.; Wang, Z.; Peng, L.; Wang, C. International scientific collaboration of China: Collaborating countries, institutions and individuals. *Scientometrics* **2013**, *95*, 885–894. [CrossRef]
93. Zhang, X.; Sun, X.; Zhang, Y. Analysis on the Characteristics of Withdrawal of International Scientific Papers. *China Sci. Technol. Resour. Rev.* **2023**, *55*, 73–81.
94. Zimmerman, A.R. Abiotic and microbial oxidation of laboratory-produced black carbon (biochar). *Environ. Sci. Technol.* **2010**, *44*, 1295–1301. [CrossRef] [PubMed]
95. International Biochar Initiative. Standardized Product Definition and Product Testing Guidelines for Biochar That Is Used in Soil (aka IBI Biochar Standards) (Version 2.1, 23 November 2015). Available online: http://www.biochar-international.org/characterizationstandard (accessed on 5 May 2025).
96. Xu, Z.; He, M.; Xu, X.; Cao, X.; Tsang, D.C.W. Impacts of different activation processes on the carbon stability of biochar for oxidation resistance. *Bioresour. Technol.* **2021**, *338*, 125555. [CrossRef]
97. Wu, M.; Han, X.; Zhong, T.; Yuan, M.; Wu, W. Soil organic carbon content affects the stability of biochar in paddy soil. *Agric. Ecosyst. Environ.* **2016**, *223*, 59–66. [CrossRef]
98. Zhang, X.; Zhang, Q.; Zhang, L.; Xu, X.; Bi, R.; Xiong, Z. Biochar addition stabilized soil carbon sequestration by reducing temperature sensitivity of mineralization and altering the microbial community in a greenhouse vegetable field. *J. Environ. Manage.* **2022**, *313*, 114972. [CrossRef]
99. Gross, A.; Bromm, T.; Polifka, S.; Fischer, D.; Glaser, B. Long-term biochar and soil organic carbon stability—Evidence from field experiments in Germany. *Sci. Total Environ.* **2024**, *954*, 176340. [CrossRef] [PubMed]
100. Cao, Y.; Shan, Y.; Wu, P.; Zhang, P.; Zhang, Z.; Zhao, F.; Zhu, T. Mitigating the global warming potential of rice paddy fields by straw and straw-derived biochar amendments. *Geoderma* **2021**, *396*, 115081. [CrossRef]
101. Nan, Q.; Xin, L.; Qin, Y.; Waqas, M.; Wu, W. Exploring long-term effects of biochar on mitigating methane emissions from paddy soil: A review. *Biochar* **2021**, *3*, 125–134. [CrossRef]
102. Liu, Y.; Weng, Z.; Han, B.; Guo, Z.; Tian, H.; Tang, Y.; Cai, Y.; Yang, Z. Recent studies on the comprehensive application of biochar in multiple environmental fields. *J. Clean. Prod.* **2023**, *421*, 138495. [CrossRef]
103. Wang, Z.; Chen, C.; Zhang, X.; Li, J.; Wu, W.; Yan, B.; Chen, W.; Zuo, X.; Li, X.; Yang, G.; et al. An operation strategy towards efficient energy recovery from integrated anaerobic digestion and pyrolysis of corn stalk. *Resour. Conserv. Recycl.* **2023**, *188*, 106710. [CrossRef]
104. He, Q.; Raheem, A.; Ding, L.; Xu, J.; Cheng, C.; Yu, G. Combining wet torrefaction and pyrolysis for woody biochar upgradation and structural modification. *Energy Convers. Manag.* **2021**, *243*, 114383. [CrossRef]
105. Cheng, N.; Wang, B.; Wu, P.; Lee, X.; Xing, Y.; Chen, M.; Gao, B. Adsorption of emerging contaminants from water and wastewater by modified biochar: A review. *Environ. Pollut.* **2021**, *273*, 116448. [CrossRef]
106. Zhang, P.; Li, Y.; Cao, Y.; Han, L. Characteristics of tetracycline adsorption by cow manure biochar prepared at different pyrolysis temperatures. *Bioresour. Technol.* **2019**, *285*, 121348. [CrossRef]
107. Zhu, X.; Li, K.; Zhang, L.; Wu, X.; Zhu, X. Comparative study on the evolution of physicochemical characteristics of biochar produced from bio-oil distillation residue under different induction atmosphere. *Energy Convers. Manag.* **2018**, *157*, 288–293. [CrossRef]

108. United Nations Environment Programme; International Waste Management Association. *Global Waste Management Outlook*; International Environmental Technology Centre: Osaka, Japan, 2015.
109. Karki, B.K. Amended biochar in constructed wetlands: Roles, challenges, and future directions removing pharmaceuticals and personal care products. *Heliyon* **2024**, *10*, e39848. [CrossRef] [PubMed]
110. Hao, L.; Zhao, Z.; Zhou, H.; Wen, L.; Liu, X.; Yu, Y.; Liu, Y. Resource utilization of tea waste in biochar and other areas: Current status, challenges and future prospects. *J. Environ. Manag.* **2025**, *377*, 124569. [CrossRef] [PubMed]
111. Yuan, S.; Zhao, L.; Meng, H.; Shen, Y. The main types of biochar and their properties and expectative researches. *J. Plant Nutr. Fertil.* **2016**, *22*, 1402–1417.
112. Li, Y. Research progress of biochar preparation and performance. *Hans J. Chem. Eng. Technol.* **2020**, *10*, 384–388. [CrossRef]
113. Li, Y.; Li, H.; Huang, W.; Sun, Y.; Cao, J.; Luo, J. Research progress on the biochar production and its applications in enhancing electron transport and catalysis performance. *Res. Environ. Sci.* **2021**, *34*, 1157–1167.
114. Liu, L.; Huang, L.; Huang, R.; Lin, H.; Wang, D. Immobilization of heavy metals in biochar derived from co-pyrolysis of sewage sludge and calcium sulfate. *J. Hazard. Mater.* **2021**, *403*, 123648. [CrossRef]
115. Bai, X.; Zhang, S.; Shao, J.; Chen, A.; Jiang, J.; Chen, A.; Luo, S. Exploring the negative effects of biochars on the germination, growth, and antioxidant system of rice and corn. *J. Environ. Chem. Eng.* **2022**, *10*, 107398. [CrossRef]
116. International Biochar Initiative. Biochar Feedstocks. Available online: https://biochar-international.org/ (accessed on 5 May 2025).
117. Chen, D.; Zhuang, X.; Gan, Z.; Cen, K.; Ba, Y.; Jia, D. Co-pyrolysis of light bio-oil leached bamboo and heavy bio-oil: Effects of mass ratio, pyrolysis temperature, and residence time on the biochar. *Chem. Eng. J.* **2022**, *437*, 135253. [CrossRef]
118. O'Boyle, M.; Mohamed, B.A.; Li, L.Y. Co-pyrolysis of sewage sludge and biomass waste into biofuels and biochar: A comprehensive feasibility study using a circular economy approach. *Chemosphere* **2024**, *350*, 141074. [CrossRef] [PubMed]
119. Márquez, A.; Ortiz, I.; Sánchez-Hervás, J.M.; Monte, M.C.; Negro, C.; Blanco, Á. Global trends of pyrolysis research: A bibliometric analysis. *Environ. Sci. Pollut. Res.* **2024**, *31*, 931–947. [CrossRef]
120. Giusitin, M.Z. Advantages of Co-Pyrolysis of Sewage Sludge with Agricultural and Forestry Waste. *Energies* **2024**, *17*, 5736. [CrossRef]
121. Beillouin, D.; Corbeels, M.; Demenois, J.; Berre, D.; Boyer, A.; Fallot, A.; Feder, F.; Cardinael, R. A global meta-analysis of soil organic carbon in the Anthropocene. *Nat. Commun.* **2023**, *14*, 3700. [CrossRef]
122. Raza, S.; Irshad, A.; Margenot, A.; Zamanian, K.; Nan, L.; Ullah, S.; Zhao, X. Inorganic carbon is overlooked in global soil carbon research: A bibliometric analysis. *Geoderma* **2024**, *443*, 116831. [CrossRef]
123. Mentges, M.I.; Reichert, J.M.; Rodrigues, M.F.; Awe, G.O.; Mentges, L.R. Capacity and intensity soil aeration properties affected by granulometry, moisture, and structure in no-tillage soils. *Geoderma* **2016**, *263*, 47–59. [CrossRef]
124. Tang, H.; Liu, Y.; Li, X.; Muhammad, A.; Huang, G. Carbon sequestration of cropland and paddy soils in China: Potential, driving factors, and mechanisms. *Greenh. Gases Sci. Technol.* **2019**, *9*, 872–885. [CrossRef]
125. Wang, J.; Wang, B.; Bian, R.; He, W.; Liu, Y.; Shen, G.; Xie, H.; Feng, Y. Bibliometric analysis of biochar-based organic fertilizers in the past 15 years: Focus on ammonia volatilization and greenhouse gas emissions during composting. *Environ. Res.* **2024**, *243*, 117853. [CrossRef]
126. Xie, Y.; Li, C.; Chen, H.; Gao, Y.; Vancov, T.; Keen, B.; Van Zwieten, L.; Fang, Y.; Sun, X.; He, Y.; et al. Methods for quantification of biochar in soils: A critical review. *Catena* **2024**, *241*, 108082. [CrossRef]
127. Gross, A.; Bromm, T.; Glaser, B. Soil Organic Carbon Sequestration after Biochar Application: A Global Meta-Analysis. *Agronomy* **2021**, *11*, 2474. [CrossRef]
128. Han, J.; Elgowainy, A.; Dunn, J.B.; Wang, M.Q. Life cycle analysis of fuel production from fast pyrolysis of biomass. *Bioresour. Technol.* **2013**, *133*, 421–428. [CrossRef] [PubMed]
129. Rasul, M.; Cho, J.; Shin, H.-S.; Hur, J. Biochar-induced priming effects in soil via modifying the status of soil organic matter and microflora: A review. *Sci. Total Environ.* **2022**, *805*, 150304. [CrossRef] [PubMed]

energies

Review

CO_2-ECBM from a Full-Chain Perspective: Mechanism Elucidation, Demonstration Practices, and Future Outlook

Yinan Cui [1,2], Chao Li [1,2], Yuchen Tian [3,4,5,*], Bin Miao [6,7], Yanzhi Liu [3,4,5], Zekun Yue [3,4,5], Xuguang Dai [3,4,5], Jinghui Zhao [2,8], Hequn Gao [1,2], Hui Li [1,2], Yaozu Zhang [1,2], Guangrong Zhang [2,8], Bei Zhang [9], Shiqi Liu [3,4,5] and Sijian Zheng [3,4,5,*]

1 Exploration and Development Research Institute, East China Oil and Gas Company, SINOPEC, Nanjing 210000, China; cuiyn1109.hdsj@sinopec.com (Y.C.); lmc937615@163.com (C.L.); gaohc.hdsj@sinopec.com (H.G.); 15050780932@163.com (H.L.); zhangyaozu.hdsj@sinopec.com (Y.Z.)
2 Key Laboratory of Deep Coalbed Methane Exploration and Development, SINOPEC, Nanjing 210000, China; zhaojh.hdsj@sinopec.com (J.Z.); zhanggr20693.hdsj@sinopec.com (G.Z.)
3 School of Resources and Geosciences, China University of Mining and Technology, Xuzhou 221116, China; liuyanzhizzz@163.com (Y.L.); yuezekun0114@163.com (Z.Y.); xuguang.dai@cumt.edu.cn (X.D.); liushiqi@cumt.edu.cn (S.L.)
4 Jiangsu Key Laboratory of Coal-Based Greenhouse Gas Control and Utilization, China University of Mining and Technology, Xuzhou 221108, China
5 Carbon Neutrality Institute, China University of Mining and Technology, Xuzhou 221008, China
6 College of Resources, Shandong University of Science and Technology, Tai'an 271019, China; miaobin@sdust.edu.cn
7 Zaozhuang Mining Group Co., Ltd., Zaozhuang 277000, China
8 Linfen Coalbed Methane Branch, SINOPEC, Linfen 041000, China
9 China Construction Eighth Engineering Division Guangxi Construction Co., Ltd., Nanning 530000, China; beizhang15578@163.com
* Correspondence: yuchentian0533@163.com (Y.T.); sijian.zheng@cumt.edu.cn (S.Z.)

Abstract: CO_2-enhanced coalbed methane recovery (CO_2-ECBM) represents a promising pathway within carbon capture, utilization, and storage (CCUS) technologies, offering dual benefits of methane production and long-term CO_2 sequestration. This review provides a comprehensive analysis of CO_2-ECBM from a full-chain perspective (Mechanism, Practices, and Outlook), covering fundamental mechanisms and key engineering practices. It highlights the complex multi-physics processes involved, including competitive adsorption–desorption, diffusion and seepage, thermal effects, stress responses, and geochemical interactions. Recent progress in laboratory experiments, capacity assessments, site evaluations, monitoring techniques, and numerical simulations are systematically reviewed. Field studies indicate that CO_2-ECBM performance is strongly influenced by reservoir pressure, temperature, injection rate, and coal seam properties. Structural conditions and multi-field coupling further affect storage efficiency and long-term security. This work also addresses major technical challenges such as real-time monitoring limitations, environmental risks, injection-induced seismicity, and economic constraints. Future research directions emphasize the need to deepen understanding of coupling mechanisms, improve monitoring frameworks, and advance integrated engineering optimization. By synthesizing recent advances and identifying research priorities, this review aims to provide theoretical support and practical guidance for the scalable deployment of CO_2-ECBM, contributing to global energy transition and carbon neutrality goals.

Keywords: CO_2-ECBM; CCUS; site suitability assessment; numerical simulation; environmental risk; energy security

1. Introduction

Global climate change represents one of the most critical challenges facing modern human society, with greenhouse gas emissions—particularly CO_2—widely recognized as a primary driver of global warming [1–4]. In light of the progress of international climate agreements, such as the Paris Agreement, achieving net-zero CO_2 emissions and global carbon neutrality has emerged as a widely shared consensus among the global community [5–7]. Against this backdrop, the central challenge facing countries worldwide is how to effectively control carbon emissions while ensuring the security of energy supply.

CCUS is regarded as a key pathway for facilitating the low-carbon transition of energy systems. CCUS enables the efficient capture of CO_2 from industrial sources and achieves significant emission reductions through secure geological storage or resource utilization [8–13] (Figure 1). CO_2-ECBM represents a synergistic pathway within the broader framework of CCUS, simultaneously targeting enhanced coalbed methane recovery and long-term CO_2 geological storage. This technology involves injecting CO_2 into coal seams, where it preferentially adsorbs onto the coal matrix due to its stronger affinity compared with methane. This competitive adsorption mechanism displaces the adsorbed CH_4, enabling its extraction, while securely storing CO_2 within the coal structure. As a result, CO_2-ECBM offers the dual benefits of energy production and greenhouse gas mitigation [14–16]. Compared with conventional coalbed methane extraction methods, CO_2-ECBM offers multiple advantages. It causes minimal disruption to the geological formation, enables efficient methane release, and exhibits strong environmental compatibility and long-term storage stability. These characteristics make CO_2-ECBM a promising technological bridge between energy development and climate mitigation. Globally, many coal-bearing regions are characterized by low permeability, complex pore structures, and limited methane desorption capacity, which often result in low recovery rates using traditional CBM extraction techniques—hindering their commercial viability [17–19]. In such contexts, CO_2-ECBM demonstrates clear technical and economic advantages, particularly in deep and complex coal seams where conventional methods fall short. This approach not only provides a sustainable pathway for CBM resource development but also offers reliable geological space for CO_2 sequestration.

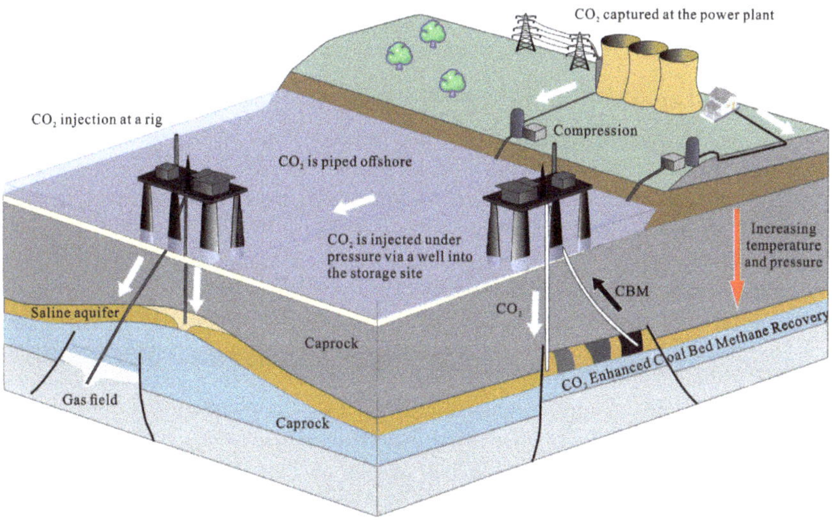

Figure 1. Schematic diagram of selected pathways for carbon dioxide capture, utilization, and storage.

In recent years, CCUS technologies have witnessed rapid advancement, with CO_2-ECBM emerging as a focal point of interest across both academia and industry. Research on this technology spans multiple scales, ranging from the fundamental mechanisms of gas adsorption and the dynamics of diffusion and seepage to field experiments and engineering demonstrations (Figure 2) [20–22]. CO_2-ECBM is currently transitioning from foundational studies to pilot-scale implementations, and is poised to move toward large-scale deployment. Nevertheless, this technology still faces numerous scientific and engineering challenges. This paper begins with a review of the fundamental mechanisms of CO_2-ECBM, and, drawing on the recent research conducted by our team, systematically summarizes the current progress and practical developments. It further explores global engineering demonstrations, key influencing factors, and potential future trends. The objective is to provide a comprehensive reference for the ongoing optimization and implementation of CO_2-ECBM, and to lay a solid foundation for its eventual large-scale application.

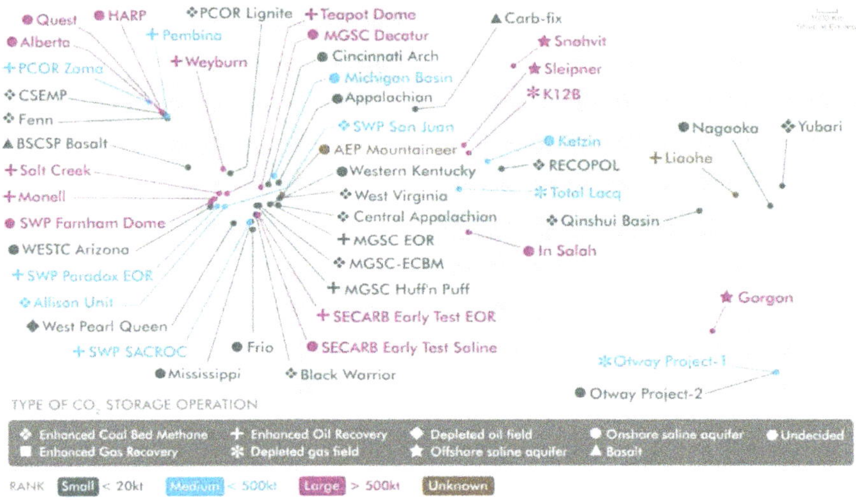

Figure 2. Global distribution of selected CCUS demonstration projects.

2. Mechanisms of CO_2-Enhanced Coalbed Methane Recovery

CO_2-ECBM enhances methane recovery and enables carbon sequestration by injecting carbon dioxide into coal seams. The core mechanisms primarily involve two interrelated processes: adsorption–desorption behavior and diffusion–seepage transport. A schematic diagram of the CO_2-ECBM mechanisms is shown in Figure 3.

The adsorption–desorption mechanism is fundamental to CO_2-ECBM. Coal exhibits a stronger adsorption affinity for CO_2 than for CH_4. Upon CO_2 injection, a distinct competitive adsorption effect occurs—CO_2 displaces CH_4 previously adsorbed on the coal surface, thereby enhancing CH_4 desorption. Concurrently, the increase in CO_2 concentration and the decrease in CH_4 concentration within the free gas phase disrupt the original adsorption–desorption equilibrium, further promoting CH_4 release. To quantitatively describe this competitive adsorption behavior, the Ideal Adsorbed Solution Theory (IAST) is widely adopted. IAST provides a thermodynamic framework for predicting the multicomponent adsorption equilibrium based on pure-component isotherms, assuming the adsorbed phase behaves as an ideal solution. This approach enables accurate estimation of CO_2 and CH_4 uptake under varying pressure and composition conditions without requiring extensive multicomponent experiments. By incorporating IAST into gas adsorption modeling, researchers can more effectively characterize selective adsorption phenomena and optimize injection strategies. The extended Langmuir model is a theoretical framework for describing the competitive adsorption of multicomponent gases on solid surfaces. It modifies the classic Langmuir single-component model to account for the complexity of gas mixtures, where each component competes for the same adsorption sites. Adsorption is influenced by both the component's own properties and the presence of other gases, with the model incorporating partial pressures and adsorption affinities to capture these interactions. This approach enables accurate prediction of gas adsorption behavior in porous media under complex conditions, such as CO_2 sequestration and methane recovery in coal seams. Overall, this process not only facilitates methane desorption and recovery, but also results in the sequestration of CO_2 in an adsorbed state within the coal matrix, establishing a synergistic "displacement–sequestration" mechanism [23–25].

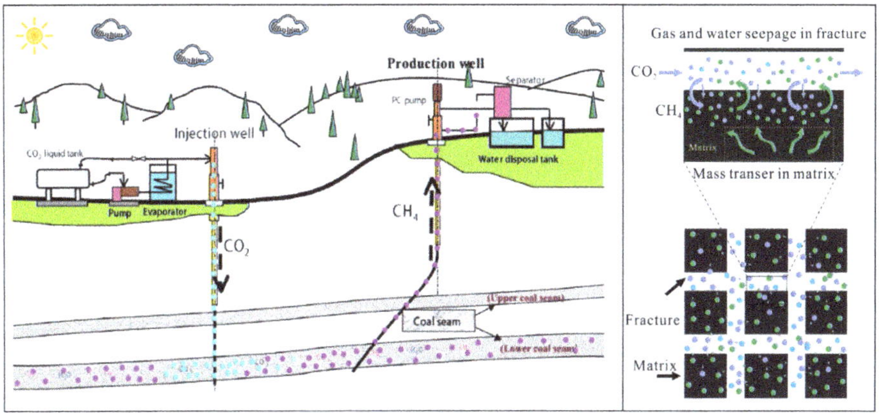

Figure 3. Mechanistic diagram of CO_2-ECBM.

In addition, the diffusion–seepage process governs the transport pathways and migration efficiency of desorbed methane and injected CO_2. The desorbed CH_4 molecules must diffuse through micropores into the cleat and fracture system, and then migrate via seepage flow toward the production wells to enable effective extraction. During this process, injected CO_2 not only functions as a displacement agent but also establishes a localized pressure gradient that helps overcome the intrinsic low permeability of coal seams, thereby improving gas mobility. Notably, the common phenomenon of "compaction-induced permeability reduction" during conventional methane extraction can be mitigated under continuous CO_2 injection. The sustained pressure around the wellbore helps preserve coal structure integrity, prevents pore closure, and ultimately enhances both permeability and flow capacity [26,27].

In summary, the CO_2-ECBM process integrates coupled mechanisms of adsorption–desorption and diffusion–seepage: adsorption–desorption enables selective gas separation and long-term CO_2 sequestration, while diffusion–seepage promotes CH_4 displacement and recovery. This synergistic mechanism typically enhances methane production by around 20–30% compared with conventional CBM methods, with CO_2 retention efficiencies after 10 years generally ranging from approximately 70% to 90%, depending on reservoir conditions.

3. Current Development of CO_2-ECBM Technology

3.1. Experimental Research Progress on CO_2-ECBM

Experimental research on CO_2-ECBM aims to elucidate the adsorption, desorption, diffusion, and seepage mechanisms of CO_2 within coal seam media at both microscopic and macroscopic scales, thereby assessing its effectiveness in enhancing coalbed methane recovery and carbon sequestration capacity. Current experimental efforts mainly focus on the characterization of reservoir properties in coal seams and surrounding rocks, simulation of CO_2 sequestration mechanisms, and the evaluation of permeability and injectivity during the injection process [28–30].

Representative coal and surrounding rock samples are used to perform comprehensive rock mechanics and petrophysical property tests. With the aid of triaxial loading apparatuses, low-field nuclear magnetic resonance (LF-NMR) instruments, various gas adsorption analyzers, mercury intrusion porosimeters, and scanning electron microscopes (SEMs), key mechanical parameters such as elastic modulus, shear modulus, and Poisson's ratio, along with physical properties including density, porosity, permeability, specific surface area, pore size distribution, pore volume, and CO_2 adsorption capacity, can be quantitatively determined. Figure 4 presents mercury intrusion porosimetry results to characterize the pore size distribution of coal samples—an effective method for evaluating pore structure. These experiments are crucial for understanding the heterogeneity of reservoir properties across different geological formations and their influence on CO_2 migration behavior.

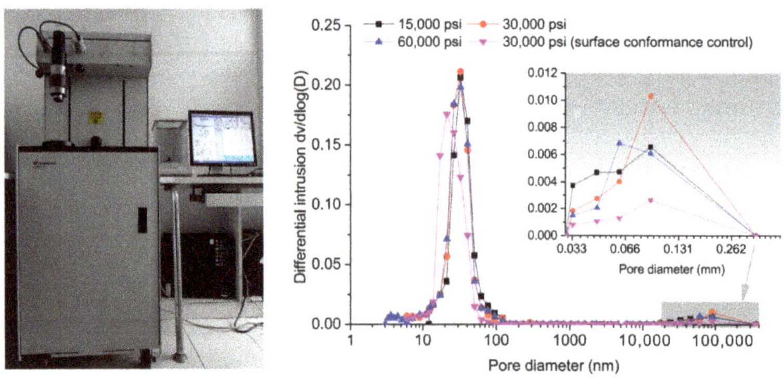

Figure 4. Pore size distribution of coal samples based on mercury intrusion porosimetry.

In terms of sequestration mechanisms, experimental studies are typically conducted under simulated reservoir conditions (e.g., temperature, pressure, and water saturation) to investigate three types of physical simulation experiments: adsorption-based, dissolution-based, and mineralization-based CO_2 sequestration. Adsorption experiments employ high-pressure gas adsorption/desorption systems to evaluate the adsorption capacity and characteristics of CO_2 on coal or rock surfaces and within pores. Dissolution experiments are conducted using CO_2–water–rock reaction vessels to examine the dissolution behavior of CO_2 in pore water and its subsequent reactions with minerals. Mineralization experiments focus on observing the formation, transformation, and stability of newly generated minerals after reactions, in order to assess the long-term sequestration potential.

Moreover, to evaluate the injectivity and evolution characteristics of CO_2 in coal seams, particular attention must be paid to geochemical effects and volumetric stress–strain responses [31–33]. For instance, CO_2 injection simulation systems are used to monitor volumetric strain, fracture development, and permeability changes in rock samples during the injection process. These observations, combined with analytical techniques such as

X-ray diffraction (XRD) and inductively coupled plasma optical emission spectrometry (ICP-OES), allow for detailed analysis of changes in fluid composition and provide a comprehensive understanding of the geochemical and mechanical responses induced by CO_2 injection. Figure 5 presents the modifications in ion concentrations and trace element compositions that are observed across different samples following geochemical interactions. The data indicate that notable alterations in both major ion concentrations and trace element profiles do occur, suggesting that geochemical interactions have a definitive impact on the chemical properties of the samples.

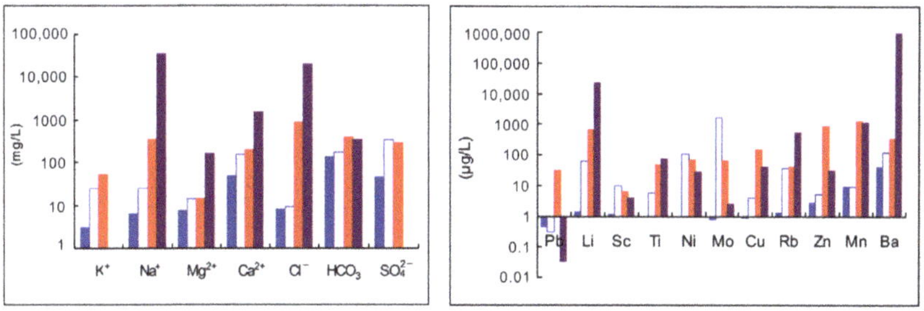

Figure 5. Combined concentration changes of major ions and trace elements before and after reactions.

In summary, these experimental studies not only elucidate the multiscale physicochemical behavior of CO_2 within coal seams, but also offer robust experimental support for the assessment of CO_2 storage capacity, suitability evaluation of CO_2-ECBM, numerical simulation research, and injectivity analysis.

3.2. Current Research on CO_2-ECBM Storage Capacity Assessment and Suitability Evaluation

In recent years, extensive research has been conducted by scholars to evaluate the storage capacity of CO_2-ECBM, leading to the development of assessment models that encompass three primary sequestration states: adsorbed, dissolved, and free-phase. These models quantitatively describe the geological sequestration potential of CO_2 in coal seams through theoretical calculations and calibration with experimental data. According to widely accepted approaches from previous studies, CO_2 sequestration in coal seams predominantly occurs in the adsorbed phase, followed by dissolved and free phases, with the adsorbed phase typically accounting for the largest proportion [34–38]. Specifically, the assessment models are generally expressed in the following form [39]:

$$M_{car} = S_{car}\rho_{car}M_{coal}\left(m_{ab} + m_d + m_f\right) \tag{1}$$

where M_{car} is the CO_2 storage capacity in t; S_{car} is the CO_2 storage coefficient in the coal reservoir; ρ_{car} is the CO_2 density under standard conditions (101.325 kPa, 273.15 K), equal to 1.977 kg/m^3; M_{coal} is the proven coal reserves in t; m_{ab} is the amount of CO_2 stored in the adsorbed state per unit mass of coal in m^3/t; m_d is the amount of CO_2 stored in the dissolved state per unit mass of coal in m^3/t; and m_f is the amount of CO_2 stored in the free state per unit mass of coal in m^3/t.

For a unit mass of coal, the CO_2 sequestration potential in the adsorbed state within the reservoir can be expressed as [37–39]:

$$m_{\text{ab}} = \frac{m_{\text{ex}}}{1 - \frac{pT_c}{8Zp_cT}} \tag{2}$$

where p is the pressure of the deep coal reservoir, corresponding to the CO_2 adsorption pressure in MPa; T_c is the critical temperature of CO_2, 304.21 K; Z is the compressibility factor of CO_2 (dimensionless), which can be estimated through multiparameter interpolation; p_c is the critical pressure of CO_2, 7.383 MPa; T is the temperature of the deep coal reservoir, also the CO_2 adsorption temperature in K; and m_{ex} is the excess adsorption capacity of CO_2 per unit mass of coal in m^3/t, which can be calculated using the Dubinin–Radushkevich (D–R) adsorption model [37–39]:

$$m_{ex} = m_0\left(1 - \frac{\rho_f}{\rho_a}\right)e^{-D[\ln(\frac{\rho_a}{\rho_f})]^2} + k\rho_f \tag{3}$$

where m_0 is the maximum CO_2 adsorption capacity per unit mass of coal determined from adsorption experiments in m^3/t; ρ_f and ρ_a are the densities of free-phase and adsorbed-phase CO_2 under the actual pressure and temperature conditions of the coal reservoir in kg/m^3; D is the adsorption constant (dimensionless); and k is a constant related to Henry's law (dimensionless).

The CO_2 density in the coal reservoir is a function of pressure and temperature, expressed as $\rho_g = f(p, T)$, and can be further expressed as [37–39]:

$$\rho_g = \frac{p}{(1 + \delta\phi_\delta^\tau)RT} \tag{4}$$

where $\delta = \rho_c/\rho_f$ is the reduced density of CO_2 (dimensionless), with ρ_c being the critical density of CO_2 (0.45×10^{-3} kg/m^3); and $\tau = T_c/T$ is the reduced temperature (dimensionless). $\phi(\delta,\tau)$ denotes the Helmholtz free energy, which is governed by both temperature and density [37–39]:

$$\phi(\delta,\tau) = \phi^0(\delta,\tau) + \phi^1(\delta,\tau) \tag{5}$$

where $\phi^0(\delta, \tau)$ represents the Helmholtz free energy of the ideal fluid; and $\phi^1(\delta, \tau)$ denotes the residual Helmholtz free energy of the real fluid.

In coal reservoirs, the solubility-based CO_2 storage potential per unit mass of coal is a function of the coal seam porosity, water saturation, coal density, and the solubility of CO_2 in water, and can be expressed as [37–39]:

$$m_d = \frac{\varphi S_w S_{car}}{\rho_{coal}} \tag{6}$$

where φ is the coal porosity in %; S_w is the water saturation of the coal seam in %; S_{car} is the solubility of CO_2 in coal seam water in mol/cm^3; and ρ_{coal} is the coal density in kg/m^3.

According to Boyle–Mariotte's law, and taking into account the gas saturation of the coal seam, the free-phase CO_2 storage potential per unit mass of coal in the reservoir can be expressed as [37–39]:

$$m_f = \frac{\varphi S_g p T_0}{\rho_{visual} Z p_0 T} \tag{7}$$

where S_g is the gas saturation of the coal seam in %; p_0 is the standard atmospheric pressure; T_0 is the temperature under standard conditions; and ρ_{visual} is the apparent density of coal in kg/m^3.

To address the issue of CO_2 geological storage suitability, researchers generally advocate for the establishment of a comprehensive evaluation index system that integrates multiple dimensions and hierarchical levels [40–45]. This system should fully consider key factors such as reservoir security, storage capacity potential, and socio-economic acceptability. Reservoir security serves as the foundation for the sustainable operation of storage projects, and is primarily determined by regional crustal stability and caprock integrity.

For crustal stability, assessments typically involve evaluating peak ground acceleration, historical seismic activity intensity, and the spatial distribution of active faults to determine the overall structural stability of the area. Caprock conditions, including lithology, thickness, continuity, and burial depth, directly affect the sealing performance and migration risk of CO_2. In terms of storage capacity, the scale of the structural unit, reservoir physical properties, and theoretical storage potential are the core evaluation components. The area and thickness of the structural unit define the spatial boundary conditions of the storage site. Reservoir development conditions—such as thickness, porosity, and permeability—determine the CO_2 injectivity and effective storage capacity. Storage potential usually refers to the estimated geological storage volume and storage capacity per unit area, reflecting the resource base of CO_2 geological storage in a given region.

In addition, the feasibility of CO_2-ECBM must also consider socio-economic conditions, particularly carbon source distribution and public acceptance. Economically, proximity to large-scale CO_2 emission sources helps reduce transportation costs and project complexity. Meanwhile, the level of public acceptance may influence the implementation and sustainability of engineering operations. Therefore, when constructing a CO_2-ECBM suitability evaluation framework, it is essential to integrate technical, geological, and economic factors into a unified system. Figure 6 illustrates the suitability evaluation index system for CO_2-ECBM constructed by the authors.

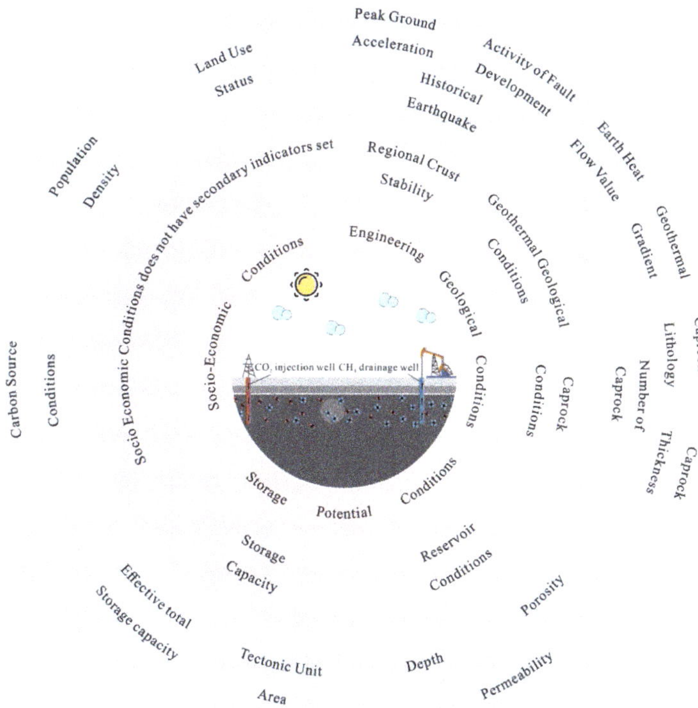

Figure 6. Evaluation indicators for CO_2-ECBM suitability assessment.

In practical suitability analyses, researchers often adopt a combination of the Analytic Hierarchy Process (AHP) and Fuzzy Comprehensive Evaluation (FCE). This approach enables the construction of a multi-level hierarchical model consisting of a target layer, a criterion layer, and an index layer. Expert judgment matrices are used to determine the weight of each indicator. Coupled with regional geological data, monitoring records, and sampling results, this method provides a quantified, comprehensive evaluation to offer a scientific basis for CO_2-ECBM suitability assessments.

3.3. Monitoring Technologies and Regulatory Frameworks for CO_2-ECBM

In response to the urgent need for CO_2 leakage and migration surveillance, numerous researchers have conducted in-depth investigations into surface deformation and environmental changes induced by CO_2 injection [46–52]. At present, a widely recognized strategy among Chinese scholars involves an integrated air–space–surface–downhole remote sensing-based monitoring system [53] (Figure 7).

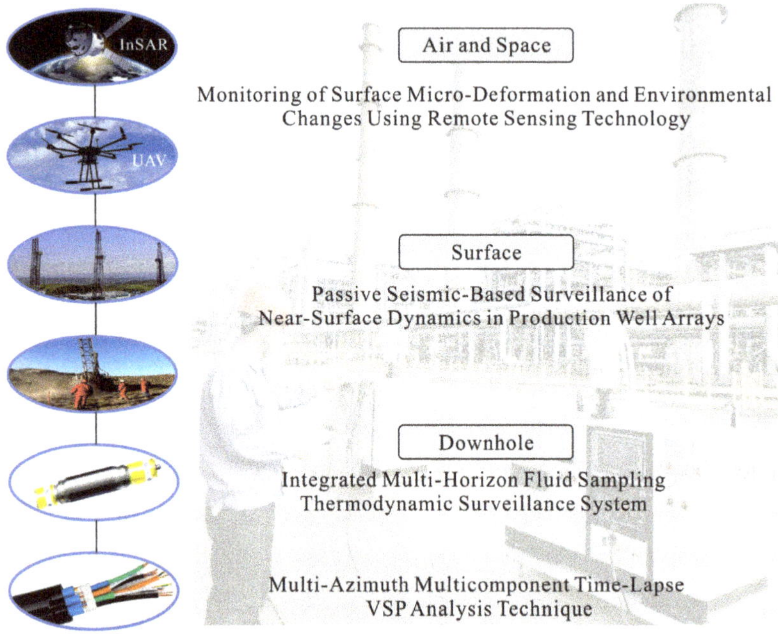

Figure 7. Air–space–surface–downhole integrated monitoring system.

Within the air and space segment, large-scale monitoring relies on optical and radar satellite platforms to acquire long-term, high-resolution InSAR datasets. These datasets enable all-weather tracking of surface micro-deformation and vegetation-related environmental dynamics, thereby revealing CO_2 migration patterns and their potential ecological impacts. Simultaneously, unmanned aerial vehicles (UAVs) equipped with optical and thermal infrared sensors are employed for targeted, rapid surveillance of the injection zone. These systems can acquire surface models and high-resolution imagery, while also detecting potential leakage hotspots and anomalous reactions in a timely manner, ensuring real-time and accurate monitoring. The integration of satellite-based remote sensing and UAV observations provides both macro and micro perspectives on the CO_2 storage site, forming a critical technological foundation for continuous and dynamic monitoring of CO_2 migration and environmental responses at the regional scale.

For surface-based monitoring, long-term surveillance of production well arrays combined with passive seismic-based surveillance of near-surface dynamics enables the effective detection of fracture propagation and surface deformation induced by CO_2 injection.

With the aid of advanced passive seismic fracture tomography, subsurface rupture events and the evolution of CO_2 plume fronts can be directly visualized. This provides crucial empirical data for understanding fluid behaviors and fault activation during the storage process. This technique also overcomes the limitations of traditional surface monitoring in terms of signal-to-noise ratio and resolution, significantly enhancing the ability to identify and provide early warnings of CO_2 leakage risks.

In the downhole monitoring system, long-term deployment of fiber-optic sensors along high-angle wells allows for real-time, stratified gas–water sampling and precise tracking of temperature and pressure changes. Additionally, a multi-azimuth multicomponent time-lapse VSP analysis technique is introduced to quantify subsurface fluid dynamics and structural deformation responses. This system not only provides real-time feedback on the in-well CO_2 storage status, but also—through the interpretation of time-lapse datasets—helps reveal the spatial distribution and evolution of CO_2 migration under multi-physics coupling conditions. Together, these insights offer robust data support and scientific guidance for the optimization of CO_2-ECBM deployment strategies.

In parallel with technological advancements, many countries, including China, the United States, and Australia, have progressively improved their policy support and management frameworks to promote the safe and orderly implementation of CO_2-ECBM. These frameworks generally emphasize the importance of long-term environmental monitoring, risk prevention and control, and accountability mechanisms throughout the project lifecycle. In China, relevant departments have begun to explore coordinated regulatory approaches that integrate environmental supervision, project approval, and data reporting, aiming to enhance oversight efficiency and project transparency. The integration of advanced monitoring technologies with a sound governance structure plays a key role in improving risk management, fostering public trust, and facilitating the sustainable and large-scale development of CO_2-ECBM projects.

3.4. Advances in Numerical Simulation of CO_2-ECBM

Coal seams are characterized by strong heterogeneity and generally low permeability, making efficient CO_2 injection and enhanced CH_4 recovery the key technical challenges of CO_2-ECBM. Numerical simulation has become a powerful tool to reproduce the production history of coalbed methane wells, quantitatively assess the potential of CO_2-ECBM, and support the optimization of engineering schemes and evaluation of CH_4 enhancement strategies. Due to its low cost and short research cycle, numerical modeling has attracted extensive attention in recent years. Research efforts have increasingly focused on developing and improving multi-physics coupled models to accurately describe the complex processes occurring within coal reservoirs, including gas adsorption, diffusion, seepage, heat transfer, geomechanical evolution, and geochemical reactions [54–59]. Early commercial simulators were primarily based on partial differential equations governing adsorption and flow fields. These models could effectively simulate isothermal adsorption and diffusion of gases within the coal matrix, as well as fluid flow in the fracture network. However, they often neglected the coupled effects of temperature fields, geostress variations, and mineral dissolution/precipitation reactions, thus limiting their applicability under the complex conditions of deep coal seams.

To address these limitations, our research team has recently developed a fully coupled adsorption–hydro–thermo–mechanical–chemical (AHTMC) model [57–60]. This type of model integrates equations of energy conservation, fluid dynamics, and geomechanics to comprehensively capture the interactions among multiple physical fields during the CO_2-ECBM process [60]. In the AHTMC framework, adsorption behavior is typically described using a modified Langmuir isotherm, while gas diffusion within the coal matrix is governed by Fick's law. Fluid transport in the fracture system is modeled using a combination of Darcy's law and gas slippage effects. To characterize thermal responses driven by temperature fluctuations, the model incorporates terms for thermal conduction and convection, alongside thermodynamic effects such as adsorption heat release, dissolution exothermicity, and endothermic deformation of coal rock. The geomechanical response is modeled using the theory of poroelasticity, fully considering reservoir deformation resulting from changes in fluid pressure, temperature, and adsorption-induced strain. Some advanced models further integrate the dynamic dissolution of CO_2 and CH_4 in formation water to refine the representation of CO_2 storage mechanisms in water-bearing coal seams.

Currently, finite element platforms such as COMSOL (Comsol multiphysics 6.2) multiphysics have emerged as important tools for simulating fully coupled CO_2-ECBM processes. Their modular and flexible multi-physics environment allows researchers to construct and solve highly complex models with ease. In parallel, advancements in finite difference and open-source simulation platforms are providing new pathways for tackling multiscale and multi-physics coupling challenges. Overall, the scope of numerical simulation in this field has evolved from single-field representations to multi-physics coupling, while the application frontier has expanded to include the quantitative evaluation of key performance indicators—such as CH_4 recovery efficiency and CO_2 adsorption capacity These efforts offer substantial data support for engineering design and risk assessment in real-world projects. Despite these advances, numerical simulation in CO_2-ECBM still faces major challenges, including high model complexity, intensive computational demands, and difficulties in obtaining accurate reservoir parameters. Future research should aim to further optimize model architecture, reduce computational costs, and validate simulations through field data, thereby enhancing their utility in guiding the engineering deployment and large-scale application of CO_2-ECBM technology.

4. Global Demonstration Projects of CO_2-ECBM

Overall, CO_2-ECBM technology remains in the exploratory engineering stage both internationally and in China. A number of pilot projects have been carried out globally, which have, to some extent, demonstrated the technical feasibility of CO_2-ECBM for both long-term CO_2 sequestration and enhanced coalbed methane recovery [61–67] (Table 1). Although research in this field began earlier in several countries, and some projects have progressed to pilot-scale or limited commercial operation, the majority are still in the field-testing phase.

As early as 1995–2001, the United States initiated the first CO_2-ECBM field trial at the Allison Unit in the San Juan Basin. The project employed multi-well co-production techniques and successfully enhanced CBM recovery. This large-scale trial covered an area of approximately 1.295×10^6 m^2, with four CO_2 injection wells and 16 production wells, and a cumulative CO_2 injection volume of 336,000 t [68,69]. Another prominent effort in the San Juan Basin, located in New Mexico, targeted the Upper Cretaceous Fruitland Formation, which has an average depth of approximately 900 m [70]. This test, initiated in 1995, aimed to assess the carbon sequestration potential of the Fruitland coal seams. The reservoir was categorized into four distinct subregions based on geological and production characteristics. The pilot area was located in a high-pressure, high-yield CBM zone. In addition to the San Juan Basin, other demonstration projects in the United States have been conducted in the Illinois and Appalachian Basins. These projects involved unmineable coal seams and lignite blocks, as well as the application of advanced technologies such as horizontal drilling and pre-fracturing, thereby contributing valuable field data and engineering experience to the development of CO_2-ECBM.

China's research and engineering application of CO_2-ECBM started relatively late, but has developed rapidly in recent years with the support of national key scientific and technological programs. Since 2013, several pilot tests have been carried out under the leadership of China United Coalbed Methane Co., Ltd., focusing on both shallow and deep coal seams, ranging from single-well trials to multi-well group operations in the Qinshui Basin (Figure 8) and the Ordos Basin. China's first pilot project was implemented in the Qinshui Basin between 2002 and 2006 through a joint initiative with the Canadian government. The trial involved intermittent CO_2 injection and production in No. 3 coal seam, where a total of 192.8 t of liquid CO_2 was injected, resulting in a significant increase in CBM recovery. From 2011 to 2012, a single-well injection-production test was conducted in a 560 m deep coal seam in Liulin, Shanxi Province, with a total CO_2 injection of 460 t. Subsequently, during 2013–2015, a deeper coal seam test at a depth of 900 m achieved a cumulative CO_2 injection of 4491 tonnes [71].

Figure 8. Field operation view and inspection/maintenance of CBM facilities in the Qinshui Basin, China.

These pilot projects illustrate the evolution of China's CO_2-ECBM development from shallow, single-well tests to deep, multi-well group experiments. The technical maturity has steadily improved, confirming the feasibility of CO_2-ECBM and revealing its promising potential for commercial-scale deployment. At the same time, these efforts have also highlighted the engineering challenges that must be overcome for widespread implementation of this technology.

Table 1. Major CO_2-ECBM projects worldwide [71].

Country	Project Location	Injection Period	CO_2 Injected (t)
USA	Allison Field, San Juan Basin	April 1995–August 2001	336,000
USA	Pump Canyon, San Juan Basin	July 2008–August 2009	16,699.00
USA	Tanquary Farm, Illinois Basin	2008	92.3
USA	Virginia, Central Appalachian Basin	January 2009–February 2009	~900.00
USA	Lignite Block, Williston Basin	2009	90
USA	Black Warrior Basin	June 2010–August 2010	225
USA	Marshall, Northern Appalachian Basin	September 2009–December 2013	4500.00
USA	Buchanan, Central Appalachian Basin	July 2015–August 2015	1470.00
China	Shizhuang South Block, Qinshui Basin	April 2004–June 2004	192.8
China	Shizhuang North Block, Qinshui Basin	April 2010–May 2010	233.6
China	Liulin Block, Eastern Ordos Basin	September 2011–March 2012	460
China	Shizhuang North Block, Qinshui Basin	2013–2015	4491.00
China	Shizhuang South Block, Qinshui Basin	June 2020–June 2021	2001.04
Canada	Fenn Big Valley, Alberta	1998	201
Canada	Alder Flats, Alberta	June 2006	Unknown
Poland	Kaniow, Silesian Basin	August 2004–May 2005	760
Japan	Ishikari Basin, Hokkaido	July 2004–September 2007	~800.00

5. Influencing Factors of CO_2-ECBM

As previously discussed, coalbed methane, as a self-sourced and self-stored unconventional natural gas within coal seams, is not only a clean energy resource but also plays a vital role in mine safety and gas explosion prevention. Its rational development can generate substantial economic benefits. Against this backdrop, CO_2-ECBM, which simultaneously enables long-term CO_2 sequestration and enhanced methane recovery, holds significant theoretical and practical value. Investigating its influencing factors is essential for optimizing production strategies and improving recovery efficiency.

During CO_2-ECBM, methane mainly exists in an adsorbed state within the micropores of the coal matrix, and the adsorption–desorption behavior is primarily controlled by gas partial pressure and temperature. After CO_2 injection, due to its stronger adsorption affinity for coal than methane, CO_2 is preferentially adsorbed onto the internal surfaces of the coal, displacing the adsorbed methane. This displacement effect significantly reduces the partial pressure of methane, promoting its desorption. The released free methane then migrates through the fracture and pore network under the pressure differential between injection and production wells, thereby enhancing the overall recovery rate [72–75]. In practical conditions, reservoir pressure, formation temperature, and CO_2 injection rate all exert significant influence on this process. Lower pressures help break the adsorption equilibrium, facilitating greater methane desorption and production. Suitable temperatures provide gas molecules with sufficient thermal energy to accelerate the desorption process. In addition, higher CO_2 injection rates can rapidly enhance the desorption rate and shorten the desorption time, playing a crucial role in improving methane recovery efficiency.

Geological conditions of the reservoir are fundamental to the feasibility and efficiency of CO_2-ECBM implementation. Optimal coal seams should possess favorable continuity and structural integrity to ensure uniform CO_2 distribution and effective contact with adsorbed methane. Specifically, laterally continuous coal seams facilitate homogeneous CO_2 migration, enhancing the overall displacement efficiency while reducing the likelihood of preferential flow paths and uneven sweep. In contrast, structurally complex formations—such as those with intense faulting or folding—can compromise sealing capacity and introduce high-permeability conduits that serve as potential leakage pathways. These not only limit the interaction between injected CO_2 and CH_4 but also elevate the risk of CO_2 escape, posing both environmental and operational safety concerns. Therefore, geologically stable, laterally extensive, and structurally simple coal reservoirs with low fault density and effective caprock integrity are considered the most suitable targets for CO_2-ECBM, as they support both enhanced methane recovery and secure long-term CO_2 storage [76–78]. In addition, permeability is a key parameter affecting injection efficiency, with higher permeability facilitating smoother fluid flow. Methane-saturated coal seams typically demonstrate better recovery outcomes, providing important guidance for the site selection and technical design of regional CO_2-ECBM projects.

Overall, the CO_2-ECBM process is governed by coupled multi-physical effects—including adsorption–desorption, diffusion and seepage, heat transfer, geomechanical evolution, and geochemical reactions—that collectively determine methane recovery and long-term CO_2 sequestration performance. These factors are not independent; rather, they interact through energy exchange, mass transport, and structural deformation in a highly complex manner. Therefore, future research should focus on elucidating the interaction mechanisms among these variables. A comprehensive understanding of these interactions is essential for revealing the fundamental nature of the CO_2-ECBM process, improving injection–production strategies, and achieving optimal technical and economic performance.

6. Discussion

6.1. Challenges in CO_2-ECBM

Although CO_2-ECBM has achieved a series of pilot-scale successes globally, numerous scientific and engineering challenges remain unresolved in practical applications. One critical issue is the development of efficient monitoring, measurement, and verification (MMV) technologies to accurately assess CO_2 migration, leakage risk, and storage capacity during sequestration, thereby ensuring long-term subsurface storage security. Despite considerable progress internationally—particularly in the United States—risk control remains at the core of CO_2-ECBM research. However, growing attention has been directed toward the associated environmental risks and sequestration costs.

One frequently discussed concern is the potential for injection-induced geomechanical responses. CO_2 injection alters the stress–pressure balance in the reservoir, potentially leading to surface uplift due to increased pore pressure and coal matrix swelling, or, conversely, surface subsidence from long-term compaction or pressure decline. Field evidence from the In Salah Krechba gas field in Algeria (Figure 9) demonstrates that sustained CO_2 injection can cause measurable surface deformation—primarily uplift—highlighting the sensitivity of overburden to subsurface pressure changes. In addition, pressure perturbations may activate pre-existing faults and induce microseismicity, posing risks to both storage security and operational safety. These deformation responses underscore the importance of geomechanical assessment, integrated monitoring, and adaptive risk management strategies during large-scale CO_2 geological sequestration [79]. Preventive measures such as optimizing injection rates, selecting structurally stable reservoirs, and implementing real-time pressure and deformation monitoring can significantly reduce the likelihood of leakage. In the event of leakage, which poses a significant safety risk during large-scale field implementations, remediation techniques—such as pressure relief through extraction wells or sealing of leakage pathways using sealants or cement—can help contain and mitigate its impact. While microfractures induced by injection may enhance reservoir permeability, they also raise concerns about seismic hazards and the integrity of geological storage.

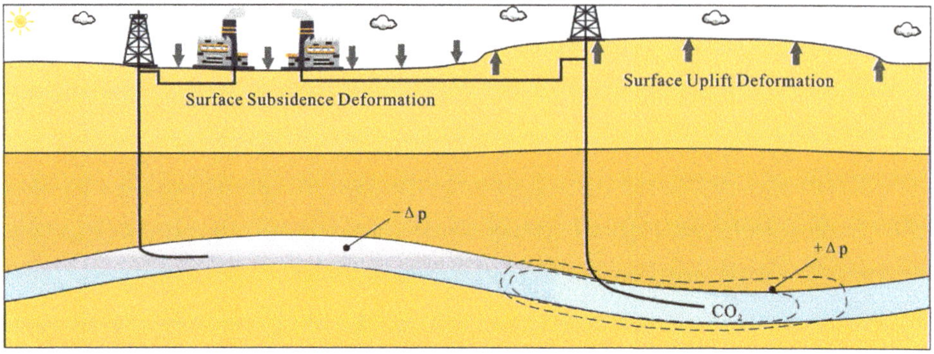

Figure 9. Surface deformation differences of the CCS injection site in Algeria In Salah Krechba gas field.

In addition, the potential environmental impacts of CO_2 geological sequestration have attracted widespread attention. Once injected into coal seams, CO_2 may react with chemical constituents in groundwater, forming acidic compounds that promote the dissolution or precipitation of minerals such as carbonates and pyrite. These geochemical reactions may alter the chemical properties of soil, vegetation, and aquifers, potentially degrading the local ecosystem. Furthermore, such reactions could adversely affect the pore structure of coal, leading to reduced gas recovery efficiency and compromised reservoir stability.

Economic cost remains another major barrier to the large-scale deployment of CO_2-ECBM. The current implementation process requires CO_2 capture, compression, and purification prior to its injection in liquid form into coal seams. The associated equipment investment and energy consumption result in high overall project costs. Although some studies propose using flue gas compression to reduce costs—while leveraging N_2 in flue gas to improve coal permeability—this approach still entails additional expenses for separation and purification at later stages. Striking a balance between technical performance and economic feasibility is a critical challenge for future development.

Another key issue is the alteration of coal seam physical properties caused by CO_2 injection. During the process, coal matrices undergo significant swelling and shrinkage, as CO_2 exhibits a much stronger adsorption capacity than CH_4. This leads to notable volumetric changes during desorption, affecting reservoir porosity and permeability, and ultimately reducing the effectiveness of gas recovery. In parallel, CO_2-induced acid generation and subsequent mineral dissolution and precipitation can further alter the pore structure of coal seams, negatively impacting gas production. Moreover, the plasticizing effect of supercritical CO_2 on coal becomes increasingly pronounced, leading to the softening of the coal's physical and mechanical properties. Significant reductions in elastic modulus and compressive strength increase the likelihood of reservoir fracturing and leakage during injection and production operations.

6.2. Development Trends of CO_2-ECBM Technology

The future development of CO_2-ECBM technology will primarily focus on in-depth investigations of multi-physics coupling mechanisms, comprehensive upgrades of monitoring technologies, and the large-scale deployment of engineering application models.

Firstly, as the decline in coal reservoir permeability after CO_2 injection becomes increasingly significant, the effects of various injection parameters—such as injection volume, pressure, and method—on coal swelling and pore structure alteration have yet to be generalized into a universally applicable theoretical framework. Therefore, future research should systematically explore the complex coupling mechanisms among these influencing factors to provide scientific guidance for site selection and the optimization of injection-production schemes, ensuring sustained and stable CO_2 injection as well as efficient methane recovery.

Uncertainties in coal cleat structure remain a key challenge for accurate risk prediction in CO_2-ECBM. As cleats govern fluid flow and pressure responses, variations in their geometry and connectivity can significantly affect model reliability. However, their fine-scale heterogeneity makes precise characterization difficult. Meanwhile, current monitoring, measurement, and verification (MMV) technologies are still inadequate for fully tracking gas transport, leakage, and reservoir dynamics within coal seams. Improving imaging, in situ monitoring, and uncertainty-informed modeling will be essential to enhance risk assessment and ensure storage safety. In addition, there is a lack of large-scale, long-term field tests—particularly in coal seams deeper than 1000 m—leading to limited understanding of optimal operational conditions and engineering control strategies. Future efforts should prioritize the development of high-precision real-time monitoring tools and the implementation of cross-regional demonstration projects to expand data diversity and support the refinement of risk management frameworks for CO_2-ECBM deployment.

Moreover, the industrial application of CO_2-ECBM continues to face challenges related to high implementation costs and insufficient market incentives. Achieving a balance between technical performance and economic viability will be a central challenge in future technological transformation and large-scale deployment. Accordingly, it is urgent for national and regional governments to enhance policies related to coalbed methane development and utilization. Support should be given to the full-process technology chain of CO_2 geological sequestration and utilization, as well as to cluster-based demonstration projects, aiming to establish industrial-scale applications with a capacity exceeding one million tons. Simultaneously, fundamental scientific research into CO_2 geological storage and utilization must be strengthened to uncover the underlying mechanisms of source-sink matching, technology integration, and system optimization under various regional energy, resource, and environmental conditions. These efforts will provide theoretical foundations and practical guidance for the development of advanced and regionally adaptive CO_2-ECBM technology models.

In summary, the future advancement of CO_2-ECBM technology will rely on deeper insights into complex coupling mechanisms, innovations in monitoring technologies, and sustained implementation of large-scale engineering trials. Through policy support and integrated technological solutions, CO_2-ECBM can achieve simultaneous improvements in storage security, methane recovery efficiency, and economic performance, ultimately promoting the healthy and widespread application of this technology on a global scale.

7. Conclusions

This study provides a comprehensive review of the fundamental mechanisms, experimental and numerical research progress, and current status of engineering demonstrations related to CO_2-ECBM technology. It highlights the synergistic advantages of this approach in enhancing coalbed methane recovery while enabling long-term CO_2 sequestration, thereby establishing an effective technological bridge between energy development and carbon emission reduction. Research indicates that the coupled multi-physical processes—including adsorption/desorption, diffusion, and seepage—play a critical role in the CO_2-driven methane displacement process. The preferential adsorption of CO_2

facilitates the displacement of adsorbed CH_4, significantly reducing the partial pressure of CH_4 in the coal matrix, promoting its desorption and migration toward production wells, and ultimately improving recovery efficiency.

Both experimental studies and numerical simulations have demonstrated that reservoir pressure, formation temperature, and CO_2 injection flow rate significantly influence CBM recovery. Meanwhile, reservoir physical properties, structural conditions, and multi-physics coupling effects collectively determine the injection efficiency and long-term storage security of CO_2. These findings provide critical references for assessing the storage capacity and site suitability of CO_2-ECBM projects.

Although a number of pilot tests and demonstration projects have been conducted globally, and some have yielded preliminary successes, CO_2-ECBM technology still faces several pressing challenges. These include insufficient monitoring and verification technologies, the risk of induced seismicity, potential environmental impacts, high economic costs, and changes in coal reservoir properties. These issues require further in-depth investigation. Future development efforts should focus on elucidating the complex multi-physics coupling mechanisms triggered by CO_2 injection, enhancing the accuracy and real-time capabilities of monitoring technologies, and promoting large-scale, long-duration, and cross-regional field demonstrations. Based on integrated technologies and full-process optimization, CO_2-ECBM can achieve coordinated improvements in storage safety, recovery efficiency, and economic viability, thus offering robust technical support for global energy transition and low-carbon development.

Funding: This research was funded by the National Natural Science Foundation of China (No. 42302194), the Natural Science Foundation of Jiangsu Province, China (No. BK20231084), the Applied Basic Research Programs of Xuzhou, China (No. KC23001), the open fund of the Key Laboratory of Deep Coalbed Methane Exploration and Development of Sinopec (No. SMKK202405), and the Postdoctoral Fellowship Program of CPSF (GZC20233002).

Conflicts of Interest: Authors Yinan Cui, Chao Li, Hequn Gao, Hui Li and Yaozu Zhang were employed by the East China Oil and Gas Company. Author Bin Miao was employed by the company Zaozhuang Mining Group Co., Ltd. Author Bei Zhang was employed by the company China Construction Eighth Engineering Division Guangxi Construction Co., Ltd. Authors Jinghui Zhao and Guangrong Zhang were employed by the company Linfen Coalbed Methane Branch, SINOPEC. The remaining authors declare that the research was conducted in the absence of any commercial or financial relationships that could be construed as a potential conflict of interest.

References

1. Yang, W.; Min, Z.; Yang, M.X.; Yan, J. Exploration of the implementation of carbon neutralization in the field of natural resources under the background of sustainable Development—An overview. *Int. J. Environ. Res. Public Health* **2022**, *19*, 14109. [CrossRef] [PubMed]
2. Li, Q.S. The view of technological innovation in coal industry under the vision of carbon neutralization. *Int. J. Coal Sci. Technol.* **2021**, *8*, 1197–1207. [CrossRef]
3. Li, Y.L.; Wei, Y.F.; Zhu, F.Q.; Du, J.Y.; Zhao, Z.M.; Ouyang, M. The path enabling storage of renewable energy toward carbon neutralization in China. *Etransportation* **2023**, *16*, 100226. [CrossRef]
4. Liu, X.H. Low-carbon utilization of coal gangue under the carbon neutralization strategy: A short review. *J. Mater. Cycles Waste Manag.* **2023**, *25*, 1978–1987. [CrossRef]
5. Chen, Q.; Zhang, H.Y.; Lau, Y.Y.; Wang, T.N.; Wang, W.; Zhang, G.S. Climate change, carbon peaks, and carbon neutralization: A bibliometric study from 2006 to 2023. *Sustainability* **2023**, *15*, 5723. [CrossRef]
6. Xu, G.W.; Bai, D.R.; Xu, C.M.; He, M.Y. Challenges and opportunities for engineering thermochemistry in carbon-neutralization technologies. *Natl. Sci. Rev.* **2023**, *10*, nwac217. [CrossRef]
7. Wang, X.P.; Zhang, Z.M.; Guo, Z.H.; Su, C.; Sun, L.H. Energy structure transformation in the context of carbon neutralization: Evolutionary game analysis based on inclusive development of coal and clean energy. *J. Clean Prod.* **2023**, *398*, 136626. [CrossRef]

8. Liu, X.Y.; Liu, X.M.; Zhang, Z.Q. Application of red mud in carbon capture, utilization and storage (CCUS) technology. *Renew. Sust. Energ. Rev.* **2024**, *202*, 114683. [CrossRef]
9. Storrs, K.D.P.; Lyhne, I.; Drustrup, R. A comprehensive framework for feasibility of CCUS deployment: A meta–review of literature on factors impacting CCUS deployment. *Int. J. Greenh. Gas Control* **2023**, *125*, 103878. [CrossRef]
10. Huang, S.P.; Liu, D.M.; Gomez-Rivas, E.; Griera, A.; Gan, Q.; Wang, M.Y.; Xing, Y.; Zhao, Y. Experimental insights into the nucleation and propagation of hydraulic fractures in anthracite coalbed methane reservoirs. *Earth Energy Sci.* **2025**. [CrossRef]
11. Li, M.X.; He, N.P.; Xu, L.; Peng, C.H.; Chen, H.; Yu, G.R. Eco-CCUS: A cost-effective pathway towards carbon neutrality in China. *Renew. Sust. Energ. Rev.* **2023**, *183*, 113512. [CrossRef]
12. Mon, M.T.; Tansuchat, R.; Yamaka, W. CCUS technology and carbon emissions: Evidence from the United States. *Energies* **2024**, *17*, 1748. [CrossRef]
13. Busch, A.; Gensterblum, Y. CBM and CO_2-ECBM related sorption processes in coal: A review. *Int. J. Coal Geol.* **2011**, *87*, 49–71. [CrossRef]
14. Li, Z.W.; Yu, H.J.; Bai, Y.S.; Wang, Y.J.; Hu, H.Q. Numerical study on the influence of temperature on CO_2-ECBM. *Fuel* **2023**, *348*, 128613. [CrossRef]
15. Fujioka, M.; Yamaguchi, S.; Nako, M. CO_2-ECBM field tests in the Ishikari Coal Basin of Japan. *Int. J. Coal Geol.* **2010**, *82*, 287–298. [CrossRef]
16. Sun, Y.F.; Zhao, Y.X.; Yuan, L. CO_2-ECBM in coal nanostructure: Modelling and simulation. *J. Nat. Gas Sci. Eng.* **2018**, *54*, 202–215. [CrossRef]
17. Mukherjee, M.; Misra, S. A review of experimental research on Enhanced Coal Bed Methane (ECBM) recovery via CO_2 sequestration. *Earth-Sci. Rev.* **2018**, *179*, 392–410. [CrossRef]
18. Zhang, L.; Li, J.H.; Xue, J.H.; Zhang, C.; Fang, X.Q. Experimental studies on the changing characteristics of the gas flow capacity on bituminous coal in CO_2-ECBM and N_2-ECBM. *Fuel* **2021**, *291*, 120115. [CrossRef]
19. Asif, M.; Wang, L.; Naveen, P.; Longinos, S.N.; Hazlett, R.; Ojha, K.; Panigrahi, D.C. Influence of competitive adsorption, diffusion, and dispersion of CH_4 and CO_2 gases during the CO_2-ECBM process. *Fuel* **2024**, *358*, 130065. [CrossRef]
20. Mwakipunda, G.C.; Wang, Y.T.; Mgimba, M.M.; Ngata, M.R.; Alhassan, J.; Mkono, C.N.; Yu, L. Recent advances in carbon dioxide sequestration in deep unmineable coal seams using CO_2-ECBM technology: Experimental studies, simulation, and field applications. *Energy Fuels* **2023**, *37*, 17161–17186. [CrossRef]
21. Zheng, S.J.; Yao, Y.B.; Liu, D.M.; Cai, Y.D.; Liu, Y. Nuclear magnetic resonance surface relaxivity of coals. *Int. J. Coal Geol.* **2019**, *205*, 1–13. [CrossRef]
22. Vishal, V.; Singh, T.N.; Ranjith, P.G. Influence of sorption time in CO_2-ECBM process in Indian coals using coupled numerical simulation. *Fuel* **2015**, *139*, 51–58. [CrossRef]
23. Liu, X.D.; Sang, S.X.; Zhou, X.Z.; Wang, Z.L.; Niu, Q.H.; Mondal, D. Modelling of geomechanical response for coal and ground induced by CO_2-ECBM recovery. *Gas Sci. Eng.* **2023**, *113*, 204953. [CrossRef]
24. Zheng, S.J.; Yao, Y.B.; Liu, D.M.; Cai, Y.D.; Liu, Y.; Li, X.W. Nuclear magnetic resonance T_2 cutoffs of coals: A novel method by multifractal analysis theory. *Fuel* **2019**, *241*, 715–724. [CrossRef]
25. Liu, S.Y.; Wang, J.L.; Li, H.Y.; Liu, J.R.; Xu, J.C.; Sun, W.Y.; Wang, X.P.; Chen, Z.H. A generalized adsorption model of CO_2-CH_4 in shale based on the improved Langmuir model. *Fuel* **2025**, *379*, 132971. [CrossRef]
26. Liu, Z.D.; Lin, X.S.; Zhu, W.C.; Hu, Z.; Hao, C.M.; Su, W.W.; Bai, G. Effects of coal permeability rebound and recovery phenomenon on CO_2 storage capacity under different coalbed temperature conditions during CO_2-ECBM process. *Energy* **2023**, *284*, 129196. [CrossRef]
27. Gong, S.L.; Zhang, L.; Zhang, T.Y.; He, W.; Hu, W.Q.; Yin, H.C.; Ma, L.S.; Hong, X.; Zhang, W.; Zhang, B. Numerical simulation of CO2-ECBM for deep coal reservoir with strong stress sensitivity. *Heliyon* **2024**, *10*, e34818. [CrossRef]
28. Yu, H.G.; Yuan, J.; Guo, W.J.; Cheng, J.L.; Hu, Q.T. A preliminary laboratory experiment on coalbed methane displacement with carbon dioxide injection. *Int. J. Coal Geol.* **2008**, *73*, 156–166. [CrossRef]
29. Zheng, S.J.; Yao, Y.B.; Liu, D.M.; Cai, Y.D.; Liu, Y. Characterizations of full-scale pore size distribution, porosity and permeability of coals: A novel methodology by nuclear magnetic resonance and fractal analysis theory. *Int. J. Coal Geol.* **2018**, *196*, 148–158. [CrossRef]
30. Zhang, C.L.; Wang, E.Y.; Li, B.B.; Kong, X.G.; Xu, J.; Peng, S.J.; Chen, Y.X. Laboratory experiments of CO_2-enhanced coalbed methane recovery considering CO_2 sequestration in a coal seam. *Energy* **2023**, *262*, 125473. [CrossRef]
31. Lu, Y.J.; Han, J.X.; Yang, M.P.; Chen, X.Y.; Zhu, H.J.; Yang, Z.Z. Molecular simulation of supercritical CO_2 extracting organic matter from coal based on the technology of CO_2-ECBM. *Energy* **2023**, *266*, 126393. [CrossRef]
32. Zhang, H.; Hu, Z.C.; Xu, Y.; Fu, X.X.; Li, W.; Zhang, D.F. Impacts of long-term exposure to supercritical carbon dioxide on physicochemical properties and adsorption and desorption capabilities of moisture-equilibrated coals. *Energy Fuels.* **2021**, *35*, 12270–12287. [CrossRef]

33. Zhang, L.S.; Shan, Y.F. Quantitative Analysis of Fracture Roughness and Multi-Field Effects for CO_2-ECBM Projects. *Energies* **2024**, *17*, 2851. [CrossRef]
34. Asif, M.; Junussov, M.; Longinos, S.; Hazlett, R.; Satibekova, S. CO_2 storage capacity of coal seams: A screening and geological review of carboniferous coal formations of Kazakhstan. *Int. J. Coal Sci. Technol.* **2025**, *12*, 18. [CrossRef]
35. Scafidi, J.; Wilkinson, M.; Gilfillan, S.M.; Heinemann, N.; Haszeldine, R.S. A quantitative assessment of the hydrogen storage capacity of the UK continental shelf. *Int. J. Hydrogen Energy* **2021**, *46*, 8629–8639. [CrossRef]
36. Zhang, K.Q.; Liu, L.R.; Huang, G.H. Nanoconfined water effect on CO_2 utilization and geological storage. *Geophys. Res. Lett.* **2020**, *47*, e2020GL087999. [CrossRef]
37. Thanh, H.V.; Sugai, Y.H.; Sasaki, K. Application of artificial neural network for predicting the performance of CO_2 enhanced oil recovery and storage in residual oil zones. *Sci. Rep.* **2020**, *10*, 18204. [CrossRef]
38. Bonto, M.; Welch, M.J.; Lüthje, M.; Andersen, S.I.; Veshareh, M.J.; Amour, F.; Afrough, A.; Mokhtari, R.; Hajiabadi, M.R.; Alizadeh, M.R.; et al. Challenges and enablers for large-scale CO_2 storage in chalk formations. *Earth-Sci. Rev.* **2021**, *222*, 103826. [CrossRef]
39. Fang, H.H.; Wang, Y.J.; Sang, S.X.; Yu, S.; Liu, H.H.; Guo, J.R.; Wang, Z.F. Potential assessment of CO_2 source/sink and its matching research during CCS process of deep unworkable seam. *Sci. Rep.* **2024**, *14*, 17206. [CrossRef]
40. Mi, Z.X.; Wang, F.G.; Yang, Y.Z.; Wang, F.; Hu, T.; Tian, H.L. Evaluation of the potentiality and suitability for CO_2 geological storage in the Junggar Basin, northwestern China. *Int. J. Greenh. Gas Control* **2018**, *78*, 62–72. [CrossRef]
41. Ye, J.; Afifi, A.; Rowaihy, F.; Baby, G.; De Santiago, A.; Tasianas, A.; Hamieh, A.; Khodayeva, A.; Al-Juaied, M.; Meckel, T.; et al. Evaluation of geological CO_2 storage potential in Saudi Arabian sedimentary basins. *Earth-Sci. Rev.* **2023**, *244*, 104539. [CrossRef]
42. Wang, J.Q.; Yuan, Y.; Chen, J.W.; Zhang, W.; Zhang, J.; Liang, J.; Zhang, Y.G. Geological conditions and suitability evaluation for CO_2 geological storage in deep saline aquifers of the Beibu Gulf Basin (South China). *Energies* **2023**, *16*, 2360. [CrossRef]
43. Zhan, J.; Su, Z.Z.; Fan, C.; Li, X.N.; Ma, X.L. Suitability evaluation of CO_2 geological sequestration based on unascertained measurement. *Arab. J. Sci. Eng.* **2022**, *47*, 11453–11467. [CrossRef]
44. Zhang, X.; Liu, W.; Chen, J.; Jiang, D.Y.; Fan, J.Y.; Daemen, J.J.K.; Qiao, W.B. Large-scale CO_2 disposal/storage in bedded rock salt caverns of China: An evaluation of safety and suitability. *Energy* **2022**, *249*, 123727. [CrossRef]
45. Carlotto, V.; Dueñas, A.; Ingol-Blanco, E.; Santa-Cruz, S. A multi-criteria approach to the selection of geological storage of CO_2 sites in Peru. *Int. J. Greenh. Gas Control* **2022**, *114*, 103600. [CrossRef]
46. Fawad, M.; Mondol, N.H. Monitoring geological storage of CO_2: A new approach. *Sci. Rep.* **2021**, *11*, 5942. [CrossRef] [PubMed]
47. Fibbi, G.; Del Soldato, M.; Fanti, R. Review of the monitoring applications involved in the underground storage of natural gas and CO_2. *Energies* **2022**, *16*, 12. [CrossRef]
48. Cao, C.; Liu, H.J.; Hou, Z.M.; Mehmood, F.; Liao, J.X.; Feng, W.T. A review of CO_2 storage in view of safety and cost-effectiveness. *Energies* **2020**, *13*, 600. [CrossRef]
49. Gholami, R.; Raza, A.; Iglauer, S. Leakage risk assessment of a CO_2 storage site: A review. *Earth-Sci. Rev.* **2021**, *223*, 103849. [CrossRef]
50. Smith, E.; Morris, J.; Kheshgi, H.; Teletzke, G.; Herzog, H.; Paltsev, S. The cost of CO_2 transport and storage in global integrated assessment modeling. *Int. J. Greenh. Gas Control* **2021**, *109*, 103367. [CrossRef]
51. Raza, A.; Glatz, G.; Gholami, R.; Mahmoud, M.; Alafnan, S. Carbon mineralization and geological storage of CO_2 in basalt: Mechanisms and technical challenges. *Earth-Sci. Rev.* **2022**, *229*, 104036. [CrossRef]
52. Zhang, T.; Zhang, W.C.; Yang, R.Z.; Liu, Y.M.; Jafari, M. CO_2 capture and storage monitoring based on remote sensing techniques: A review. *J. Clean Prod.* **2021**, *281*, 124409. [CrossRef]
53. Zhang, S.R.; Sang, S.X.; Wu, J.; Zhou, X.Z.; Zhang, B.; Yang, R.Z.; Guo, J.C.; Liu, X.D.; Zhang, W.Q.; Li, Y. Progress and application of key technologies for CO_2 enhancing coalbed methane. *J. China Coal Soc.* **2022**, *47*, 3952–3964.
54. Li, Z.W.; Yu, H.J.; Bai, Y.S. Numerical simulation of CO_2-ECBM based on multi-physical field coupling model. *Sustainability* **2022**, *14*, 11789. [CrossRef]
55. Fang, H.H.; Xu, H.J.; Sang, S.X.; Liu, S.Q.; Song, S.L.; Liu, H.H. 3D reconstruction of coal pore network and its application in CO_2-ECBM process simulation at laboratory scale. *Front. Earth Sci.* **2022**, *16*, 523–539. [CrossRef]
56. Ma, Q.Q.; Li, H.; Ji, K.; Huang, F.J. Thermal-hydraulic-mechanical coupling simulation of CO_2 enhanced coalbed methane recovery with regards to Low-Rank but relatively shallow coal seams. *Appl. Sci.* **2023**, *13*, 2592. [CrossRef]
57. Cheng, X.Y.; Gong, X.P.; Cheng, C.; Li, Q.G.; Li, Z.Q. Molecular Dynamics Simulation of CO_2-ECBM Under Different Moisture Contents. *Energies* **2025**, *18*, 239. [CrossRef]
58. Liu, M.Y.; Wen, H.; Fan, S.X.; Wang, Z.P.; Fei, J.B.; Wei, G.M.; Chen, X.J.; Wang, H. Experimental study of CO_2-ECBM by injection liquid CO_2. *Minerals* **2022**, *12*, 297. [CrossRef]
59. Wang, Z.L.; Sang, S.X.; Zhou, X.Z.; Liu, X.D. Numerical study on CO_2 sequestration in low-permeability coal reservoirs to enhance CH_4 recovery: Gas driving water and staged inhibition on CH_4 output. *J. Pet. Sci. Eng.* **2022**, *214*, 110478. [CrossRef]
60. Liu, X.D.; Sang, S.X.; Zhou, X.Z.; Wang, Z.L. Coupled adsorption-hydro-thermo-mechanical-chemical modeling for CO_2 sequestration and well production during CO_2-ECBM. *Energy* **2023**, *262*, 125306. [CrossRef]

61. Damen, K.; Faaij, A.; Bergen, F.V.; Gale, J.; Lysen, E. Identification of early opportunities for CO_2 sequestration—Worldwide screening for CO_2-EOR and CO_2-ECBM projects. *Energy* **2005**, *30*, 1931–1952. [CrossRef]
62. Godec, M.; Koperna, G.; Gale, J. CO_2-ECBM: A review of its status and global potential. *Energy Proc.* **2014**, *63*, 5858–5869. [CrossRef]
63. Gale, J.; Freund, P. Coal-bed methane enhancement with CO_2 sequestration worldwide potential. *Environ. Geosci.* **2001**, *8*, 210–217. [CrossRef]
64. Su, X.B.; Lin, X.Y.; Liu, S.B.; Zhao, M.J.; Song, Y. Geology of coalbed methane reservoirs in the Southeast Qinshui Basin of China. *Int. J. Coal Geol.* **2005**, *62*, 197–210. [CrossRef]
65. Liu, D.M.; Jia, Q.F.; Cai, Y.D.; Gao, C.J.; Qiu, F.; Zhao, Z.; Chen, S.Y. A new insight into coalbed methane occurrence and accumulation in the Qinshui Basin, China. *Gondwana Res.* **2022**, *111*, 280–297. [CrossRef]
66. Su, X.B.; Lin, X.Y.; Zhao, M.J.; Song, Y.; Liu, S.B. The upper Paleozoic coalbed methane system in the Qinshui basin, China. *AAPG Bull.* **2005**, *89*, 81–100. [CrossRef]
67. Jiang, L.; Song, Y.; Zhao, W.; Bo, D.M.; Liu, S.B.; Hao, J.Q. Main controlling factor of coalbed methane enrichment area in southern Qinshui Basin, China. *J. Pet. Explor. Prod. Technol.* **2024**, *14*, 165–173. [CrossRef]
68. Clarkson, C. The Allison Unit CO_2-ECBM Pilot: A Reservoir Modeling Study. *Retrieved April* **2003**, *7*, 1–56.
69. Shi, J.Q.; Durucan, S. A numerical simulation study of the Allison unit CO_2-ECBM pilot: The impact of matrix shrinkage and swelling on ECBM production and CO_2 injectivity. *Greenh. Gas Control Technol. 7* **2005**, *I*, 431–439.
70. Weber, M.; Wilson, T.H.; Akwari, B.; Wells, A.W.; Koperna, G. Impact of geological complexity of the Fruitland Formation on combined CO_2 enhanced recovery/sequestration at San Juan Basin pilot site. *Int. J. Coal Geol.* **2012**, *104*, 46–58. [CrossRef]
71. Liu, S.Q.; Huang, F.S.; Du, R.B.; Chen, S.H.; Guan, Y.T.; Liu, Y.H.; Wang, T. Progress and typical case analysis of demonstration projects of the geologic al sequestration and utilization of CO_2. *Coal Geol. Explor.* **2023**, *51*, 158–174.
72. Massarotto, P.; Golding, S.D.; Bae, J.S.; Iyer, R.; Rudolph, V. Changes in reservoir properties from injection of supercritical CO_2 into coal seams—A laboratory study. *Int. J. Coal Geol.* **2010**, *82*, 269–279. [CrossRef]
73. Ranathunga, A.S.; Perera, M.S.A.; Ranjith, P.G.; Rathnaweera, T.D.; Zhang, X.G. Effect of coal rank on CO_2 adsorption induced coal matrix swelling with different CO_2 properties and reservoir depths. *Energy Fuels* **2017**, *31*, 5297–5305. [CrossRef]
74. Reisabadi, M.Z.; Sayyafzadeh, M.; Haghighi, M. Stress and permeability modelling in depleted coal seams during CO_2 storage. *Fuel* **2022**, *325*, 124958. [CrossRef]
75. Cao, B.; Fu, X.H.; Kang, J.Q.; Lu, J.L.; Tang, P.; Xu, H.; Huang, M. Mini-Review on Influence of CO_2-Enhanced Coalbed Methane Recovery and CO_2 Geological Storage on Physical Properties of Coal Reservoir. *Energy Fuels* **2024**, *38*, 23268–23280. [CrossRef]
76. Shang, Z.; Wang, H.F.; Li, B.; Cheng, Y.P.; Zhang, X.H.; Wang, Z.Y.; Geng, S.K.; Wang, Z.Y.; Chen, P.; Lv, P.F.; et al. The effect of leakage characteristics of liquid CO_2 phase transition on fracturing coal seam: Applications for enhancing coalbed methane recovery. *Fuel* **2022**, *308*, 122044. [CrossRef]
77. Gordon, D.; Reuland, F.; Jacob, D.J.; Worden, J.R.; Shindell, D.; Dyson, M. Evaluating net life-cycle greenhouse gas emissions intensities from gas and coal at varying methane leakage rates. *Environ. Res. Lett.* **2023**, *18*, 084008. [CrossRef]
78. Zhang, M.J.; Gao, L.; Wang, Q.D.; Xie, D.L.; Gao, J.; Wang, S.X.; Lu, X. Methane leakage measurement of natural gas heating boilers and greenhouse gas emissions accounting of "coal-to-gas" transition for residential heating in rural Beijing. *Environ. Sci. Technol. Lett.* **2022**, *10*, 93–97. [CrossRef]
79. Rutqvist, J.; Birkholzer, J.T.; Tsang, C.F. Coupled reservoir-geomechanical analysis of the potential for tensile and shear failure associated with CO_2 injection in multilayered reservoir-caprock systems. *Int. J. Rock Mech. Min. Sci.* **2008**, *45*, 132–143. [CrossRef]

energies

MDPI

Article

Numerical Studies of Advanced Methane Drainage Employing Underground Long-Reach Directional Drilling

Wiesław Szott [1,*], Małgorzata Słota-Valim [2], Piotr Ruciński [1], Krzysztof Miłek [1] and Piotr Łętkowski [1]

[1] Department of Hydrocarbon Reservoir and UGS Simulation, Oil and Gas Institute—National Research Institute, 31-503 Krakow, Poland; rucinski@inig.pl (P.R.); milek@inig.pl (K.M.); letkowski@inig.pl (P.Ł.)

[2] Department of Geology and Geochemistry, Oil and Gas Institute—National Research Institute, 31-503 Krakow, Poland; slota-valim@inig.pl

* Correspondence: szott@inig.pl; Tel.: +48-606-431-704

Abstract: This paper presents the procedures and results of the numerical modelling and simulations performed to analyse an innovative method of advanced methane drainage employing underground long-reach directional drilling (LRDD) technology. The analysis involved the implementation of geomechanical and dynamic reservoir models to simulate processes in coal seams and the surrounding rocks during coal mining and concurrent methane drainage, in accordance with the proposed technology. The analysis aimed to quantitatively assess the effectiveness of the technology, evaluate its sensitivity to the geological and geomechanical properties of the rocks, and identify the potential for optimisation of its technological and operational parameters in the proposed strategy. The works presented in this paper include the following key tasks: the construction of a system of geological, geomechanical, and dynamic simulation models; the analysis of the geomechanical effects of various types and regions of occurrence; the implementation of the correlation between the geomechanical states of the rocks and their transport properties; and the performance of the effectively coupled geomechanical and reservoir fluid flow simulations. The proposed approach was applied to the specific conditions of the multi-seam Murcki–Staszic Coal Mine operated by Jastrzębska Spółka Węglowa, Poland.

Keywords: coal seam; methane drainage; directional drilling; geomechanical modelling; dynamic modelling

1. Introduction

The problem of methane emissions from coal mines has become more severe due to the increasing depth of coal seam exploitation, which is associated with higher methane content in deeper-lying seams. The presence of methane in these seams creates significant challenges during exploitation, as it can lead to hazards such as explosions and sudden rock outbursts [1,2]. Due to the risks associated with methane emissions during mining processes, draining this gas is crucial for managing methane hazards in Polish coal mines [3]. Additionally, the methane captured serves as a cleaner energy source and helps reduce emissions of this highly potent greenhouse gas into the atmosphere [4,5]. Diverse technologies have been implemented to mitigate methane emissions from coal mining operations; however, these practices are still not standard across the industry. Among the mitigation methods is ventilation of air methane (VAM) and the application of methane drainage systems [5]. Two fundamental types of methane drainage methods can be distinguished based on the timing of the drainage process relative to mining operations: pre-drainage and post-drainage methods. Pre-drainage involves extracting methane from the coal seam in high-gas-content coal mines using various borehole-drilling techniques, including vertical boreholes, horizontal in-seam boreholes drilled directionally from the ground surface, and in-seam or superjacent boreholes originating from within the mine workings before mining operations begin. The effectiveness of pre-drainage methods varies significantly depending on geological conditions, including coal seam burial depth and specific coal properties that determine both methane storage capacity and gas flow characteristics. Key factors affecting methane storage capacity include adsorbed methane content and coal porosity, while gas flow rate is controlled by permeability, reservoir pressure, and coal diffusivity. Particularly high methane capture rates are achieved during pre-drainage in fractured rock formations with high gas permeability [6].

In contrast, post-drainage focuses on capturing methane and other gases released from surrounding coal seams and rock strata due to longwall-induced subsidence, stress relaxation, and enhanced permeability caused by mining activities [7]. Post-mining drainage technologies aim to capture methane released during and after mining operations through vertical goaf wells, superjacent boreholes, and cross-measure in-seam boreholes [8] drilled into the strata overlying the coal seam targeted for production. This method, frequently complemented by ventilation systems, has been proven to achieve methane drainage efficiency in the range of 41–64% [9].

Methane pre-drainage in Polish coal mining conditions exhibits low effectiveness due to the inherently low gas permeability of coal seams. Consequently, drainage technology in Poland relies primarily on post-mining methods in which methane is captured from zones where mining activities have increased permeability, specifically in areas surrounding active coal panels during extraction [9].

To meet the increasing demand for more efficient drainage in gassy coal mines, a drainage technique utilising long-reach directional drilling (LRDD) was developed and has been tested both as a pre-drainage and post-drainage configuration in coal mines in the U.S.A. [10], China [11], Australia [10,12], Turkey [8], and more recently in Poland [13,14].

In general, any quantitative analysis of methane production from coal structures applies a modern numerical modelling and simulation approach. It refers to both methane recovered from coal unmined beds (CBM—coal bed methane) and methane removed from active or abandoned coal mines (CMM—coal mine methane, AMM—abandoned mine methane). In particular, various aspects of methane drainage from coal mines were studied with this approach.

The use of numerical modelling and simulations is recommended as a standard for the analysis of methane drainage in coal mines by various investigation centres devoted to coal mining issues (e.g., [15]). In particular, this approach is recommended by international agencies such as the Global Methane Initiative, a partnership of 44 member countries and the European Commission that focuses on cost-effective, near-term methane recovery and use as a clean energy source (e.g., [16]). Various theoretical gas flow and simulation models have been developed by research groups from academia and research institutes. Hao Xu et al. [17] developed an integrated numerical model of desorption and diffusion during borehole gas extraction for a dual-porosity structure. They found the relative gas permeability function of gas saturation in fractures to mainly control the gas flow rate in the early extraction stages, while the coal matrix size and matrix-to-fracture diffusion coefficient mainly control the gas flow rate in the late extraction stages. Adel Taheri et al. [18] investigated methane flow in a coal structure to calculate the pressure of gas and its molecular velocity for methane gas drainage operations. They used a coal piece surrounded by cleats for geometrical modelling and numerical simulation and successfully simulated movements of fluid and gas molecules in a porous medium by applying COMSOL Multiphysics software. C. Özgen Karacan et al. [19] reviewed the technical aspects of coal mine methane capture in and from coal mines and the main factors affecting coal mine methane accumulations in underground mines. They also summarized the following issues: (i) various methods for capturing methane using boreholes as well as specific borehole designs for effective methane capture, (ii) aspects of removing that gas from abandoned mines and from sealed/active goafs of operating mines, (iii) benefits of capturing and controlling coal mine methane for mine safety, and (iv) benefits for energy production and greenhouse gas reduction. They also emphasised the significance of simulation studies in the quantitative analysis of methane drainage processes. Q. Haisheng and W. Kuijun [20] proposed a mathematical model of gas flow around bedding drainage boreholes that is a modification of the existing gas flow models with the free gas and adsorbed gas considered separately and the free gas flowing in the coal fractures with adsorbed gas being the positive quality source of the fracture. The governing equations were implemented and solved using COMSOL Multiphysics. Peng Liu et al. [21] developed a mechanism-based model to describe gas desorption/diffusion/flow from/between/within the coal matrix/fractures around the drainage borehole and studied the effects of parametric sensitivity on drainage performance using numerical simulations and mine site measurements. Shadi Jamalan et al. [22] found numerical methods to be powerful tools for studying the gas flow inside a coal block from their study of the gas drainage from coal seams using COMSOL software. They determined the appropriate pattern for drilling drainage holes and inferred from the model that the methane outlet pressure decreases over time, depending on the porosity of the coal and the number of boreholes. Tomasz Janoszek [23] presented a methodology for predicting the influence of the longwall shearer's control parameter on methane flow rate to the working of a longwall panel based on computational fluid dynamics methods. Kazem Oraee and Arash Goodarzi [24] constructed a two-dimensional mathematical model based on the finite element approach to simulate a methane drainage system where the resultant risk is linked to the coefficient of diffusion, coal permeability, hole pattern, and the duration of drainage. Mingyun Tang et al. [25] established a mathematical model of gas–solid interaction in coal rock based on the Kozeny–Carman equation, where the coal rock is considered a fracture-porosity dual medium. Yi Zhang et al. [26] carried out a numerical simulation of methane emission under coal mining conditions to optimise coalbed methane extraction. They used the dynamic coupling of a permeability field and virtual well with the mining face to identify the rapid-pressure high-speed decrease effect on the adjacent layer under the influence of mining. According to the authors, their method provided effective guidance for future coal bed methane capacity prediction under mining conditions.

Other groups of studies on methane drainage from coal seams refer to more specific issues. Xinguang Zhu et al. [27] proposed a fracture–pore equivalent model to establish the relationship between the fracture aperture under hydraulic fracturing and the permeability coefficient of coal in the process of gas extraction. They quantitatively analysed the change in the cracks' distribution with the injection pressure and indicated the significance of optimal design of gas extraction activity under hydraulic fracturing. Xiaoyan Li et al. [28] studied the impact of borehole diameter and drainage time on gas drainage volume using numerical simulations. The simulation results revealed that the gas drainage volume increased along with the borehole diameter increase and the gas pressure decrease. Accordingly, the effective pre-drainage radius varied as a power function of the pre-drainage time. In another paper, Xiaojun Li et al. [29] studied the problem of gas overlimits in the upper corner of a hard roof of a selected coal mine working face. Their findings offered guidance and technical references for gas control in the upper corners of hard roofs. Ali Bagherzadeh et al. [30] addressed the optimizing problem of the operational parameters of the borehole pattern in the coal seam to maximise gas drainage with the least amount of drilling, investment, and drainage time. They performed numerical modelling of the boreholes with different parameters and three drainage patterns (rectangular, parallelogram, and triangular) and showed the triangular pattern to be more suitable than the other two. The presented model was found to be applicable in coal mines where gas drainage operations are necessary and helpful for the design of drainage borehole patterns to maximise their gas drainage efficiency. D.I. Blokhin et al. [31] studied the effect of high-rate mining of methane-bearing coal seams on gas drainage performance in longwalls under difficult geological conditions. They identified an increase in local gas release and the stability of gas drainage wells due to the rise in the reservoir characteristics of rocks under the impact of swelling. They performed numerical experiments to model the stress–strain changes in the stratified coal–rock mass of a selected coal mine with gas drainage wells at different positions of the longwall face. The justification was provided for the expedience of correcting gas drainage patterns occurring in similar mining conditions as in the presented research. C.Ö. Karacan et al. [32] investigated different horizontal methane drainage borehole patterns, borehole lengths, and degasification times before and during panel extraction and evaluated their effectiveness using dynamic 3D reservoir modelling of a selected longwall panel operating in the Pittsburgh coalbed. They found dual and tri-lateral boreholes to be more effective in decreasing emissions compared to fewer, shorter, cross-panel, horizontal boreholes parallel to the longwall face. The dual and tri-lateral boreholes are also more effective in shielding the entries. Hua Guo et al. [33] studied a system of horizontal boreholes for longwall goaf gas drainage using computational fluid dynamics simulations. Their results showed that horizontal drainage boreholes significantly improved gas management through controlling goaf gas pressure distribution and flow pattern. Consequently, gas-related longwall production delays were largely eliminated. Moreover, the fugitive emissions from the mine were significantly reduced. Qingquan Liu et al. [34] found a method of improving sealing performance and borehole space maintenance for high-efficiency methane drainage. They proposed a visco-elastic plastic model including the plastic softening and dilatancy features of soft coal. The model included deformation, shrinkage, and fracture characteristics of the coal-surrounded boreholes and roadways to determine the proper sealing depth and the borehole spacing that provides favourable conditions for maximum methane pre-drainage from outburst-prone coal and maximum utilisation of the in-seam boreholes. Liang Zhang et al. [35] constructed a fully coupled thermo-hydro-mechanical model to simulate methane flow in coal and used this model and data from a coal mine to investigate various effects. They included the effect of multiple tree-type borehole layouts, tree-type borehole spacing, different Langmuir volumes and different Langmuir pressure constants, and initial coal permeabilities on methane drainage. The results showed that the total pre-drainage time increases as a power function of the tree-type borehole spacing. The effective drainage zone expands with the increase in the Langmuir pressure constant, the fracture permeability, and the matrix permeability. Xincheng Hu et al. [36] established a numerical model of surface boreholes draining methane from overlying layers and mining goafs in underground gassy mines. They studied the influence of surface drainage boreholes on coal spontaneous combustion. The simulation results showed the occurrence of an elliptic combustion zone around the borehole located in a deep goaf. It was found that the advancement of drainage boreholes enlarges the width of the combustion zone in the middle goaf. Zixi Guo et al. [37] conducted a feasibility analysis and optimal design of acidising drainage wells in a selected block of a Chinese coal bed. By combining their method with the current numerical model of acidising, they presented an optimal design that optimises the volume and injection rate of the acid. It turned out that acidising is an effective stimulation technology for some specific drainage wells. The feasibility analysis and the optimal design could thus improve the effect of acidising drainage wells. Pan Wei et al. [38] constructed a fluid–solid coupling mathematical model considering the creep properties of coal. The properties were determined from a permeability evolution equation, considering the effects of matrix shrinkage and effective stress. The results showed that the reduction in gas pressure around the boreholes was larger than that from a single borehole. The borehole spacing should be more than twice that of the effective drainage radius. The effect of gas extraction after optimising parameters conformed to industry standards. Ali Hosseini et al. [39] developed a 3D numerical model for methane drainage from an underground coal mine to determine the suitable distance between methane drainage stations in a cross-measure borehole. The model simulation results showed that the gas flow rate and the amount of gas produced by the boreholes increased to an acceptable level when the distance between methane drainage stations was reduced. Numerical simulation in that study provided basic support for the design of drainage boreholes in underground coal mines. Ying-ke Liu et al. [40] investigated a double-seam mining operation at a selected coal mine in China. In particular, they, studied the deformation and the resultant pressure relief of highly gassy seams. They simulated the extraction process by constructing a scaled physical model and a computational model. The simulation results showed the features of deformation resulting from the double-seam extraction to be different from those in the case of single-seam extraction. They found that the goaf below the gassy seam provided a buffering effect for the protective seam to be mined. They also found that the depressurisation effect was largely weakened. Shujun Ma et al. [41] proposed a method of combining gas content field data and numerical simulation calculation to accurately determine the effective drainage radius of a drainage borehole. The results showed that the effective drainage radius increases rapidly at the initial stage and then gradually decreases, eventually reaching a limit value. According to the authors, their research provided references for determining and laying out the effective drainage radius of a drainage borehole in a mining coal seam. Wenjie Xu et al. [42] used the finite element method and reservoir simulation model to study gas drainage by high-pressure air injection. They established a fluid–structure model for gas drainage by air injection. They found an optimal spacing between the extraction and injection holes as a function of extraction time. According to the authors, their approach provides a basis for guiding gas drainage by injecting high-pressure air on-site. Dangyu Zhang et al. [43] established a coal damage permeability evolution model of a selected coal mine. They simulated the influence of various factors, such as the initial coal permeability, negative pumping pressure, and pumping time, on the pre-drainage borehole effective pumping radius. The simulation results showed the significant powerful effects of the factors on the effective extraction radius. According to the authors, their research results provided theoretical references for the pre-drainage of gas in the selected mine. Xingkui Liu et al. [44] established 3D models to investigate different borehole positions to drain methane in low-protective coal seams. The models covered three kinds of drill holes. The model results showed that the high-level crossing holes contribute to better methane drainage than the other drill holes. Accordingly, the research findings provided theoretical grounds for methane drainage from coal mine workings. Szott et al. [45] presented procedures and results of drainage borehole stimulation methods. They performed numerical simulations under the specific conditions of a selected coal mine in Poland. The simulations were carried out for hydraulic fracturing, high-pressure jet slotting, and the use of explosives. The influence of the basic technological parameters of the stimulation methods upon the drainage process characteristics was determined. In another paper, Szott et al. [46] addressed the problem of initial conditions before commencement of methane drainage. They used a numerical model to simulate the influence of the broad surroundings of the analysed coal seam on initial conditions for the simulation of coal extraction from the seam. They analysed the influence of spatial and temporal effects of the prior mining activities on detailed distributions of various parameters, including pore pressure, the direction of reservoir fluid flows, fluid saturation in natural fractures, and methane desorption from the coal matrix.

The studies presented hereafter address specific issues referring to long-reach directional drilling (LRDD) for methane drainage purposes, a problem not covered by the previous studies referenced above and, to the authors' knowledge, not addressed by any research group worldwide. In particular, the reported studies aimed to assess the effectiveness of coal mine methane drainage in a multi–seam coal mine situated within the Upper Silesian Coal Basin (USCB) in Poland using LRDD. The studies also included a sensitivity analysis of the methane drainage effects to basic geological and geomechanical parameters of the studied system and partial optimisation analysis of the studied method for the most significant operational parameters. Determination of the above aims was carried out using numerical modelling methods with coupled reservoir fluid flow simulations and geomechanical simulations. The basis of both geomechanical and fluid flow models was developed from geological structural and parametric models of the coal seams and surrounding rocks. As a result of a comprehensive analysis and data integration, complex numerical models of the Polish hard coal mines from the USCB were developed. The models were used to perform the simulations of processes taking place in coal seams and the surrounding rocks during coal mining and concomitant methane drainage according to the proposed technology of LRDD. The thorough analysis of the simulation results provided the means with which to quantitatively assess the above fundamental aims, that is, the effectiveness of the technology, its sensitivity to the geological and geomechanical properties of the rocks, and its potential optimisation with respect to the technological and operational parameters of the proposed strategy.

2. Geological Setting

The study area lies within the central part of the Upper Silesian Coal Basin (USCB) in southern Poland, as shown in Figure 1A. The USCB, the largest coal basin in the country and one of the largest in Europe, originated as a foredeep of the Moravo-Silesian fold belt. It is characterised as a deep molasse basin with a polygenetic origin.

The Carboniferous coal-bearing deposits within the USCB exhibit a distinct dual zonation. The lower part of the Upper Mississippian coal-bearing lithostratigraphic succession, representing a paralic depositional system dated to Namurian A, unconformably lies over marine siliciclastic sediments. The upper part consists of continental (terrestrial) deposits, including the Upper Silesian Sandstone Series (belonging to the Upper Namurian A, B, and C), the Mudstone Series, and the Kraków Sandstone Series (of Westphalian age). The zone of highest interest in the study area includes the sediments of the Upper Silesian Sandstone Series, dated to the Namurian B and C. These deposits are developed predominantly as sandstones, with sandy shales and shales, interbedded with thick coal seams [47]. The continental deposits mentioned above lie concordantly on the paralic series with a sedimentation gap. Individual lithostratigraphic units significantly thin towards the east and southeast [48,49]. The main structural features of the USCB are related to the block structure of the basement, formed primarily during the Asturian phase. Three zones with distinct structural features can be distinguished within the USCB: fold tectonics, disjunctive tectonics, and fold–block tectonics. The fold tectonics zone occurs in the western part of the USCB and forms a narrow SSW-NNE belt where the Carboniferous formations are strongly folded into mostly asymmetrical brachysyncline types and also exhibit minor strata disturbances. The block tectonics zone encompasses most of the basin, with the predominant structural elements being WNW-ESE faults that have formed horst and graben structures. Carboniferous strata in the study area are gently inclined, with dips generally not exceeding 15 degrees, forming anticlinal highs, domes, synclinal zones, and troughs.

Fault throws typically range from 100 to 150 m. The fold–block tectonics zone covers the relatively narrow eastern part of the USCB, characterised by structures with N-E vergence, intersected by faults of meridional direction [50,51]. Additionally, the Alpine orogeny significantly influenced the structure of the basin, leading to the rejuvenation of numerous faults that were originally formed during the Variscan orogeny.

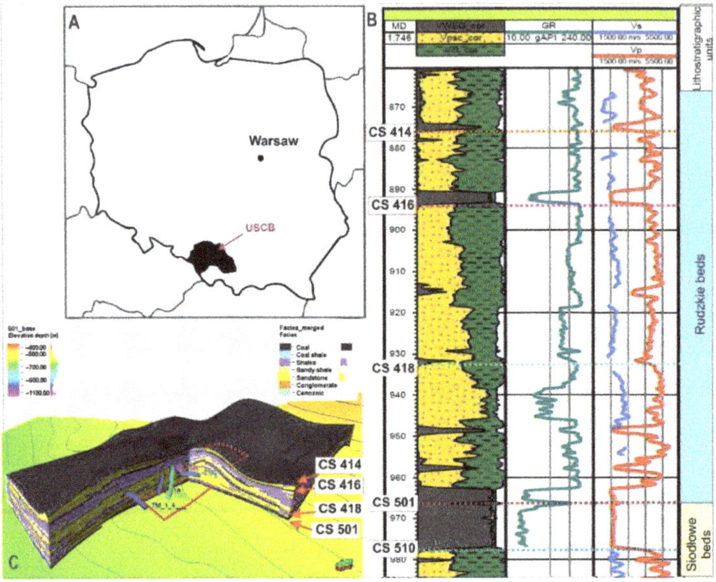

Figure 1. Location of the study area: (**A**) model setup; (**B**) lithological profile in the area of interest with well W-1 log data used for interpretation, including gamma ray (GR), compressional wave velocity (Vp), and shear wave velocity (Vs), all plotted versus measured depth (MD); (**C**) location of the IC panel with relevant coal seams (CS) selected for further detailed analysis and the boreholes (LRDB) trajectories.

3. Geological Model

The data required for structural and parametric modelling were provided by the PGG Mining Group of the Department of Staszic-Wujek Coal Mine, as well as the PIG (Polish Geological Institute—NRI), as the owners of geological information from the W-1 borehole. The dataset consisted of results from laboratory measurements of various rock and fluid properties, well log data, borehole sedimentological profiles from 18 wells, and structural maps and cross-sections capturing the structural information of coal seams observed in the modelled mine. The wells included 3 deep surface research boreholes—B-10 (1958), W-1—PIG1 (2013), and W-PIG-2H (2014); 10 downholes—G-876 (2011), G-25 (1961), G-945 (2019), G-25 (2019), G-2 (2019), G-944 (2019), G-4 (2019), G-8 (2019), G-16 (2019), and G-17 (2019); and 5 LRDD draining boreholes drilled in 2019—TM-2, TM-3, TM-4, TM-5, TM-1/4, and TM-1A.

The maps included the geometry of the bottom and information regarding the thickness of the 510, 501, 418, 416, and 414 coal seams of interest. The use of general scale cross-sections aided the reproduction of the structure of more shallow-lying coal seams.

The well-log data from the W-1 borehole, approximately 1000 m long, was particularly important as the data set from this well contained a wide spectrum of geophysical profiling results to be used in a subsequent stage of petrophysical properties interpolation. The lithological profile in the area of interest, with well log data of the W-1 borehole used for interpretation, is shown in Figure 1B. Despite this well not being located directly in the vicinity of the project's area of interest, due to the drilled geological profile, set of measured logs, and laboratory data of the petrophysical parameters, it was found useful for drawing conclusions on a regional scale and extrapolating the results to the study area being modelled.

Two-scale geological modelling was implemented for the C sector in the Staszic-Wujek coal mine, using both a regional model and higher-resolution local models focused on drainage test areas. Both were able to capture the geometry of the 510, 501, 418, 416, and 414 coal seams and surrounding formations (Figure 1C) as well as petrophysical/geomechanical properties from the overlying 610 coal seam to the surface.

The regional model incorporated all data, including data from the W-1 well, reducing structural uncertainty and enabling stress field determination affected by multi-seam mining activity. This model used a 25 × 25 m horizontal grid with vertical resolution varying from 10 through 20 m in overburden to approximately 5 m in drainage areas.

Local models used 1 × 1 m horizontal resolution with approximately 1 m vertical resolution in the 501 coal seam and surrounding strata to capture mine infrastructural details and wall face movement. These models incorporated all available structural and parametric data for subsequent coupled geomechanics and fluid flow numerical modelling.

The visualisation of data used for regional-scale geological model development with the well data location is shown in Figure 2.

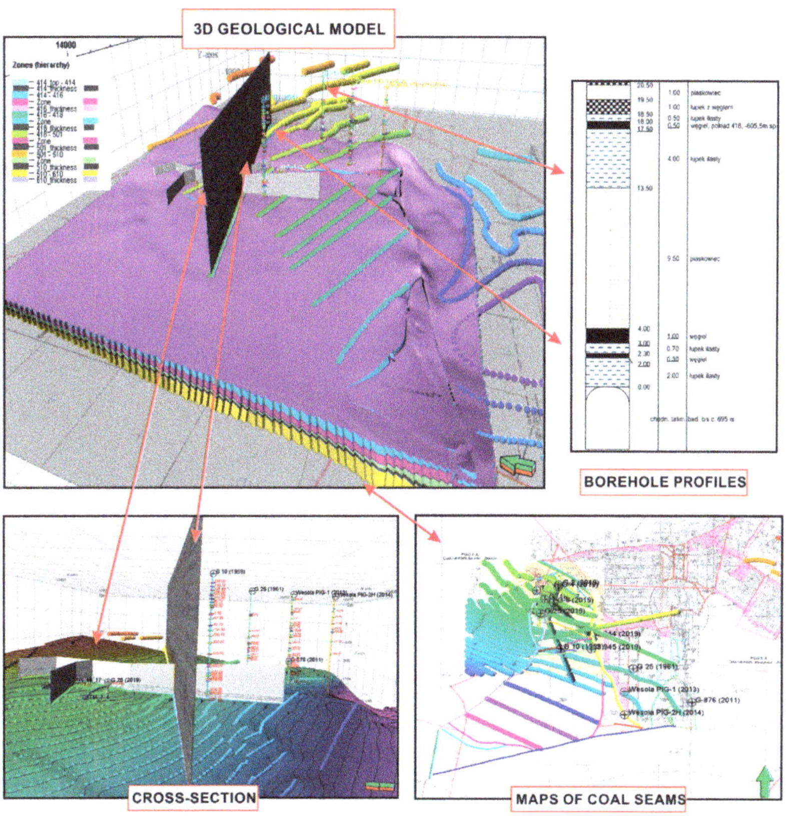

Figure 2. Data type used for 3D geological model development.

Development of 3D Parametric Models

The structural model presenting the geometry of the C sector of the Staszic-Wujek Coal mine was parametrised with petrophysical and geomechanical parameters, which is necessary in stress field calculations and the modelling of fluid flow in the area of methane drainage. The parametrisation was preceded by the modelling of lithotypes occurring in the area of interest.

The 3D lithotype model development was primarily motivated by the value of understanding the spatial distribution of lithotypes for geomechanical property modelling. The strong correlation between rock lithology and mechanical characteristics [52] enabled more effective assignment of geomechanical parameters to specific lithotypes, which was particularly valuable given the limited geomechanical data available for the Staszic-Wujek coal mine area. The model of lithotypes' distribution was constructed based on lithofacies profiles from 17 boreholes. In the borehole profiles (Figure 1C), 7 main lithotypes were distinguished, further marked in the Figure with a specified color and numerical code as follows: 0—coal (gray), 1—coal shale (blue), 2—shale (purple), 3—sandy shale (beige), 4—sandstone (yellow), 5—conglomerate (orange), 6—sediments of Cenozoic age (green).

In the initial phase, digitised lithotype profiles from boreholes underwent upscaling based on lithotype representativeness in each interval. Geostatistical analysis using Petrel 2019 software determined lithotype proportions and variogram parameters defining spatial correlation in vertical and horizontal directions. The spatial distribution of lithofacies was estimated using sequential indicator simulation (SIS), with computations performed individually for each distinguished lithotype within the 3D model. The developed spatial facies distribution was then used to guide the modelling of both petrophysical and geomechanical properties in 3D space.

The models of density, porosity and permeability were constructed using data from two key sources: the profile of these petrophysical parameters from the W-1 borehole (derived from well-log interpretation) and laboratory measurements from 36 samples collected from the 501 coal seam near the drainage area [53,54]. Input data were averaged to match the vertical resolution of the 3D grid through an up-scaling procedure then subjected to statistical analysis performed individually for each lithotype.

Statistical analysis of the input data determined the parameters controlling modelled parameter interpolation within the 3D structural model space. A stochastic algorithm—Gaussian random function simulation—was employed to model the spatial distribution of density, porosity, and permeability. The modelling process was executed 25 times, generating equally probable 3D porosity distributions. These distributions were then averaged to create the final 3D models of density, porosity and permeability.

4. Geomechanical Model

The models of stress and strain fields were developed based on the static parametric models of petrophysical and mechanical properties of lithotypes occurring in the Staszic-Wujek coal mine. The parametric models presenting the spatial distribution of geomechanical properties were developed based on well log data recorded in the W-1 reference borehole, as well as laboratory measurements of rock samples collected from the drainage zone.

The 1D models of elastic properties (including Young's modulus and Poisson's ratio) and strength properties (such as unconfined compressive strength (UCS) and tensile strength (T)) were developed using well log data from the W-1 borehole. For the estimation of dynamic elastic properties, we used compressional wave velocity (v_p), shear wave velocity (v_s), and rock density (ρ), utilising the following Formulas (1) and (2) [55]:

$$v_{dyn} = \left[{v_p}^2 - 2{v_s}^2\right] / \left[2\left({v_p}^2 - {v_s}^2\right)\right] \quad (1)$$

$$E_{dyn} = \rho {v_s}^2 \left[\left(3{v_p}^2 - 4{v_s}^2 \right) / \left({v_p}^2 - {v_s}^2 \right) \right] \quad (2)$$

where v_{dyn} stands for the dynamic Poisson's ratio, E_{dyn} denotes the dynamic Young's modulus, v_p and v_s are the compressional shear wave velocities, respectively, and ρ is the rock density.

To determine static equivalents of elastic and strength properties for particular lithotypes occurring in the W-1 borehole profile, we used empirical models reported in the literature that describe the relationship between dynamic and static elastic properties for specific lithotypes [56–58].

In Figure 3, track 3 displays the developed static geomechanical properties alongside lithological interpretation (track 1) and input data (track 2) including compressional wave velocity (v_p), shear wave velocity (v_s), rock density (ρ), and gamma ray (GR).

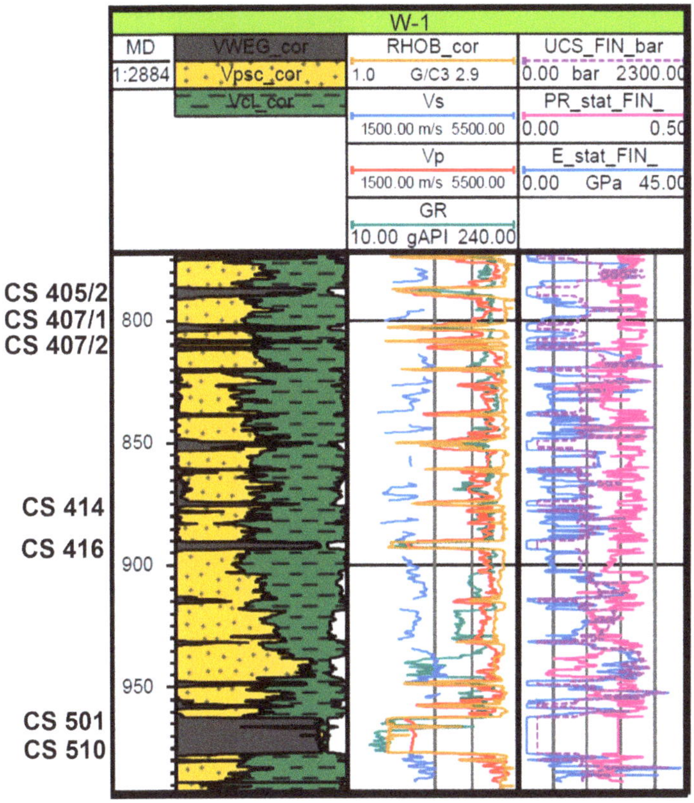

Figure 3. Diagram presenting the lithological model with major coal seams (CSs) in the interval of the interest (track 1), input data including gamma ray (GR), compressional wave velocity (Vp), shear wave velocity (Vs), bulk density (RHOB_cor) (track 2), and calculated static mechanical parameters: unconfined compressive strength (UCS_stat), Young's modulus (E_stat) and Poisson's ratio (PR_stat) (track 3) plotted versus measured depth (MD).

Each of the three distinguished lithotypes was assigned an appropriate model [56–58] to calculate static equivalents of elastic properties (track 3 in Figure 3). For UCS estimation, relationships between this parameter and Young's modulus were used for both sandstones and sandy shales [58], while a relationship between UCS and porosity was utilised for shales [56]. Tensile strength was determined using the formula proposed by Hoek [59] that describes the relationship between tensile strength and unconfined compressive strength.

Three-dimensional models of geomechanical parameters were developed using the estimated static geomechanical properties (Young's modulus, Poisson's ratio, and UCS) from the W-1 borehole profile. During modelling of geomechanical properties near the drainage area, results from geomechanical measurements of samples collected in the drainage area, archival laboratory measurements shared by the mine operator, and data from the literature for other strata were incorporated. The value ranges for specific parameters and their source references are provided in Table 1 [60–65].

Table 1. The mean values of mechanical properties in the Staszic-Wujek coal mine area.

Parameter [unit]	Q + III	Conglomerate	Coal	Goaf	Coal Shale	Shale			SandyShale			Sandstone		
						Westfal A CS300	Namur C CS400	Namur B CS500	Westfal A CS300	Namur C CS400	Namur B CS500	Westfal A CS300	Namur C CS400	Namur B CS500
E [GPa]	0.8 [60]	41 [61]	2.08	1.77 [62]	5 [63]	1.42–6.79 av 4.69 (coal mine archival data)			3.94–6.89 av 5.14 (coal mine archival data)			4.09–8.18 av 5.73 (coal mine archival data)		
PR [-]	0.35 [60]	0.25 [61]	0.29	0.27 [62]	0.28 [63]	0.07–0.37 av 0.16 (coal mine archival data)			0.06–0.3 av 0.14 (coal mine archival data)			0.07–0.2 av 0.12 (coal mine archival data)		
UCS [MPa]	6.9 [60]	40 [61]	24.6	12.3 [62]	28.78 [63]	22.6–51.4 av. 42.6 (coal mine archival data)	31.7–61.8 av. 49.4 (coal mine archival data)		8–31.7 av. 24.5 (coal mine archival data)	24.8–47.6 av. 38.7 (coal mine archival data)		14–36.1 av. 29.1 (coal mine archival data)	39–56.3 av. 46.3 (coal mine archival data)	
TENSILE [MPa]	0.69 [60]	4 [61]	0.6	1.23 [62]	2.88 [63]	0.29–4.42 av 1.47 (coal mine archival data)			1.23–3.82 av 2.15 (coal mine archival data)			1.15–4.55 av 2.82 (coal mine archival data)		
DENSITY [g/cm^3]	Well log	Well log	2.14	1.14	Well log	Well log			Well log			Well log		
POROSITY [m^3/m^3]	Well log	Well log	Well log	0.3	Well log	Well log			Well log			Well log		
FA [°]	28 [60]	35 [64]	22 [62]	30 [62]	33.5 [65]	46.5 [65]	33.5 [65]	46.5 [65]	37.5 [65]	46 [65]	37.5 [65]	57 [65]	53 [65]	55.5 [65]
BIOT [-]	1	1	1	1	1	1			1			1		

Before data analysis, the input data were averaged to the vertical scale of the 3D model. During the geostatistical analysis, variogram parameters were defined, and spatial relationships between input data were determined for calculating the distribution of geomechanical parameters in 3D space. This procedure was performed individually for each lithotype.

The modelling of geomechanical properties utilised the Gaussian random function simulation algorithm, calculating 25 equally probable realisations, which were then averaged to create the final 3D distribution of mechanical properties (Figure 4). The developed models of geomechanical properties, similar to the petrophysical 3D models, took into account the effects of the mining works carried out so far, manifesting in local changes in the properties of the rock material in the mining zone.

Models were developed on a regional scale to capture the entire scope of data and local scale models of higher resolution in the area of the IC-panel in the zone of drainage.

The models constructed on a local scale, visualised in Figure 4, were further used as an input for modelling of fluid flows as well as geomechanical simulations to calculate the changes in stress and strain field with the progression of the mining works. The stress and strain field were calculated using the finite element geomechanical simulator Visage (Schlumberger). The initial effective stress conditions within a reservoir and its overlying rock formations are described using the following Formula (3) applicable to isotropic rocks [66,67].

$$\sigma_h = \frac{\nu}{1-\nu}(\sigma_v - \alpha P_P) + \alpha P_P + P_T \quad (3)$$

where σ_h is the minimum horizontal stress, ν is the Poisson's ratio, σ_v is the vertical stress, α is the Biot's constant, P_P is the pore pressure, and P_T is the tectonic pressure.

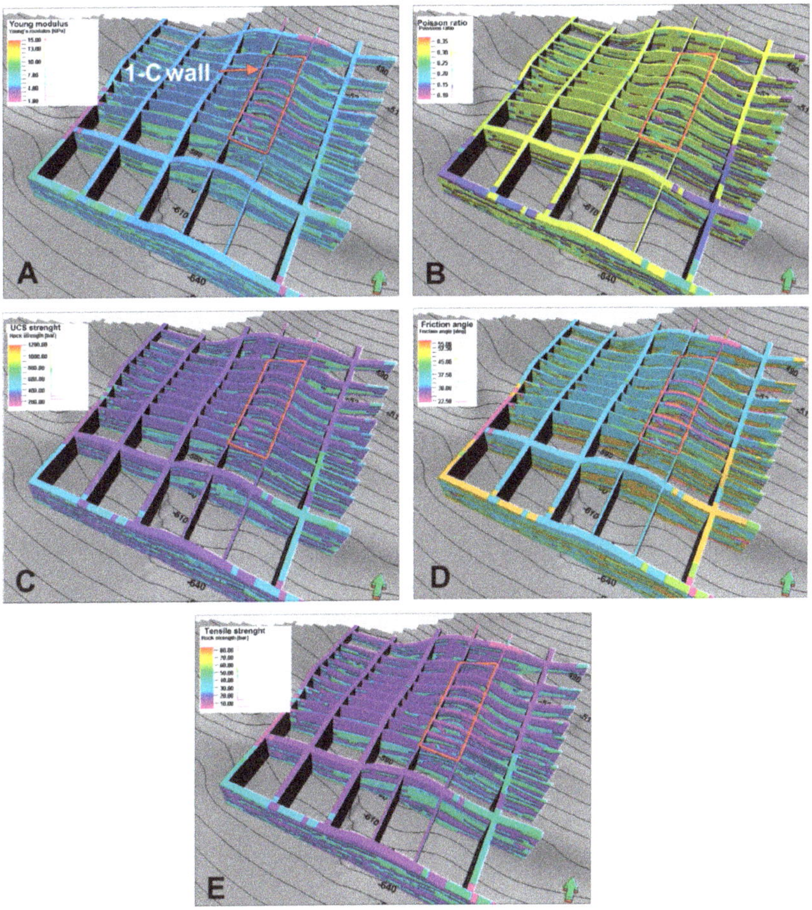

Figure 4. Visualisations of 3D parametric models on a local scale presenting 3D distributions of the (**A**) Young's modulus, (**B**) Poisson's ratio, (**C**) unconfined compressive strength, (**D**) friction angle, and (**E**) tensile strength.

To reproduce changes in stress and strain fields resulting from mining activity, specifically the IC-panel initialisation and its longwall movement, initial pore pressure distribution and boundary conditions were determined.

A pore pressure model was developed to calculate effective stresses using Terzaghi's principal Formula (4) [68]:

$$\sigma' = \sigma_{tot} - P_P, \tag{4}$$

where σ' stands for the effective stress, σ_{tot} is the total stress, and P_P is the pore pressure.

The distribution of the initial pore pressure in the Staszic-Wujek coal mine was determined based on the work by Różkowski et al. [69]. In their research, due to the different content of solids dissolved in the underground waters within the USCB, the authors distinguished five 200 m thick zones in the profile (Table 2). This solid content, causing an increase in water mineralisation with depth and corresponding rise in the hydrostatic pressure gradient, was used to determine the pore pressure gradients in these zones and, consequently, the 3D distribution of pore pressure (P_P) in the research area.

Table 2. The content of solids dissolved in underground water in Carboniferous productive intervals (based on [69]).

Depth [m]	Solid Average Content [g/L]	Pore Pressure Gradient [bar/m]
0–200	3.35	0.10002
200–400	13.94	0.10108
400–600	51.84	0.10487
600–800	76.89	0.10737
800–1000	117.38	0.11142

The hydrostatic pressure at the level of the 501 coal seam in the vicinity of the IC panel was determined to be 94 bars (at true vertical subsea depth of −585.5 m). The visualisation of the calculated 3D distribution of pore pressure in the study area is presented in Figure 5. The estimated pore pressure was then used in the geomechanical simulations to obtain the initial stress field.

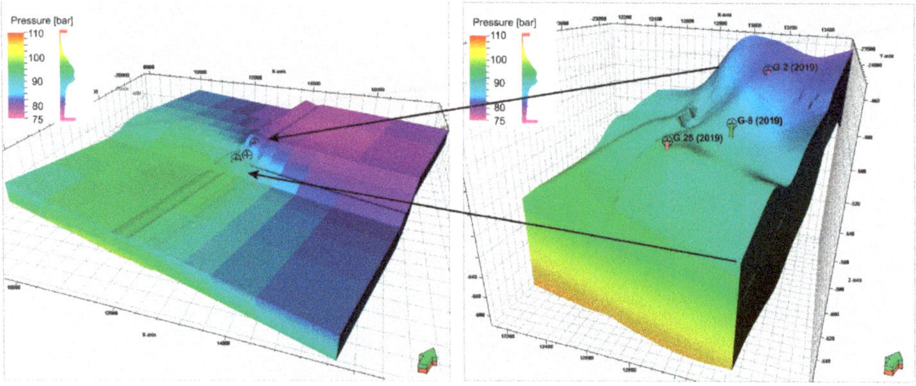

Figure 5. Three-dimensional distribution of pore pressure in the large-scale model (**left**) and the local-scale model of the area subjected to methane drainage (**right**).

Geomechanical Boundary Conditions

To determine the initial stress conditions, we applied the load of the overlying rocks and tectonic stresses as boundary conditions. Based on the interpretation of borehole wall images available in the W-1 borehole capturing the drilling-induced damage observed in the wellbore wall, such as breakouts and drilling-induced fractures (DIFs), the direction of present maximum horizontal stress σ_H was determined [70]. The azimuth of the DIFs was between 90 and 110° (ESE). These directions are consistent with the findings of Dubiński et al. [71], who stated that the maximum horizontal stress azimuth in the close vicinity of the study area is 99° (ESE).

The gradient of the minimum horizontal stress σ_h was determined to be approx. 0.1707 bar/m based on mini-frac tests in the nearest available borehole location [72]. The stress regime in the study area was assumed to be normal faulting according to the studies carried out in the area of interest [71,73]. The anisotropy between principal horizontal stresses σ_H/σ_h was also determined to be 1.25, based on the findings from the literature using the reference borehole [72]. The initial stress conditions in the Staszic-Wujek coal mine are listed in Table 3.

Table 3. Initial principal stresses acting in the Staszic-Wujek coal mine.

Parameter [Unit]	Value
Gradient σ_h [bar/m]	0.1707
σ_H / σ_h [-]	1.25
Azimuth σ_h [°]	99.00

During geomechanical simulations coupled with dynamic reservoir fluid flow modelling, pressure changes induced by the advancing longwall front led to the redistribution of stress and strain fields. This, in turn, caused both elastic and plastic deformations. When the evolving stresses exceeded the critical strength of the rock mass (as listed in Table 1), failure occurred according to the Mohr–Coulomb criterion used in simulations, ultimately affecting the transport properties of the reservoir, as discussed below.

5. Dynamic Model

Based on the geological and geomechanical characterisations employed to generate the corresponding geological and geomechanical models, dynamic models of the selected IC panel of the 501 coal seam and its surroundings were constructed. A standard double-component system was assumed in coal seams of the analysed structure; this comprised cleats containing free fluids (gas and water) and described by multiphase viscous flow and matrices containing adsorbed gas and described by the diffusion type of gas transport. The other regions of the model (clastic rocks) assumed conventional single-porosity systems.

To complete the dynamic model construction, additional components of the models were determined as follows:

(a) Thermodynamic model of the reservoir gas:
the thermodynamic properties of the reservoir gas were determined by the Peng–Robinson equation of state (gas formation volume factor) and by the Lorentz–Bray–Clark correlation (gas viscosity) from gas composition, which are presented in Table 4.

(b) Capillary pressure:
the dependence of capillary pressure upon water saturation was implemented from the provided measurements—the capillary pressure determined the distribution of pressure and water/gas saturations for initial hydrostatically balanced conditions.

(c) Relative permeabilities of reservoir fluids:
due to the lack of proper measurements, relative permeability curves ($k_{rg/w}$ vs. $S_{g/w}$) were assumed to be of conventional types, i.e., linear for the naturally fractured coal and power-like for the porous clastic rocks.

(d) Adsorption/desorption isotherm of methane:
the Langmuir adsorption/desorption isotherm was determined as an average dependence from the provided measurement data described by the Formula (5):

$$C(P) = \frac{PC_0}{P + P_0} \tag{5}$$

where C(P) is the concentration of the adsorbed methane as a function of pressure, P, parametrized by asymptotic concentration C_0 = 11.25 Sm^3 of gas/m^3 of coal and characteristic pressure P_0 = 7.3 b,
The above Langmuir isotherm was obtained as an average of 23 measurements on samples obtained from different depths. They covered the pressure range of 0–18 bar.

(e) Trajectories of drainage boreholes;
trajectories and completions of both conventional and directional methane drainage boreholes were implemented from the geological model of the structure and other

technical parameters provided by the Murcki-Staszic mine operator, Jastrzębska Spółka Węglowa, Jastrzębie-Zdrój, Poland.

(f) Schedule of gate-road making:
a detailed schedule of the gate-road excavation around the IC panel of the 501 coal seam was implemented into the dynamic simulation model to estimate the road's influence upon the pressure and reservoir fluid distributions and to obtain the state of the analysed panel before the mining longwall activities.

Table 4. Composition of reservoir gas from the 501 coal seam.

CH_4	C_2H_6	H_2	CO_2	CO
[%]	[%]	[%]	[%]	[%]
92.867	0.436	0.035	6.593	0.069

Note: In the remaining part of the paper, the name "methane" is used in place of "reservoir gas".

Initial Conditions

Based on the information concerning the state of the structure elements surrounding the analysed coal seam, the dynamic model was applied to simulate their impact upon the IC panel of the seam. In particular, two factors were analysed in detail, as described below.

1. Neighbouring excavated coal seams.
The closest exploited coal seam was found to be the 349 coal seam, located about 350 m above the analysed seam and mined in the years 1988–1992. The pressure distribution in the structure as obtained from the simulation in 2019 (the beginning of the mining activities in the IC panel of the 501 coal seam) is presented in Figure 6.

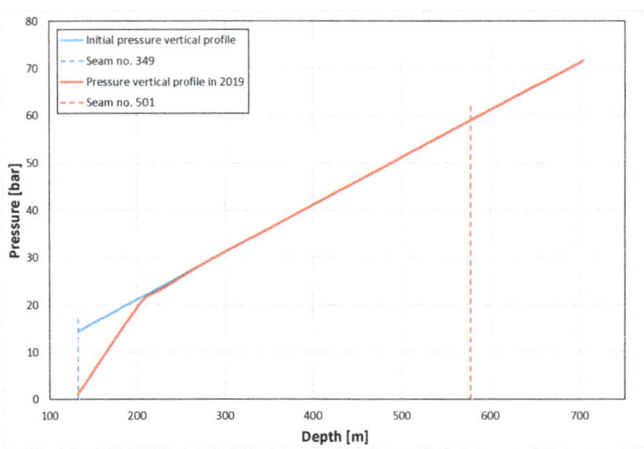

Figure 6. Vertical pressure profile of the analysed structure.

The results of the simulations proved the negligible influence of the 349 coal seam upon the 501 coal seam.

2. Gate-roads:
The constructed simulation model of the structure was used to determine the influence of the gate-roads around the IC panel on pressure distribution and desorbed methane content in the analysed coal seam and its surroundings. It was assumed that the gate-roads were excavated gradually according to the provided schedule and ventilated proportionally to the length of the completed roads.

In both the static (geological and geomechanical) and dynamic modelling described above, the effects of temperature variations were neglected due to the following reasons:

(a) The depth variations of the models and, consequently, the initial temperature variations were small;
(b) Dynamic temperature variations during the IC panel excavation were also negligible due to the very small gas (injected air) heat capacity compared to the heat capacity of the rocks.

6. Effective Coupling of Geomechanical and Fluid Flow Simulations

To study geomechanical effects, numerical modelling of fluid flow coupled with a geomechanical analysis of the medium [74–77] was applied. This coupling was achieved by partially coupled numerical simulations based on an external coupling between separate numerical simulations [74,76,78–80].

The fluid flow simulations resulted in variations in the pressure and fluid saturation distributions that, together with general rock and fluid characteristics, determined the geomechanical properties of the system. Based on those results and properties, the geomechanical simulations resulted in variations in the geomechanical state (stress and strain), thus affecting the transport properties of the system to be used in the flow simulations [75,81,82]. The two simulations were iteratively coupled at each time step. The schematic of this approach is shown in Figure 7.

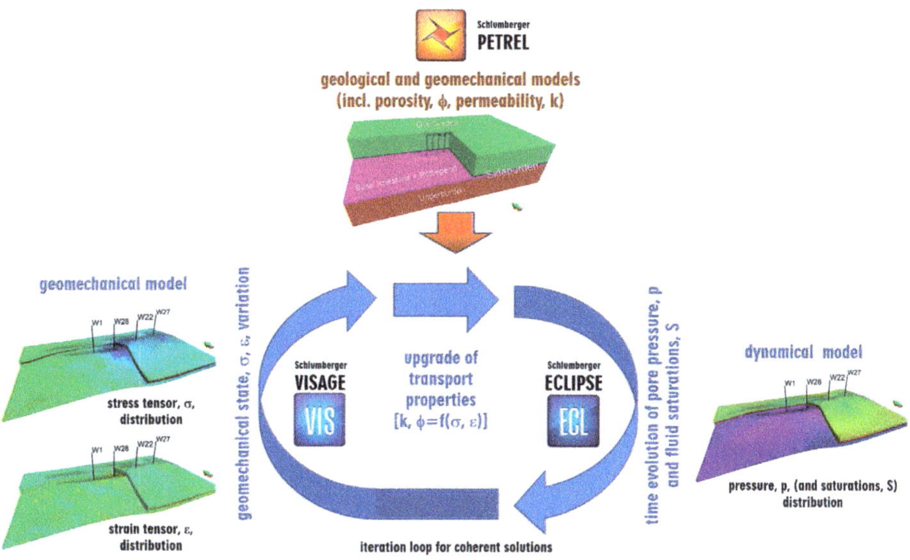

Figure 7. Dynamic and geomechanical simulation coupling.

A detailed description of the simulation coupling method used in this study was given in an earlier paper [83].

7. Geomechanical Effects

Three types of rock deformations were recognised: elastic, plastic, and failure. The elastic ones are strictly dependent on pressure variations and are different for each type of rock. Their range is not limited; however, their amplitude is relatively small. Although precisely included in the general procedure, it will not be discussed in detail here. Instead, plastic and failure-type rock deformations are presented thoroughly below. Based on the analysed results of geomechanical simulations, significant changes were observed in the diagonal components of stress and strain tensors to occur within the limited-range vicinity of the coal workings in all directions, confirming the plastic and failure type of the deformations within those regions.

The blocks around the gate-roads were grouped into several regions according to the major spatial directions following the route of the gate-roads around the analysed panel. Exemplary Regions no. 6 and no. 7 are shown in Figure 8.

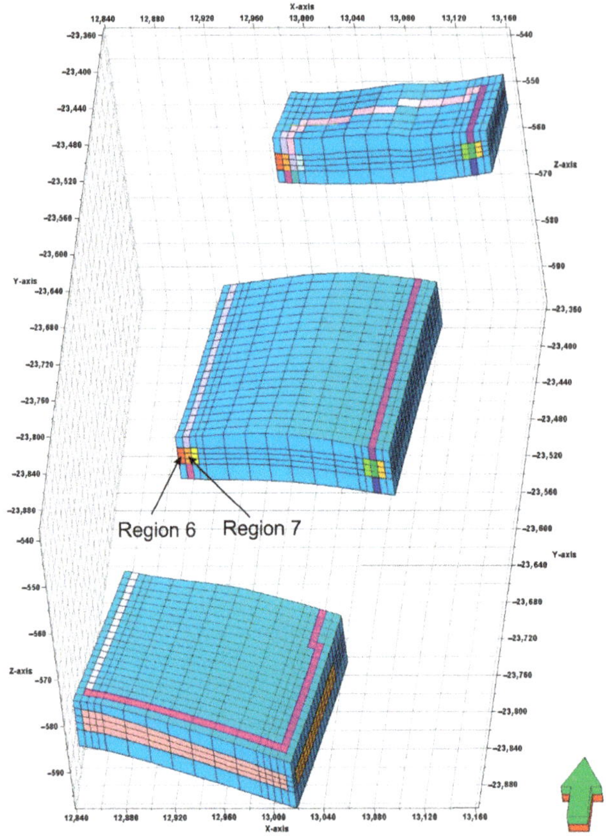

Figure 8. Location of Regions no. 6 and no. 7 in the simulation model of the 501 coal seam and its surroundings.

Changes in the stress tensor components were correlated with the changes in pressures in individual regions (Figure 9).

Next, transport properties in terms of inter-block transmissibilities were analysed for successive time steps. To relate changes in the components of the stress tensor with the changes in the transmissibilities, the relationship (6) proposed by J.Q. Shi and S. Durucan [84] was employed:

$$T_i = T_{i0} \cdot e^{-c \cdot \sum_{j=1}^{3} \Delta\sigma_{jj} \cdot (1-\delta_{ij})} \quad (6)$$

where

$T_{i/i0}$—stands for the modified/original i-th component of the transmissibility,

c—stands for pore compressibility,

$\Delta\sigma_{jj}$—stands for the change in the diagonal stress component, $j = 1(x),\ 2(y),\ 3(z)$,

δ_{ij}—stands for Kroncker's delta function.

Effective dependence of transmissibilities (transmissibility multiplication factors) on pressure was found as presented in Figure 10 for the example of Region no. 6. During the flow simulations, the effective changes of the inter-block transmissibilities, resulting from the pressure changes, were implicitly performed, taking into account appropriate simulator functions for the multiplicative transmissibility coefficients.

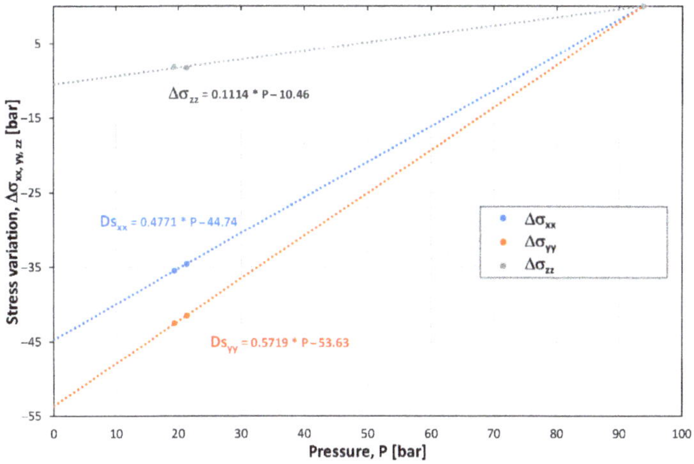

Figure 9. Variations in diagonal stress tensor components with pressure changes from the simulation results of selected analysed time steps. The example is from Region No. 6.

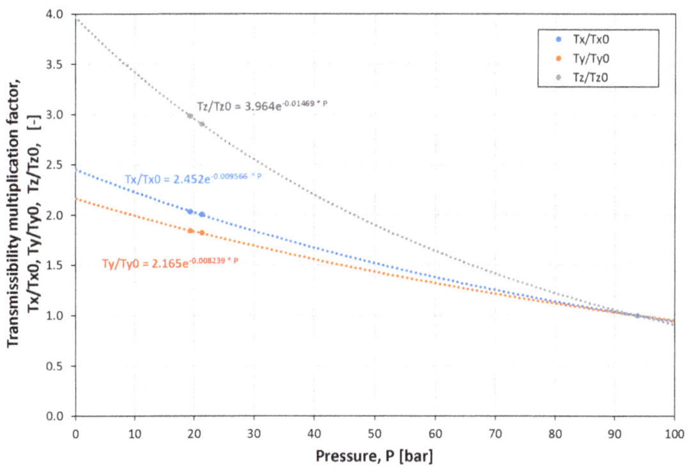

Figure 10. Variations of inter-block transmissibilities with pressure changes. Th example comes from Region No. 6.

Excavation of the analysed coal seam caused further significant modifications of the geomechanical state of the unmined coal seam and rocks surrounding the excavated part of the seam. Consequently, the modifications affected the transport properties of those rocks. Examples of the distributions of the geomechanical state variations are shown in Figures 11–14 for two selected excavation advancements: Figures 11 and 12 for partial (50%) excavation progress and Figures 13 and 14 for the complete excavation of the analysed panel of the 501 coal seam. Those distributions included three principle components of the effective stress tensor (σ'_h, σ'_H, σ'_v) within the mid-depth surface of the 501 coal seam

(Figures 11 and 13) and 3D views (Figures 12 and 14) of those distributions on the side walls of the model with front walls along the cross-section lines A-A′, B-B′, and C-C′ defined in Figure 11.

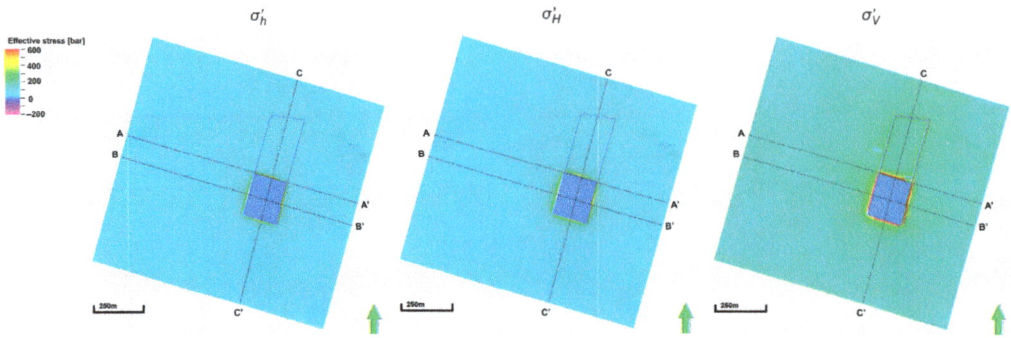

Figure 11. Distributions of the effective principal stresses (σ'_h, σ'_H, and σ'_v) within the mid-depth surface of the 501 coal seam upon partial excavation of the IC panel and location of the cross-section lines A-A′, B-B′, and C-C′.

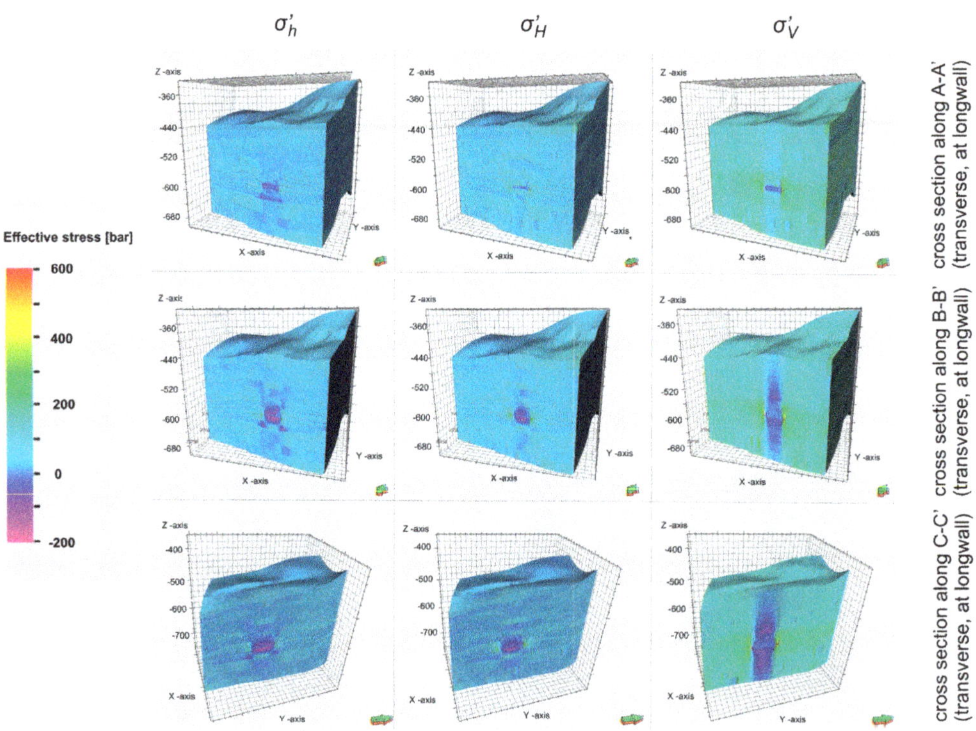

Figure 12. Three-dimensional distributions of the effective principal stresses (σ'_h, σ'_H, and σ'_v) with front walls along the vertical cross-sections A-A′, B-B′, and C-C′ of the 501 coal seam upon partial excavation of the IC panel.

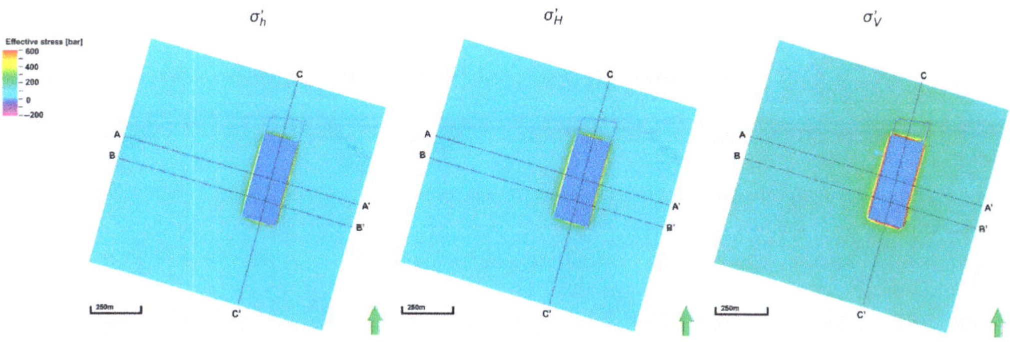

Figure 13. Distributions of the effective principal stresses (σ'_h, σ'_H, and σ'_v) within the mid-depth surface of the 501 coal seam at complete excavation of the IC panel, with provided A-A′, B-B′, and C-C′ vertical cross-section lines geometry.

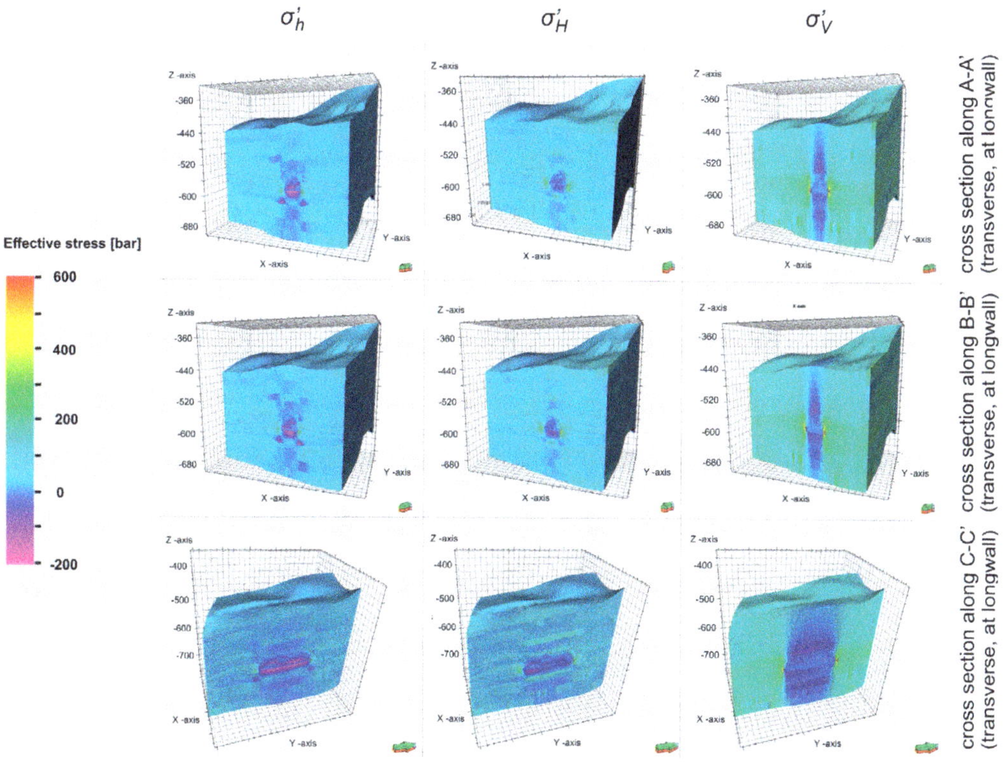

Figure 14. Three-dimensional distributions of the effective principal stresses (σ'_h, σ'_H, and σ'_v) with front walls along the vertical cross-sections A-A′, B-B′, and C-C′ of the 501 coal seam upon complete excavation of the IC panel.

An analysis establishing relations between pressure distribution, geomechanical states (stress tensor), and transport properties was made to apply the effective method of coupled geomechanical and flow simulations. Stress tensor variations resulted in changes in rock transport properties (transmissibilities) according to the Formula (6) by J.Q. Shi and S. Durucan given above.

To draw conclusions concerning the geomechanical effects of the unmined coal seam and rocks surrounding the excavated part of the IC panel upon transport properties (transmissibilities), common trends of the transmissibility multiplication factors were identified and found to depend on the distance from excavated regions of the IC panel. Detailed distributions of the transmissibility multiplication factors are shown for various regions in Figure 15 upon partial (50%) excavation and in Figure 16 upon complete excavation as a function of depth expressed by the model layer numbers.

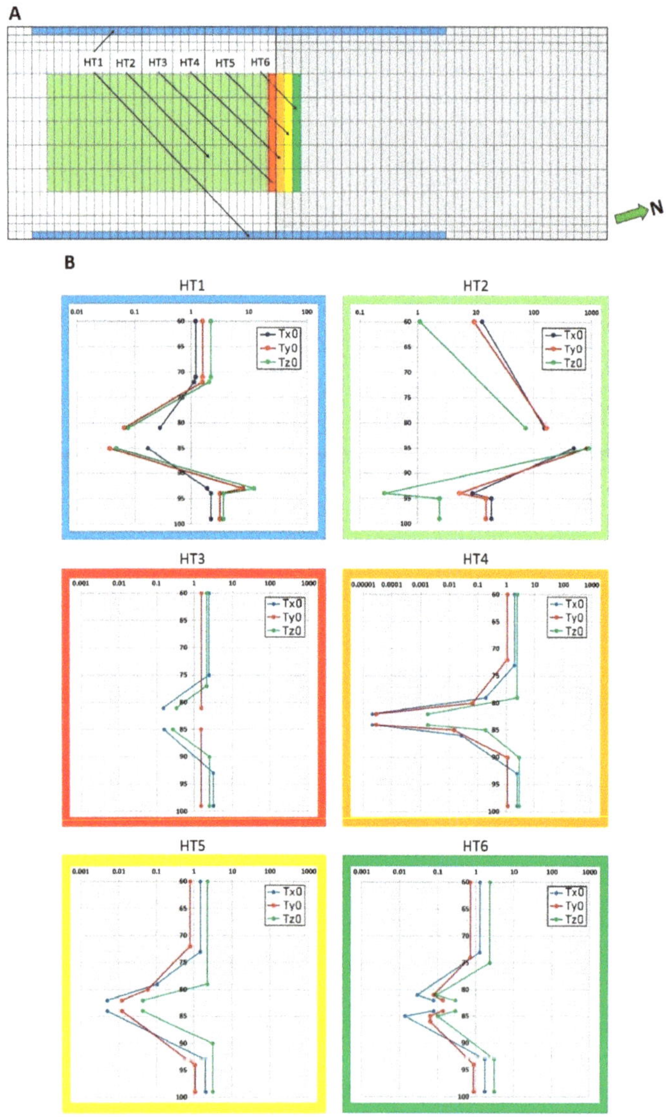

Figure 15. Common trends of the transmissibility multiplication factors in selected regions surrounding the excavated part of the IC panel upon partial (50%) excavation, (**A**) region (HT1, HT2,. . ., HT6) definitions in seam layer, (**B**) transmissibility multiplication factors vs. depth in regions HT1, HT2,. . ., HT6.

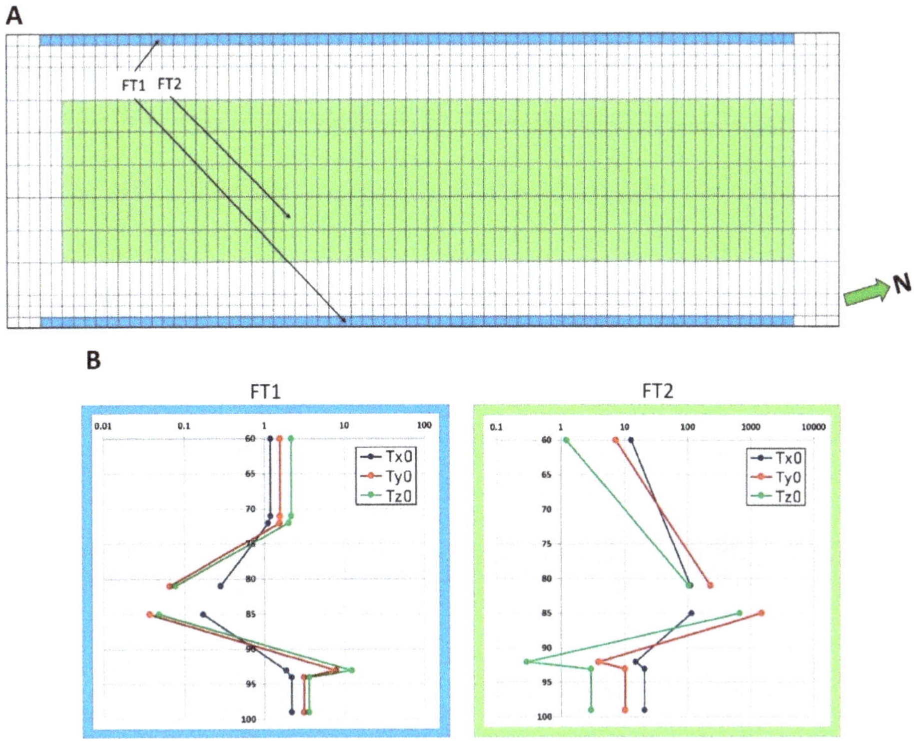

Figure 16. Common trends of the transmissibility multiplication factors in selected regions surrounding the IC panel upon the complete excavation of the panel, (**A**) region (HT1, HT2) definitions in seam layer, (**B**) transmissibility multiplication factors vs. depth in regions HT1, HT2.

The above results of the analysis concerning the influence of geomechanical states upon the transport properties of unmined parts of the 501 coal seam and rocks surrounding the seam were applied to the calibration procedure of the analysed simulation model described below.

8. Model Calibration

The calibration procedure of the model was based on the following list of historical data related to the coal production from the IC panel of the seam.

1. Schedule and measurement data from the gate-road excavation and ventilation before the coal excavation from the panel, including air injection, total gas (combined air and methane) production, and methane production of the ventilation system vs. time (from 29 January 2018 to 31 August 2019),
2. Schedule and measurement data from the panel coal production, including air injection to the ventilation system and total gas and methane production by three elements of the methane drainage system: the ventilation system, the standard methane drainage boreholes, and the directional methane drainage boreholes vs. time (from 1 September 2019 to 28 February 2020).

In addition, the historical data included the pressure conditions of the ventilation system and the methane drainage boreholes. The schedule of the IC panel coal production is presented in Figure 17 as the longwall advancement.

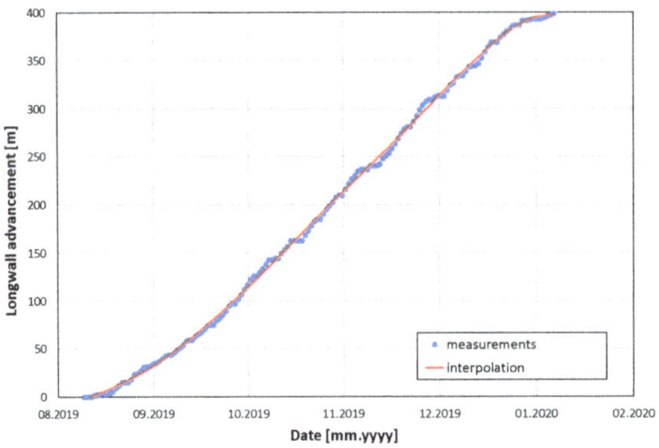

Figure 17. The longwall advancement during the IC panel coal production.

Similarly, the schedule of the standard clusters of drainage borehole activities (the cluster production intervals) is shown in Figure 18 for the boreholes grouped into 19 clusters.

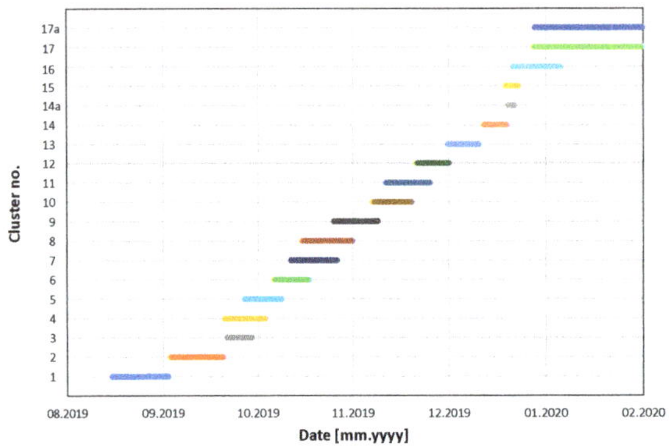

Figure 18. The production intervals of standard methane drainage borehole clusters. Colours are used for better readability only.

Furthermore, the schedule of the directional borehole activities (the borehole production intervals) is shown in Figure 19.

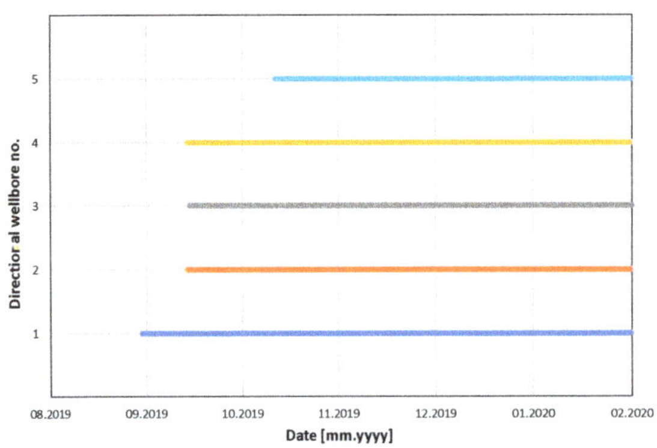

Figure 19. The production intervals of directional methane drainage boreholes. Colours are used for better readability only.

The basic measured data of the methane ventilation and drainage processes during the coal production from the IC panel included the following:

1. the volume of the injected air,
2. the volume of the total gas produced by the ventilation system (Figure 20),
3. the methane volume in the total gas produced by the ventilation system (Figure 21),
4. the volume of the total gas produced by the standard drainage system (Figure 22),
5. the methane volume in the total gas produced by the standard drainage system (Figure 23),
6. the volume of the total gas produced by the directional drainage boreholes (Figure 24),
7. the methane volume in the total gas produced by the directional drainage boreholes (Figure 25).

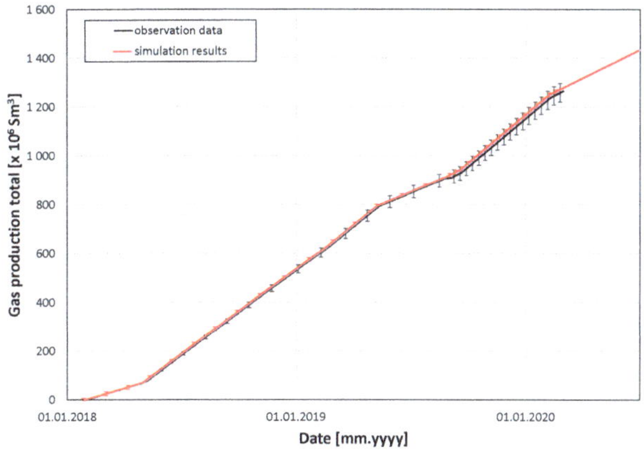

Figure 20. Total production of ventilation gas (combined air and methane): observed data vs. simulation results. Error bars in the data series are explained in the text.

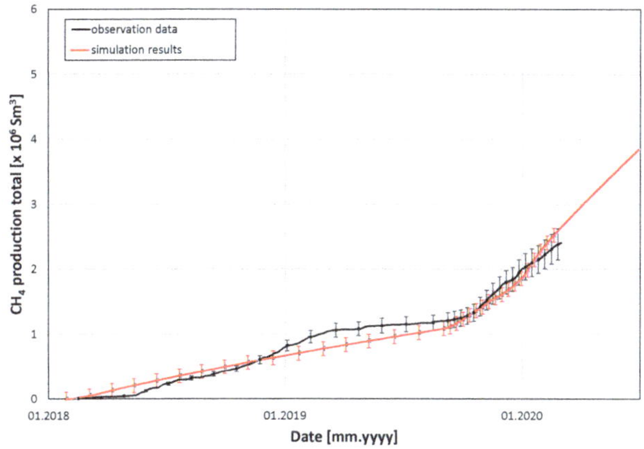

Figure 21. Total methane production in the ventilation gas: observed data vs. simulation results. Error bars in the data series are explained in the text.

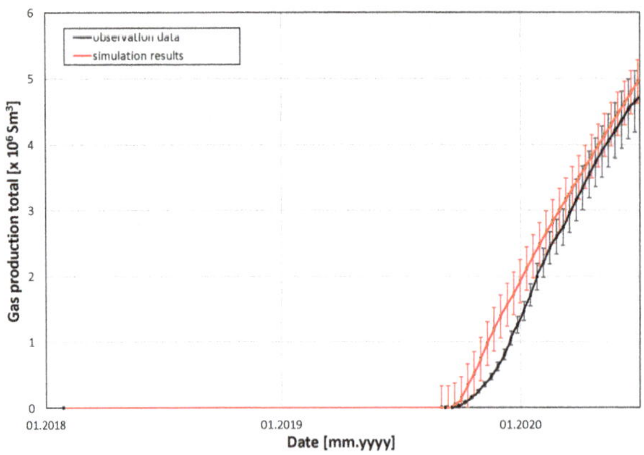

Figure 22. Total gas production by the standard drainage boreholes: simulation results vs. observed data. Error bars in the data series are explained in the text.

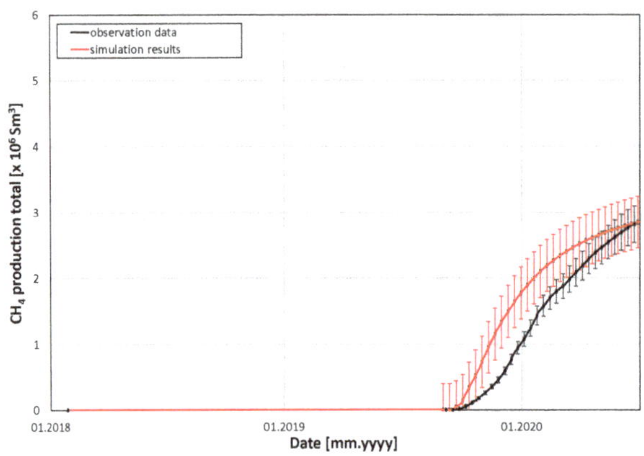

Figure 23. Total methane production by the standard methane drainage boreholes: observed data vs. simulation results. Error bars in the data series are explained in the text.

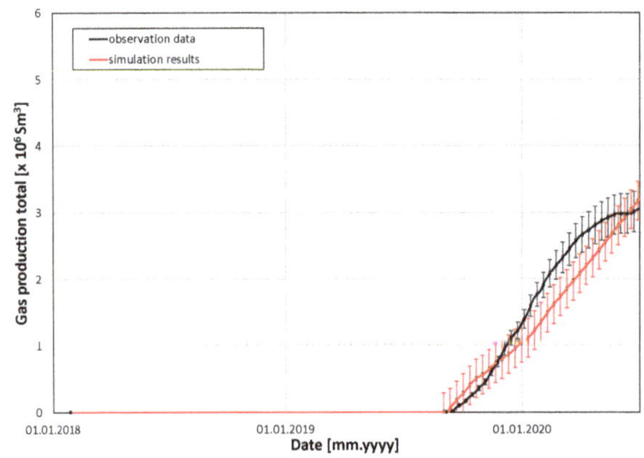

Figure 24. Total gas production by the directional boreholes: observed data vs. simulation results. Error bars in the data series are explained in the text.

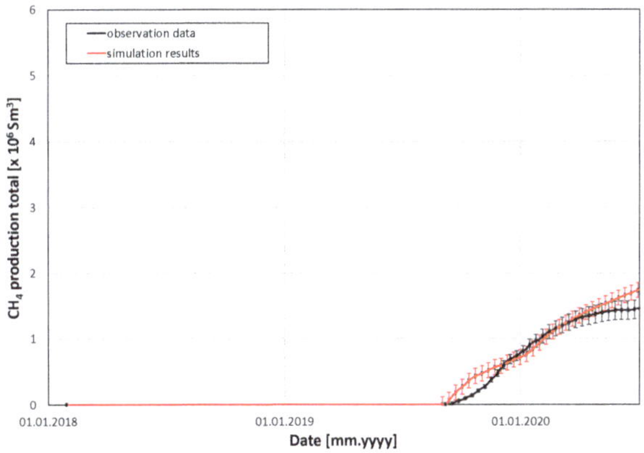

Figure 25. Total methane production by the directional boreholes: observed data vs. simulation results. Error bars in the data series are explained in the text.

Among the above seven items, the first was treated as a control variable, and the other six were history-matched to calibrate the model through multiple simulation runs.

The final results of the calibration procedure are shown in Figures 20–25 for items 2 through 7, respectively, as a comparison of the measurement (observed) data with the simulation results.

The main data series in these Figures are supplemented with error bars attached to both measurement data and simulation results. The former indicates typical errors of measurements (with a relative error of 10%), and the latter corresponds to estimated mean absolute deviations (MADs) as measures of goodness of fit for the above six items. The precise values of MAD, defined as

$$MAD = \frac{1}{k}\sum_{i=1}^{k}|m_i - d_i| \tag{7}$$

are included in Table 5 together with average measurement error (AME) and their ratio (MAD/AME). In the above formulas, the following symbols are used:

k—total number of data points,

m_i—simulation result, $i = 1,\ldots, k$,

d_i—measurement data, $i = 1,\ldots, k$.

Table 5. Calibration goodness-of-fit. Measures of goodness of fit: mean absolute deviation (MAD) and MAD to average measurement error (AME) ratio.

Fitted Parameter for Total Production of [Component Type]	Figure No. [-]	MAD [$\times 10^6$ Sm^3]	AME [$\times 10^6$ Sm^3]	MAD/AME [-]
ventilation gas	Figure 20	7.678	39.05	0.197
methane in ventilation gas	Figure 21	0.105	0.120	0.876
gas by standard drainage boreholes	Figure 22	0.329	0.210	1.567
methane by standard drainage boreholes	Figure 23	0.394	0.139	2.835
gas by directional boreholes	Figure 24	0.289	0.169	1.713
methane by directional boreholes	Figure 25	0.115	0.087	1.321

The above results of the goodness of fit show a satisfactory agreement between the measurement data and the simulation results, confirming good reliability of the model to correctly describe the methane drainage taking place during the coal production from the IC panel of the 501 coal seam.

9. Effectiveness of Methane Drainage by the Directional Boreholes

The calibrated model of the 501 coal seam and its surroundings, as described above, was used to simulate a hypothetical scenario that did not use directional boreholes for methane drainage. The fundamental results of that simulation in terms of total methane production by the ventilation system, with analogous results for the scenario with directional boreholes, are shown in Figure 26.

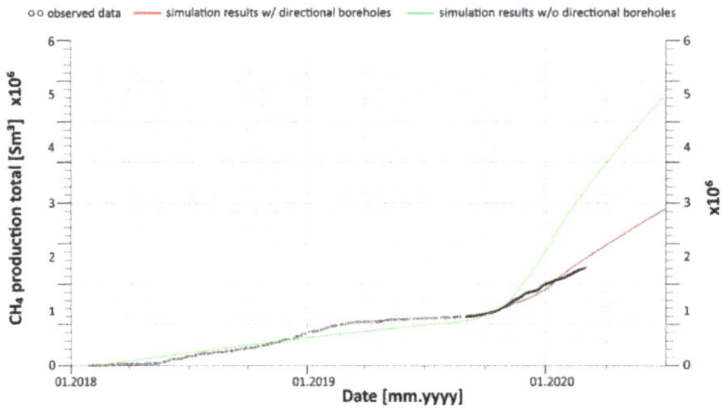

Figure 26. Total methane production of the ventilation system. Scenario with vs. without directional boreholes.

It is worth noting that other basic simulation results, including total gas production by the ventilation system and standard drainage borehole system, as well as total methane production by the standard drainage borehole system, did not vary much when the directional boreholes were used.

In particular, the comparison of methane concentration in the gas outflow to the ventilation system for the cases with and without the directional boreholes is shown in Figure 27.

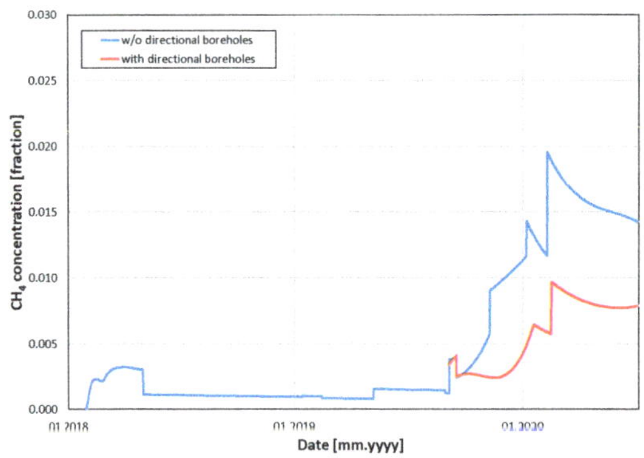

Figure 27. Methane concentration in the gas outflow to the ventilation system.

In addition, the results for the volume of methane left in the IC panel of the 501 coal seam are shown in Figure 28 for both scenarios.

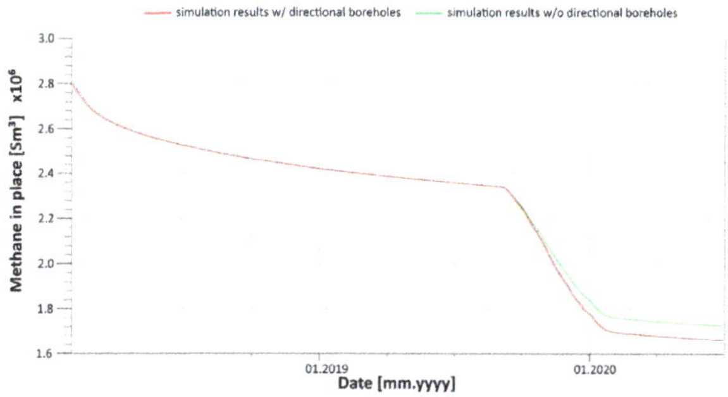

Figure 28. The decline of methane volume within the analysed panel of the 501 coal seam. Scenarios with vs. without directional boreholes.

The distributions of methane adsorbed in coal matrices of the coal seams within the model are presented in Figure 29. It includes four cross-sections for four rows defined as the mid-depth surface of the 501 coal seam, the vertical transverse cross-section along line A-A′, the vertical transverse cross-section along line B-B′, and the vertical longitudinal cross-section along line C-C′. Every set of four cross-sections corresponds to three situations (in three columns): the initial state before the panel excavation commencement, the final state after the panel excavation completion with directional boreholes, and the final state after the panel excavation completion without directional boreholes.

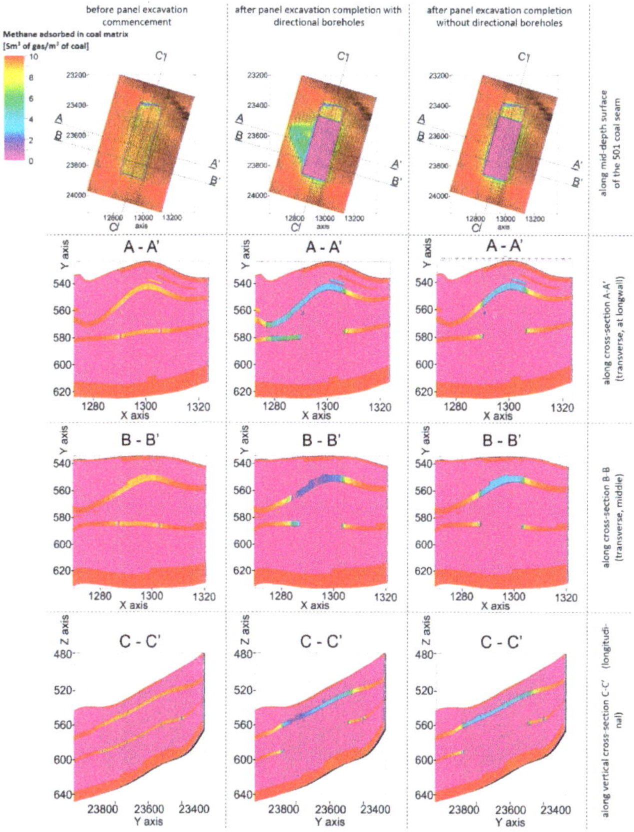

Figure 29. Distribution of methane volume adsorbed in coal seams.

To summarise, the application of the long-reach directional boreholes proved the effectiveness of the analysed methane drainage method, as can be concluded from all the results presented above. In particular, the method significantly (by 52%) decreased the methane concentration in the ventilation gas (Figure 26) and remarkably reduced the methane content in the coal matrix of the analysed coal seam (Figure 28).

10. Sensitivity Analysis

The results of methane drainage from the analysed coal seam were dependent on various parameters, including the geological parameters of the structure. Multiple simulations of the coal excavation process, accompanied by the studied methane drainage strategy, provided the data required for the appropriate sensitivity analysis. Below, its results are presented for the basic parameters, including

- two parameters characterising the methane adsorption/desorption isotherm in the coal matrix—the asymptotic concentration, C_0, and the characteristic pressure, P_0, of the standard Langmuir isotherm;
- the overall amplitude of the rock transmissibility factors, T/T_0, due to geomechanical elastic perturbations;
- the effective amplitude of the coal transmissibility multiplicative factor, T_c/T_{c0}, due to geomechanical inelastic and failure-type perturbations;
- uncertainty of the original porosity, ϕ, vertical permeability, k_z, and horizontal permeabilities, k_x, k_y,+;
- the matrix–fracture coupling coefficient, σ, in the coal seams;
- the gas diffusion coefficient, D, for the methane migration from the coal matrix to coal fractures;
- the residual gas saturation, S_{gcr}, in the coal fractures.

The influence of the above parameters for their relative modifications by $\pm$10% upon four significant methane drainage effectiveness criteria is shown as tornado diagrams in Figures 30 and 31 for the total gas and methane drainage by directional drainage boreholes, respectively.

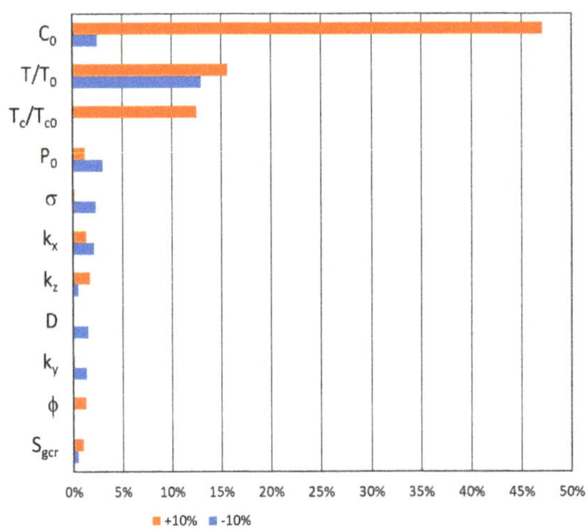

Figure 30. Sensitivity of the total gas drainage by the directional drainage boreholes to eleven selected parameters.

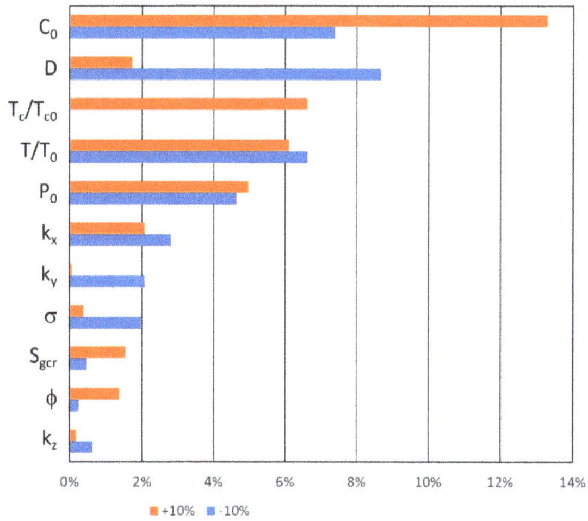

Figure 31. Sensitivity of the total methane drainage by the directional drainage boreholes to eleven selected parameters.

As far as the total gas drainage by the directional drainage boreholes (Figure 30) is concerned, the prevailing parameters are those determining the amount of adsorbed methane, namely the asymptotic concentration, C_0, and, to a lesser extent, the characteristic pressure, P_0, of the standard Langmuir isotherm. The other dominant parameters are the overall amplitude of the rock transmissibility factors, T/T_0, due to geomechanical elastic perturbations and the effective amplitude of the coal transmissibility multiplicative factor, T_c/T_{c0}, due to geomechanical inelastic and failure-type perturbations. Regarding the total methane drainage by the directional drainage boreholes (Figure 31), it is primarily sensitive to the above four parameters and to the gas diffusion coefficient, D, for the methane migration from the coal matrix to coal fractures.

It is worth noting that the results of the sensitivity analysis show asymmetry between positive and negative variations of the key simulation results with respect to some parameters. The most probable reason for this behaviour is a non-linear correlation between the parameters and the results. Nevertheless, the relative significance of various parameters, as discussed above, is valid and can be treated as reliable. Hence, this observation does not affect the main conclusions stated below and cannot reduce the practical values of the presented sensitivity analysis.

Based on the sensitivity analysis results presented above, the following summary remarks concerning the total volume of methane drainage by the directional boreholes may be formulated.

- It is proportional to the original methane reserves adsorbed in the coal matrix—parameter C_0 of the adsorption isotherm.
- It is also proportional to the methane volume available for the drainage at the effective pressure in the coal matrix—parameter P_0 of the adsorption isotherm.
- It increases with the diffusive properties of the coal matrix—diffusion coefficient D of the coal matrix.
- It increases with the effective transmissibility coefficients of the coal matrix due to geomechanical perturbations.
- It also increases with the effective transmissibility coefficients of the clastic rocks neighbouring the coal seam.

Based on our analysis, the following conclusions have been drawn regarding the estimates of methane available for the drainage process and the corresponding drainage requirements:

- the reserves of methane adsorbed in the coal matrix, as well as the degree of its desorption at a given drainage pressure, must be determined through the identification and parametrisation of the adsorption isotherm,
- the significance of the diffusion transport properties of the coal matrix must be ascertained to accurately estimate the effective migration of methane from the coal matrix to the coal cleats,
- the detailed geomechanical characterisation of both the coal matrix and the surrounding clastic rocks must be acknowledged to assess the modification of their geomechanical state—specifically, the distributions of stress and strain—resulting from pressure variations,
- the correlation between the modified geomechanical state and the transport properties—namely, effective transmissibilities and diffusion coefficients—of the coal and clastic rocks needs to be established.

All the above parameters are of a geological and geomechanical type and should be analysed before the commencement of the coal excavation process and implementation of the method of methane drainage and its operation parameters. The influence of the latter on the effectiveness of the methane drainage process is presented in the next chapter.

11. Drainage Process Optimisation

The calibrated model of the analysed carbon seam and the reliable simulations of its excavation process, accompanied by methane drainage, provide an opportunity to optimise the proposed strategy of methane drainage by a set of directional boreholes. According to practical aspects of the analysed underground long-reach directional drilling technology, the details of the directional borehole trajectories and their completion are the most significant and controllable. Among several parameters describing borehole trajectories, the most significant ones seem to be their depths and separation. Below, the results of the trajectory depths optimisation analysis are presented first. The total amount of methane drained by the directional boreholes and the total recovery of methane originally present in the excavated panel of the coal seam are selected as the basic criteria for the effectiveness of the analysed methane drainage method.

Two groups of different scenarios of directional borehole trajectories were considered. The first one included the original trajectories modified for their depth relative to the IC panel of the 501 coal seam, while they preserved their original localisation in the x-y plane. As an example, Figure 32 presents the modifications of the original trajectory of the borehole TM3 denoted TM3a, TM3b, and TM3c for the alternative depths realised in Scenario 1a, Scenario 1b, and Scenario 1c, respectively. Analogous modifications of the other directional borehole trajectories were applied to the above scenarios. The most significant simulation results of Scenarios 1a, 1b, 1c, and the original are presented in Figures 33 and 34 in terms of the total methane production by the directional boreholes from the analysed panel and the methane remaining in place of that panel vs. time, respectively.

Simulation results of those scenarios may be summarised by the following conclusion: the closer the trajectories to the excavated coal seam, the higher the degree of methane recovered from the seam within the IC panel. The additional recovery in terms of total methane production may exceed 200% (Figure 33).

The other significant parameter describing borehole trajectories is their separation scheme. The corresponding group of analysed scenarios, denoted 2a, 2b, and 2c, included borehole trajectories covering the excavated panel with a horizontally uniform scheme.

These scenarios differed in the depth of the trajectories in comparison with the previous group.

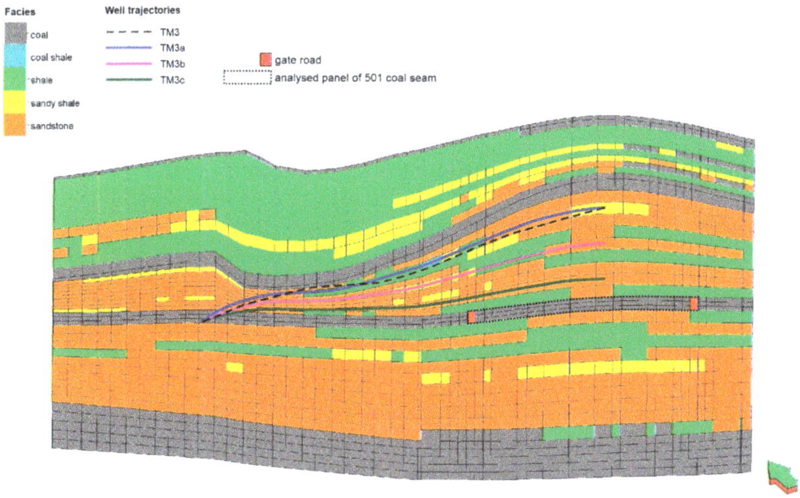

Figure 32. The vertical cross-section along the TM3 borehole trajectory and the original and alternative TM3 trajectories.

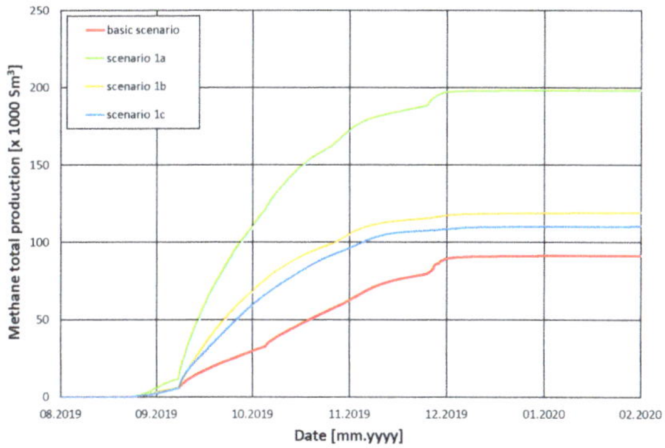

Figure 33. Total methane production by the directional boreholes from the analysed panel and the results of Scenarios 1a, 1b, and 1c vs. the basic scenario.

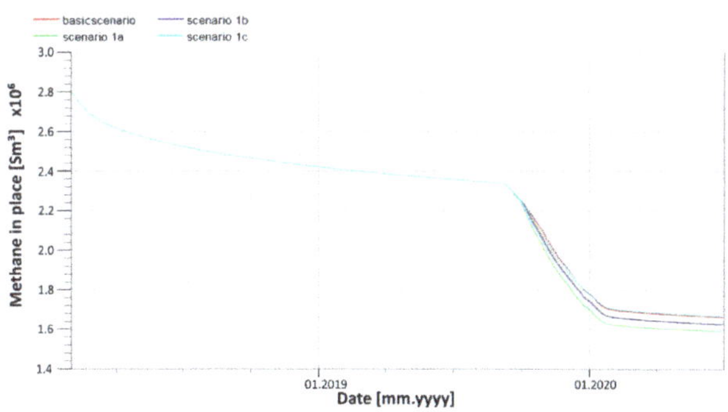

Figure 34. The decline of methane volume within the analysed panel. The results of Scenarios 1a, 1b, and 1c vs. the basic scenario.

The simulation results of Scenarios 2a, 2b, 2c, and the original one are presented in Figures 35 and 36 in terms of the total methane production by the directional boreholes from the IC panel of the 501 coal seam and the methane remaining in place of that panel vs. time, respectively.

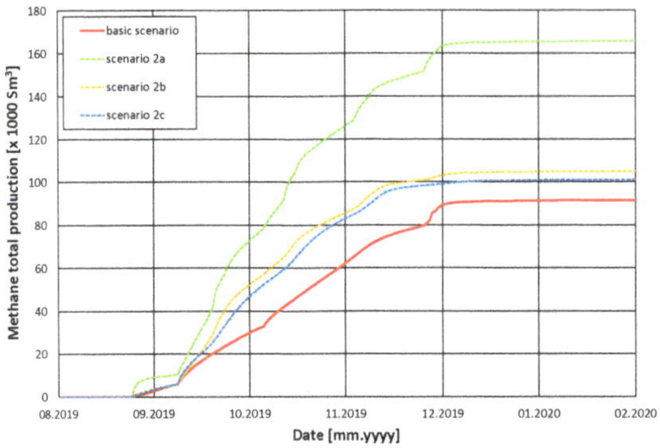

Figure 35. Total methane production by the directional boreholes from the analysed panel. The results of Scenarios 2a, 2b, 2c vs. the basic scenario.

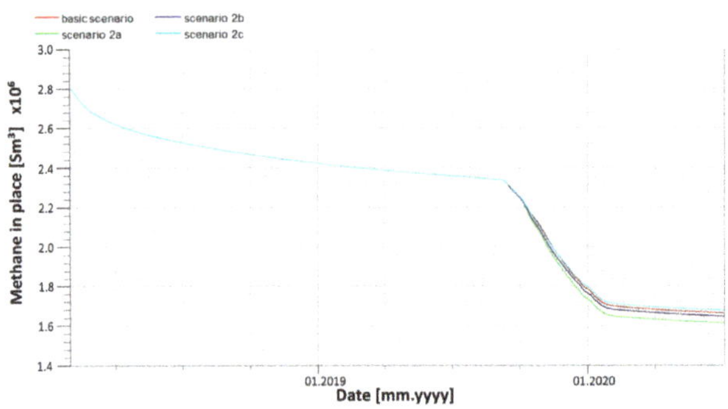

Figure 36. The decline of methane volume within the analysed panel. The results of Scenarios 2a, 2b, and 3c vs. the basic scenario.

Similar to the previous group of scenarios, the closer the trajectories to the 501 seam, the larger the methane total production (Figure 35) and the higher the recovery (Figure 36) of the methane originally adsorbed in the analysed panel. It is worth noting that the results of the latter group of Scenarios 2a, 2b, and 2c are systematically more advantageous than those of the first group. This result is caused by the difference in the total length of directional borehole completions draining methane from the analysed panel. This length equals 476 m for Scenarios 1a, 1b, and 1c vs. 304 m for Scenarios 2a, 2b, and 2c.

However, it should be noted that there is another factor limiting that conclusion, namely, wellbore instability. Openings of the directional boreholes may be partially located in the coal seam, and they usually pass through mudstone and sandstone when entering the fracture zone. Due to the relatively large diameters of the boreholes, they are susceptible to loss of stability, resulting in their deformation and collapse. Therefore, the stability of the long-reach directional boreholes may directly affect their drainage performance. The influencing factors of borehole stability mainly include rock strength, drainage pressure, stress and strain, burial depth, lateral pressure coefficient, and anisotropic permeability, as indicated by [85–88].

In particular, X. Yao et al. [85] constructed a local model of a long drilling borehole through the soft coal seam of Dingji Coal Mine, China and performed numerical simulations to obtain stress and displacement fields and plastic failure zone. They concluded that the stress concentration of the surrounding rock becomes higher with the increase in the side pressure coefficient. For the same reason, the displacement of drilling increases and the radius of the plastic zone decreases. Moreover, the drilling becomes more unstable with this variation in the side pressure coefficient. E. Karatela and A. Taheri [86] constructed a three-dimensional hydro-mechanical model of a borehole in fractured rock mass. They simulated overbalanced drilling conditions and validated the model against log measurements of the caliper log and found the strength of rock to be a governing factor in controlling the borehole stability. In addition, they observed that high mud flow rates as well as high pore pressure increased the instability around the borehole. Z. Wang et al. [87] performed studies on the instability of directional boreholes localized on coal seam roofs. They adopted the geological and geomechanical conditions of a coal seam in the Tingnan Coal Mine, Shaanxi Province, China. They performed numerical simulations and obtained stress, strain, and plastic deformations of the rocks surrounding the boreholes with different diameters. The simulation results showed that the borehole stability depended on rock type. It was lower for coal and mudstone than for sandstone. They also concluded that the larger the borehole diameter, the lower the borehole stability. According to the authors, their research results provided reliable technical support for the protection of directional boreholes on coal seam roofs. X-B Zhang et al. [88] studied the dependence of gas extraction in deep mining soft coal seams on deformation and instability of drainage boreholes. They analysed deformation instability mechanisms and dynamic changes of borehole suction pressures for three deformation instability modes: (1) complete holes, (2) collapse holes, and (3) plug holes. The resultant suction pressure loss of the drainage system was found to be very small for case (1) and much bigger in case (2). In case (3), a continuous medium formed at the plug segment, and the plug segment drainage pressure turned into coal bed gas pressure. According to the authors, their study provided important guidelines for improving gas drainage and effectively preventing gas disasters.

From the above results, it follows that to study directional borehole stability, it is necessary to use a high-resolution and appropriate (non-cartesian) symmetric type of numerical model of all the borehole vicinities combined with a large-scale model. This kind of study goes beyond the scope of the research presented in this paper.

12. Summary and Conclusions

The studies reported in this paper address specific issues referring to long-reach directional drilling (LRDD) for methane drainage purposes; such issues have not been covered by other research activities in the area of methane emission in coal mines. These innovative studies are based on a comprehensive numerical modelling approach. They include the following tasks: (i) the construction of structural and parametric geological models, (ii) the construction of dynamic simulation models of a selected coal mine structure, (iii) the implementation of the initial conditions of the structure, (iv) the analysis of the geomechanical effects of various types and regions of occurrence, (v) the implementation of the correlation between the geomechanical state of the rocks and their transport properties, (vi) the performance of the effectively coupled geomechanical and fluid flow simulations, (vii) the calibration of the models based on historical coal seam exploitation and methane drainage data, (viii) the analysis of the effectiveness of methane drainage by the proposed long-reach directional drilling technology, (ix) sensitivity analysis of the drainage results to geological and geomechanical properties of the rocks, and (x) the limited optimization of the drainage process with respect to the technological and operational parameters of the proposed strategy.

The results of the above simulation studies lead to the following main conclusion: the effectiveness of the analysed methane drainage method with the use of long-reach directional drilling is strongly confirmed and, in the analysed case, results in a 52% decrease in the methane content in the ventilation gas. It also leads to a concomitant reduction of the methane content in the coal matrix of the excavated coal seam. Thus, the studies, whose results are presented and discussed above, provide evidence for numerical modelling methods with coupled reservoir fluid flow simulations and geomechanical simulations to as a very useful and feasible approach to analysing the very specific method of methane drainage employing underground long-reach directional drilling technology.

In addition to the assessment of the method's effectiveness, this approach allows one to quantitatively determine the sensitivity of the methane drainage to both geological and geomechanical parameters (reserves of methane adsorbed in the coal matrix, degree of desorption, diffusive migration of methane from coal matrix to cleats, etc.) and thus also allows one to estimate the uncertainty and risks of the analysed method. Moreover, practical optimisation of the analysed method is feasible for the application of the approach used in this study by the proper selection of operational parameters, such as trajectories and completions of drainage boreholes.

It is worth noting that an analysis of the possible risks or limitations of the proposed method is provided in the report of a previous project [89], where the studies reported in this paper were performed. Particular issues concerning environmental risk assessment in coal mining with the proposed methane degassing approach are presented in [90].

In general, to achieve an analogous conclusion for any specific case of multi-seam coal systems, it is necessary to apply a similar approach taking into account all the key phenomena determining the complete gas migration route, including gas desorption from coal matrix, its diffusion to coal cleats, and viscous multiphase flow through porous and fractured media of coal multi-seams and other clastic rocks. Moreover, the variations in the geomechanical states of all the rocks present in the studied system have to be taken into account along the fluid flow, as they may significantly influence rock transport properties and thus reciprocally affect the flow.

To this aim, complex models describing both the transport properties of an appropriate fluid–rock system and the mechanical properties of both its separate components have to be developed to be used in coupled simulations of the key phenomena; they can also be used to reliably assess the effectiveness of various methane drainage methods, as shown in these studies, in the particular case of underground long-reach directional drilling method.

Only then can one resolve the fundamental question of the method's feasibility, the sensitivity of the methane drainage effects to both geological and geomechanical parameters, and the method's uncertainty and application risks, as well as practical optimisation with respect to the method's operational parameters.

Author Contributions: Conceptualization, W.S.; methodology, M.S.-V., P.R. and P.Ł.; validation, W.S. and M.S.-V.; formal analysis, W.S.; investigation, M.S.-V., P.Ł. and P.R.; resources, P.R. and K.M.; data curation, P.R., P.Ł. and K.M.; writing—original draft preparation, W.S. and M.S.-V.; writing—review and editing, W.S. and M.S.-V.; visualization, P.R. and K.M.; supervision, W.S.; project administration, W.S. All authors have read and agreed to the published version of the manuscript.

Funding: This report is part of the project entitled "Advanced methane drainage strategy employing underground directional drilling technology for major risk prevention and greenhouse gases emission mitigation" (acronym: DD-MET) that has received funding by the European Union's Research Programme of the Research Fund for Coal and Steel, Technical Group Coal 1 TGK1 research and innovation programme under grant agreement number 847338-DD-MET-RFCS-2018/RFCS-2018 and from the Polish Ministry of Science and Higher Education (contract no. 5038/FBWiS/2019/2).

Data Availability Statement: The original contributions presented in the study are included in the article; further inquiries can be directed to the corresponding author.

Acknowledgments: The authors thank the team from the Department of Geology and Geochemistry of Oil and Gas Institute—NRI, for conducting laboratory studies and analyses of the well logging geophysical data used in the construction of the static geological models.

Conflicts of Interest: The authors declare no conflicts of interest.

References

1. Swolkień, J.; Szlązak, N. The Impact of the Coexistence of Methane Hazard and Rock-Bursts on the Safety of Works in Underground Hard Coal Mines. *Energies* **2021**, *14*, 128. [CrossRef]
2. Krause, E. *Zagrożenie Metanowe W Kopalniach Węgla Kamiennego*; Wydawnictwo GIG: Katowice, Poland, 2019; pp. 162–163, ISBN 978-83-65503-16-9.
3. Przystolik, A.; Krause, E.; Borowski, M.; Jura, B.; Skiba, J. Distribution of the volume of methane intake by classical drainage and long directional boreholes in longwalls 1C and 2C in seam 501 in the Staszic-Wujek coal mine. In Proceedings of the GZN2023: XXX Jubileuszowa Międzynarodowa Konferencja Naukowo-Techniczna z Cyklu Górnicze Zagrożenia Naturalne, Wisła, Poland, 14–16 November 2023; Główny Instytut Górnictwa: Katowice, Poland, 2023; S.52.
4. IPCC Contribution of Working Groups I, II and III to the Fifth Assessment Report of the Intergovernmental Panel on Climate Change. In *Climate Change 2014: Synthesis Report*; IPCC: Geneva, Switzerland, 2014; p. 151.
5. Global Methane Tracker 2023, IEA, Paris. Available online: https://www.iea.org/reports/global-methane-tracker-2023 (accessed on 30 November 2024).
6. Nawrat, S.; Kuczera, Z.; Łuczak, R.; Życzkowski, P.; Napieraj, S.; Gatnar, K. *Utylizacja Metanu Z Pokładów Węgla W Polskich Kopalniach Podziemnych*; Wydawnictwa AGH: Kraków, Poland, 2009; ISBN 978-83-7464-213-2.
7. UNECE. *Best Practice Guidance for Effective Methane Drainage and Use in Coal Mines*, 2nd ed.; United Nations Publications ECE Energy Series: No. 47; United Nations, Economic Commission for Europe: Geneva, Switzerland, 2016; ISBN 978-92-1-117121-1. ISSN 1014-7225.
8. U.S. EPA, Coalbed Methane, Outreach Program. Application of Longhole Directional Drilling for Methane Drainage at the Amasra Hard Coal Mine: Amasra, Turkey. U.S. Environmental Protection Agency, February 2015. Available online: https://www.epa.gov/sites/default/files/2016-03/documents/turkey_pre-feas_amasra.pdf (accessed on 21 March 2025).
9. Szlązak, N.; Swolkień, J. Possibilities of Capturing Methane from Hard Coal Deposits Lying at Great Depths. *Energies* **2021**, *14*, 3542. [CrossRef]
10. Wang, F.; Ren, T.X.; Hungerford, F.; Tu, S.; Aziz, N. Advanced directional drilling technology for gas drainage and exploration in Australian coal mines. *Procedia Eng.* **2011**, *26*, 25–36.
11. Wang, F.; Ren, T.X.; Tu, S.; Hungerford, F.; Aziz, N. Implementation of underground longhole directional drilling technology for greenhouse gas mitigation in Chinese coal mines. *Int. J. Greenh. Gas Control.* **2012**, *11*, 290–303.
12. Black, D.J.; Aziz, N.J. Improving UIS Gas Drainage in Underground Coal Mines. In Proceedings of the Coal Operators' Conference, University of Wallongong, Wallongong, Australia, 14–15 February 2008; pp. 186–196.
13. Skiba, J.; Topór, T.; Cicha-Szot, R.; Jura, B. Application of long reach directional drilling for gas drainage of adjacent seams in coal mines with severe geological conditions. In Proceedings of the Methane in the Context of the Transition of the Coal Sector. XXX Szkoła Eksploatacji Podziemnej, Kraków, Poland, 27–28 September 2021.
14. Leśniak, G.; Brunner, D.J.; Topór, T.; Słota-Valim, M.; Cicha-Szot, R.; Jura, B.; Skiba, J.; Przystolik, A.; Lydall, B.; Plonka, G. Application of long-reach directional drilling boreholes for gas drainage of adjacent seams in coal mines with severe geological conditions. *Int. J. Coal Sci. Technol.* **2023**, *10*, 39. [CrossRef]
15. *Best Practice Guidance for Effective Methane Drainage and Use in Coal Mines*; ECE Energy Series, No. 31; United Nations, Economic Commission for Europe: Geneva, Switzerland, 2010; ISBN 13: 9789211170184.
16. *Pre-Feasibility Study for Methane Drainage and Utilisation at the TengHui Coal Mine Shanxi Province People's Republic of China*; U.S. Environmental Protection Agency: Washington, DC, USA, 2019; Publication No. 430R19005.
17. Xu, H.; Qin, Y.; Wu, F.; Zhang, F.; Liu, W.; Liu, J.; Guo, M. Numerical modelling of gas extraction from coal seam combined with a dual-porosity model: Finite difference solution and multi-factor analysis. *Fuel* **2022**, *313*, 122687. [CrossRef]
18. Taheri, A.; Sereshki, F.; Ardejani, F.D.; Mirzaghorbanali, A. Numerical modelling of gas flow in coal pores for methane drainage. *J. Sustain. Min.* **2016**, *15*, 95–99. [CrossRef]

19. Karacan, C.O.; Ruiz, F.A.; Cote, M.; Phipps, S. Coal mine methane: A review of capture and utilisation practices with benefits to mining safety and to greenhouse gas reduction. *Int. J. Coal Geol.* **2011**, *86*, 121–156. [CrossRef]
20. Haisheng, Q.; Kuijun, W. Mathematical model of gas drainage through bedding boreholes in a coal seam and numerical simulation. *Electron. J. Geotech. Eng.* **2014**, *19*, 3723–3731.
21. Liu, P.; Fan, J.; Jiang, D.; Li, J. Evaluation of underground coal gas drainage performance: Mine site measurements and parametric sensitivity analysis. *Process Saf. Environ. Prot.* **2021**, *148*, 711–723, ISSN 0957-5820. [CrossRef]
22. Jamalan, S.; Sereshki, F.; Ataei, M.; Najafi, M. Numerical modelling of coal gas drainage in a three-dimensional framework. *Arab. J. Geosci.* **2022**, *15*, 826. [CrossRef]
23. Janoszek, T. Numerical Simulation of Methane Distribution at the Longwall Working with Various Stages of Shearer Advance. *Arch. Min. Sci.* **2024**, *69*, 353–373. [CrossRef]
24. Oraee, K.; Goodarzi, A. Mathematical Modelling of Coal Seam Methane Drainage in Longwall Mining. In Proceedings of the 100 Years of Mining Research Symposium and SME Annual Meeting & Exhibit, Phoenix, AZ, USA, 28 February–3 March 2010; Society for Mining, Metallurgy, and Exploration; pp. 548–552.
25. Tang, M.; Zheng, P.; Zheng, C.; Hu, Z.; Wang, J.; Dai, G. Numerical Analysis and Prediction of Coal Mine Methane Drainage Based on Gas–Solid Coupling Model. *Teh. Vjesn.* **2019**, *26*, 752–761. [CrossRef]
26. Zhang, Y.; Wang, R.; Zhang, B.; Liu, B.; Zhang, Q.; Wang, E.; Shen, L. Numerical simulation of coalbed methane under the mining conditions based on dynamic virtual well and variable permeability field. *Energy Explor. Exploit.* **2022**, *41*, 170–186. [CrossRef]
27. Zhu, X.; Pengda, C.; Chun, F. A numerical method of hydraulic fracturing in coal seams for enhancing gas drainage. 11th Conference of Asian Rock Mechanics Society. *IOP Conf. Ser. Earth Environ. Sci.* **2021**, *861*, 032040. [CrossRef]
28. Li, X.; Zheng, J.; Liu, J. Numerical Simulation of Borehole Parameters for Coal Seam Gas Pre-Drainage. *Int. J. Heat Technol.* **2012**, *39*, 1328–1334. [CrossRef]
29. Li, X.; Guo, S.; Mu, B. Numerical simulation of gas drainage from goaf via high-level borehole in gently inclined seams. *AIP Adv.* **2025**, *15*, 035037. [CrossRef]
30. Bagherzadeh, A.; Najafi, M.; Mohammad, F.; Noroozi, M. A proper borehole pattern design for coal seam methane drainage in Tabas coal mine using Comsol Multiphysics. *J. Sustain. Min.* **2022**, *21*, 5. [CrossRef]
31. Blokhin, D.I.; Zakorshmenniy, I.M.; Kubrin, S.S.; Kobylkin, A.S.; Pozdeev, E.E.; Pushilin, A.N. Numerical research of effect of stress strain changes on stability of gas drainage wells in coal-rock mass. *MIAB. Min. Inf. Anal. Bull.* **2023**, *11*, 17–32. [CrossRef]
32. Karacan, C.O.; Diamond, W.P.; Schatzel, S.J. Numerical analysis of the influence of in-seam horizontal methane drainage boreholes on longwall face emission rates. *Int. J. Coal Geol.* **2007**, *72*, 15–32. [CrossRef]
33. Guo, H.; Todhunter, C.; Qu, Q.; Kerr, H.; Qin, J. An Innovative Drainage System for Coal Mine Methane Capture Optimisation and Abatement Maximisation. In Proceedings of the 2016 Coal Operators' Conference, Mining Engineering, University of Wollongong, Wollongong, Australia, 18–20 February 2016; Aziz, N., Kininmonth, B., Eds.; pp. 380–393. Available online: https://ro.uow.edu.au/coal/624 (accessed on 15 February 2025).
34. Liu, Q.; Zhou, H.; Cheng, Y.; Shu, L.; Ullah, B. An improved method for high-efficiency coal mine methane drainage: Theoretical analysis and field verification. *Energy Sci. Eng.* **2018**, *6*, 739–748. [CrossRef]
35. Zhang, L.; Qi, Q.; Deng, K.; Zuo, S.; Liu, Y. Optimum Layout of Multiple Tree-type Boreholes in Low-Permeability Coal Seams to Improve Methane Drainage Performance. *Front. Energy Res.* **2021**, *9*, 732827. [CrossRef]
36. Hu, X.; Yang, S.; Liu, W.V.; Cai, J.; Zhou, X. Numerical modelling and rescaled range analysis on spontaneous combustion under surface methane drainage in a Chinese coal mine. In Proceedings of the 3rd International Symposium on Mine Safety Science and Engineering, Montreal, QU, Canada, 13–19 August 2016. Paper No. 89.
37. Guo, Z.; Chen, Y.; Yao, S.; Zhang, Q.; Liu, Y.; Zeng, F. Feasibility Analysis and Optimal Design of Acidising of Coalbed Methane Wells. *ASME J. Energy Resour. Technol.* **2019**, *141*, 082906. [CrossRef]
38. Wei, P.; Huang, C.; Li, X.; Peng, S.; Lu, Y. Numerical simulation of boreholes for gas extraction and effective range of gas extraction in soft coal seams. *Energy Sci. Eng.* **2019**, *7*, 1632–1648. [CrossRef]
39. Hosseini, A.; Najafi, M.; Morshedy, A.H. A 3D Numerical Model for Improving Methane Drainage Efficiency in Underground Coal Mines Using the Cross-Measure Borehole Method. *Int. J. Energy Res.* **2024**, *2024*, 19. [CrossRef]
40. Liu, Y.-K.; Zhou, F.-B.; Liu, L.; Liu, C.; Hu, S.-Y. An experimental and numerical investigation on the deformation of overlying coal seams above double-seam extraction for controlling coal mine methane emissions. *Int. J. Coal Geol.* **2011**, *87*, 139–149. [CrossRef]
41. Ma, S.; Wang, Z.; Qi, L.; Wang, L.; Chen, H. Determination of Effective Drainage Radius of In-Seam Borehole Based on Gas Content Test and Numerical Simulation Calculation Method. *ACS Omega* **2024**, *9*, 36485–36496. [CrossRef]
42. Xu, W.; Zheng, X.; Liu, C.; Li, P.; Li, B.; Shayanowako, K.M.; Wang, J.; Guo, X.; Lai, G. Numerical Simulation Study of High-Pressure Air Injection to Promote Gas Drainage. *Sustainability* **2022**, *14*, 13699. [CrossRef]
43. Zhang, D.; Zhang, M.; Zhang, S.; Wang, Z.; Jin, Y.; Liu, R. Numerical Simulation of Effective Extraction Radius of Pre-Drainage Borehole Based on Coal Damage Model. *Sustainability* **2023**, *15*, 4446. [CrossRef]

44. Liu, X.; Yang, S. Three-dimensional numerical simulation of methane drainage by high-level drill holes in a lower protective coal seam with a "U" type face. *Int. J. Coal. Sci. Technol.* **2014**, *1*, 434–440. [CrossRef]
45. Szott, W.; Słota-Valim, M.; Gołąbek, A.; Sowiżdżał, K.; Łętkowski, P. Numerical studies of improved methane drainage technologies by stimulating coal seams in multi-seam mining layouts. *Int. J. Rock Mech. Min. Sci.* **2018**, *108*, 157–168. [CrossRef]
46. Szott, W.; Miłek, K.; Łętkowski, P.; Gołąbek, A. Study of the impact of exploitation structures on pressure distribution and adsorbed methane content in coal seams using dynamic flow models. A case study. *Min. Sci.* **2020**, *27*, 133–153. [CrossRef]
47. Jureczka, J.; Dopita, M.; Gałka, M.; Krieger, W.; Kwarciński, J.; Martinec, P. *Geological Atlas of Coal Deposits of the Polish and Czech Parts of the Upper Silesian Coal Basin*; Polish Geological Institute—National Research Institute: Warsaw, Poland, 2005. (In Polish)
48. Opluštil, S.; Laurin, J.; Jureczka, J.; Nadłonek, W.; Naglik, B.; Horák, J.; Kedzior, A.; Lojka, R.; Nádaskay, R. Depositional Setting and Sequence Stratigraphy of Upper Mississippian Coal-Bearing Paralic Cyclothems in Upper Silesian Foreland. *Int. J. Coal Geol.* **2024**, *287*, 104516. [CrossRef]
49. Kotarba, M.J.; Bilkiewicz, E.; Jurek, K.; Waliczek, M.; Więcław, D.; Zych, H. Thermogenic gases generated from coals and carbonaceous shales of the Upper Silesian and Lublin Coal basins: A hydrous pyrolysis approach. *Geol. Q.* **2021**, *65*, 26. [CrossRef]
50. Kotas, A. Ważniejsze cechy budowy geologicznej Górnośląskiego Zagłębia Węglowego na tle pozycji tektonicznej i budowy głębokiego podłoża utworów produktywnych. W: Problemy geodynamiki i tąpań. *Probl. Geodyn. I Tąpań* **1972**, *56*, 5–55.
51. Jureczka, J.; Kotas, A. Coal deposits–Upper Silesian Coal Basin. *Carbonif. Syst. Pol.* **1995**, *148*, 164–173.
52. Abdlmutalib, A.; Abdullatif, O.; Korvin, G.; Abdulraheem, A. The relationship between lithological and geomechanical properties of tight carbonate rocks from Upper Jubaila and Arab-D Member outcrop analogue. *Cent. Saudi Arabia. Arab. J Geosci.* **2015**, *8*, 11031–11048. [CrossRef]
53. DD-MET 1st Periodic Technical Report. Advanced Methane Drainage Strategy Employing Underground Directional Drilling Technology for Major Risk Prevention and Greenhouse Gases Emission Mitigation. Project Number: 847338, 1–115. Available online: https://www.uniovi.es/documents/39158/78881/DD-MET.pdf/d6999636-2446-4248-33da-f887db8b519f?t=1676391577278 (accessed on 15 February 2025).
54. Topór, T.; Grzegorz, L.; Cicha-Szot, R.; Słota-Valim, M.; Budak, P.; Plonka, G.; Surma, R. Application of Long-Rich Directional Drilling for Gas Drainage of Adjacent Seams in Coal Mines with Severe Geological Conditions. In Proceedings of the 38th Annual Virtual International Pittsburgh Coal Conference, Virtual, 20–23 September 2021.
55. Herwanger, J.; Koutsabeloulis, N. *Seismic Geomechanics: How to Build and Calibrate Geomechanical Models Using 3D and 4D Seismic Data*; European Association of Geoscientists & Engineers: Houten, The Netherlands, 2011; ISBN 9789462820005, ISBN 9789073834101. [CrossRef]
56. Chang, C.; Zoback, M.; Khaksar, A. Empirical relations between rock strength and physical properties in sedimentary rocks. *J. Pet. Sci. Eng.* **2016**, *51*, 223–237. [CrossRef]
57. Słota-Valim, M. Static and dynamic elastic properties, the cause of the difference and conversion methods–case study. *Nafta-Gaz* **2015**, *71*, 816–826. [CrossRef]
58. Xu, H.; Zhou, W.; Xie, R.; Lina, D.; Xiao, C.; Shan, Y.; Zhang, H. Characterisation of Rock Mechanical Properties Using Lab Tests and Numerical Interpretation Model of Well Logs. *Math. Probl. Eng.* **2016**, *2016*, 13. [CrossRef]
59. Hoek, E. Rock Mechanics–An Introduction for the Practical Engineer Parts I.; II and III. *Min. Mag.* **1966**, *144*, 236–243.
60. Zhu, Y.; Chen, L.; Zhang, H.; Zhou, Z.; Chen, S. Physical and Mechanical Characteristics of Soft Rock Tunnel and the Effect of Excavation on Supporting Structure. *Appl. Sci.* **2019**, *9*, 1517. [CrossRef]
61. Jacobsen, L.; Rosenfeld, M.A.; Ferm, J.C. A comparison of sieve and thin section techniques for size analysis. *J. Geol.* **1943**, *61*, 114–132.
62. GasDrain Development of Improved Methane Drainage Technologies by Stimulating Coal Seams for Major Risks Prevention and Increased Coal Output. RFCS Project RFCR-CT-2014-00004. Available online: https://research-and-innovation.ec.europa.eu/document/download/2a4fb946-0b0e-4412-aae9-5e2429e050e3_en?filename=summaries-rfcs_en_2003-2014.pdf (accessed on 1 December 2024).
63. Małkowski, P.; Niedbalski, Z.; Majcherczy, T. Konwergencja wyrobisk chodnikowych na podstawie wyników obliczeń numerycznych i ich weryfikacja pomiarami in situ. *Górnictwo I Geoinżynieria* **2008**, *32*, 199–215.
64. Rodríguez Ortiz, J.M.; Serra Gesta, J.; Oteo Mazo, C. *Curso Aplicado de Cimentaciones*, 6th ed.; Colegio Oficial de Arquitectos de Madrid: Madrid, Spain, 1995.
65. Godula, T. Własności mechaniczne skał karbońskich w południowo zachodniej części górnośląskiego zagłębia węglowego. *Przegląd Geol.* **1984**, *32*, 445–448.
66. Zoback, M.D. *Reservoir Geomechanics*; Cambridge University Press: Cambridge, UK, 2010; ISBN 978-0-521-77069-9.
67. Belyadi, H.; Fathi, E.; Belyadi, F. Chapter Thirteen: Rock Mechanical Properties and In Situ Stresses. In *Hydraulic Fracturing in Unconventional Reservoirs*; Gulf Professional Publishing: Oxford, UK, 2017; pp. 207–224, ISBN 9780128498712.
68. Terzaghi, K. *Theoretical Soil Mechanics*; John Wiley & Sons, Inc.: New York, NY, USA, 1943.

69. Różkowski, A.; Chmura, A.; Gajowiec, B.; Wagner, J. Określenie warunków hydrogeologicznych głębokich poziomów karbonu Górnośląskiego Zagłębia Węglowego. In *Opracowanie Warunków Geologicznogórniczych Górotworu Karbonu Górnośląskiego Zagłębia Węglowego*; Archiwum PIG Odział Górnośląski: Sosnowiec, Poland, 1990; CPBR nr 1.8: Budowa Geologiczna Polski i Poszukiwania Złoże Surowców Mineralnych.
70. Tingay, M.; Reinecker, J.; Birgit, M. Borehole Breakout and Drilling-Induced Fracture Analysis from Image Logs; World Stress Map Project–Guidelines: Image Logs. Helmholtz Vent. Potsdam, GFZ. 2008, pp. 1–18. Available online: https://www.researchgate.net/profile/Mark-Tingay/publication/242115376_Borehole_breakout_and_drilling-induced_fracture_analysis_from_image_logs/links/00b4953a17ea376cfc000000/Borehole-breakout-and-drilling-induced-fracture-analysis-from-image-logs.pdf (accessed on 20 March 2025).
71. Dubiński, J.; Stec, K.; Bukowska, M. Geomechanical and tectonophysical conditions of mining-induced seismicity in the Upper Silesian Coal Basin in Poland: A case study. *Arch. Min. Sci.* **2019**, *64*, 163–180. [CrossRef]
72. Jarosiński, M. Recent tectonic stress regime in Poland based on analyses of hydraulic fracturing of borehole walls. *Przegląd Geol.* **2006**, *53*, 863–872.
73. Zuberek, W.; Teper, L.; Idziak, A.; Sagan, G. Seismicity and tectonics in the Upper Silesian Coal Basin, Poland. *Papers Pol. Geol. Inst.* **1997**, *157*, 199–207.
74. Settari, A.; Mourits, F.M. A Coupled Reservoir and Geomechanical Simulation System. *SPE J.* **1998**, *3*, 219–226.
75. Settari, A.; Walters, D.A. Advances in Coupled Geomechanical and Reservoir Modelling with Applications to Reservoir Compaction. *Spe J.* **2001**, *6*, 334–342.
76. Rutqvist, J.; Wu, Y.-S.; Tsang, C.-F.; Bodvarsson, G. A modelling approach for analysis of coupled multiphase fluid flow, heat transfer, and deformation in fractured porous rock. *Int. J. Rock Mech. Min. Sci.* **2002**, *39*, 429–442. [CrossRef]
77. Vidal-Gilbert, S.; Nauroy, J.-F.; Bross, E. 3D geomechanical modelling for CO_2 geologic storage in the Dogger carbonates of the Paris Basin. *Int. J. Greenh. Gas Control.* **2009**, *3*, 288–299. [CrossRef]
78. Longuemare, P.; Mainguy, M.; Lemonnier, P.; Onaisi, A.; Gérard, C.; Koutsabeloulis, N. Geomechanics in Reservoir Simulation: Overview of Coupling Methods and Field Case Study. *Oil Gas Sci. Technol.* **2002**, *57*, 471–483.
79. Settari, A.; Mourits, F.M. Coupling of Geomechanics and Reservoir Simulation Models. In *Computer Methods and Advances in Geomechanics*; Siriwardane, H.J., Zaman, M.M., Eds.; CRC Press: Morgantown, WV, USA, 1994; Volume 3, pp. 2151–2158.
80. Tsang, C.-F. Linking Thermal, Hydrological, and Mechanical Processes in Fractured Rocks. *Annu. Rev. Earth Planet. Sci.* **1999**, *27*, 359–384.
81. Minkoff, S.E.; Stone, C.M.; Arguello, J.G.; Bryant, S.; Eaton, J.; Peszynska, M.; Wheeler, M.F. Staggered in Time Coupling of Reservoir Flow Simulation and Geomechanical Deformation: Step 1-One-Way Coupling. In Proceedings of the SPE Reservoir Simulation Symposium. Society of Petroleum Engineers, Houston, TX, USA, 14–17 February 1999; p. 2.
82. Minkoff, S.E.; Stone, C.M.; Bryant, S.; Peszynska, M.; Wheeler, M.F. Coupled fluid flow and geomechanical deformation modelling. *J. Pet. Sci. Eng.* **2003**, *38*, 37–56.
83. Szott, W.; Ruciński, P.; Słota-Valim, M.; Sowiżdżał, K. Investigation of the Impact of Natural Fracture Geomechanics on the Efficiency of Oil Production and CO_2 Injection from/to a Petroleum Structure: A Case Study. *Energies* **2023**, *16*, 4219. [CrossRef]
84. Shi, J.Q.; Durucan, S. Drawdown induced changes in the permeability of coalbeds: A new interpretation of the reservoir response to primary recovery. *Transp. Porous Media* **2004**, *56*, 1–16.
85. Yao, X.; Cheng, G.; Shi, B. Analysis on gas extraction borehole instability and control method of pore-forming in deep surrounding rock with weak structure. *J. China Coal Soc.* **2010**, *35*, 2073–2081.
86. Karatela, E.; Taheri, A. Three-dimensional hydro-mechanical model of borehole in fractured rock mass using discrete element method. *J. Nat. Gas Sci. Eng.* **2018**, *53*, 263–275.
87. Wang, Z.; Yang, X.; Wang, G.; Gong, H. Study on Instability Characteristics of the Directional Borehole on the Coal-Seam Roof: A Case Study of the Tingnan Coal Mine. *Processes* **2023**, *11*, 1675. [CrossRef]
88. Zhang, X.-B.; Shen, S.-S.; Feng, X.-J.; Ming, Y.; Liu, J.-J. Influence of Deformation and Instability of Borehole on Gas Extraction in Deep Mining Soft Coal Seam. *Hindawi Adv. Civ. Eng.* **2020**, *2021*, 11. [CrossRef]
89. DD-MET. RFCS Research Project. Available online: https://dd-met.inig.pl/docs/DD-MET_BEST_PRACTICE_GUIDELINES_1.pdf (accessed on 30 April 2025).
90. Fidalgo-Valverde, G.; Menéndez-Díaz, A.; Krzemień, A.; Riesgo-Fernández, P.; Marques Sierra, A. Environmental risk assessment in coal mining with methane degassing. *MATEC Web Conf.* **2024**, *389*, 00039. [CrossRef]

energies

MDPI

Article

Development and Characterization of High-Strength Coalbed Fracturing Proppant Based on Activated Carbon Skeleton

Kai Wang [1,2,*], Chenye Guo [2], Qisen Gong [3], Gen Li [2], Xiaoyue Zhuo [3], Peng Zhuo [3] and Chaoxian Chen [3]

[1] College of Safety Science and Engineering, Xi'an University of Science and Technology, Xi'an 710048, China
[2] Southwest Branch of China National Coal Group Corp, Chongqing 400025, China; guocheny@chinacoal.com (C.G.); ligen@chinacoal.com (G.L.)
[3] Chongqing Energy Investment Group Technology Co., Ltd., Chongqing 401420, China; gong_qisen@126.com (Q.G.); 13996289571@163.com (X.Z.); 18512390497@163.com (P.Z.); 13883227679@163.com (C.C.)
* Correspondence: kiz1024@outlook.com

Abstract: To address the challenges of low permeability, high gas adsorption, and a fragile structure in coalbed methane reservoirs, this study developed a high-strength composite proppant with an activated carbon skeleton via nitric acid pretreatment, silica–alumina sol coating, and calcination. Orthogonal experiments optimized the preparation conditions: 30–40 mesh activated carbon, Si/Al molar ratio of 4:1, calcination at 650 °C for 2 h. The resulting proppant exhibited an excellent performance: a single-particle compressive strength of 55.5 N, porosity of 33.2%, crushing rate of only 2.3% under 50 MPa closure pressure, and permeability 48.5% higher than quartz sand. In simulated acidic coalbed environments (pH 3–5), its acid corrosion rate was <2.8%, and it enhanced methane desorption by 16.2% compared to pure coal. Additionally, the proppant showed a superior transport performance in fracturing fluids, with better distribution uniformity in fractures than ceramsite, and its hydrophobic surface (contact angle 115.32°) improved fracturing fluid flowback efficiency. This proppant integrates high strength, good conductivity, gas desorption promotion, and corrosion resistance, offering a novel material solution for efficient coalbed methane extraction.

Keywords: activated carbon skeleton proppant; coalbed methane fracturing; coalbed adaptability; silica–alumina coating

1. Introduction

In the context of the global energy transition toward low-carbon sources, coalbed methane has emerged as a pivotal clean and efficient unconventional natural gas resource, playing a critical role in ensuring energy security and mitigating greenhouse gas emissions [1–3]. China, endowed with abundant coalbed methane reserves, boasts recoverable reserves reaching 36.8 trillion cubic meters, ranking among the world's top reserves [4]. However, the commercial exploitation of coalbed methane faces formidable challenges due to the unique geological characteristics of most reservoirs [5,6]. These reservoirs are typically characterized by ultra-low permeability (<0.1 mD), attributed to the dense and poorly connected pore–fracture networks in coal matrices. A high gas adsorption capacity stems from the extensive microporous structure of coal, which tightly binds methane molecules [7,8]. Fragile coal rock is prone to collapse under the disturbance of fracturing operations [6,9]. These factors collectively result in low single-well production and hinder the large-scale development of coalbed methane resources [10,11].

Fracturing technology has long been recognized as the cornerstone for unlocking the potential of low-permeability coalbed methane reservoirs [12,13]. By injecting high-pressure fracturing fluids, artificial fractures are created in the coalbed, and proppants are introduced to prop open these fractures, maintaining their conductivity for sustained gas flow [14]. As such, proppants are indispensable components that directly determine the success of fracturing operations and the long-term productivity of coalbed methane wells. Traditional proppants, however, fail to meet the complex demands of coalbed environments. Quartz sand, despite its low cost and widespread use, exhibits insufficient compressive strength (typically <30 MPa), leaving it susceptible to crushing under the high-geostress conditions prevalent in deep coalbeds (often exceeding 30 MPa) [15]. This crushing leads to the blockage of fractures and a drastic decline in conductivity. Ceramsite, another commonly used proppant, offers higher strength but suffers from high density (>3.0 g/cm^3), causing rapid sedimentation in fracturing fluids [16,17]. This uneven distribution within fractures results in inadequate propping of the fracture tips and excessive accumulation at the fracture mouths, compromising the overall fracturing effect. Traditional proppants such as quartz sand and ceramsite, in their unmodified state, lack active interactions with coal matrices to effectively promote gas desorption. While some studies have explored surface modification of these proppants to partially enhance gas desorption, their intrinsic inert surface properties still limit such promotion compared to carbon-based materials with intrinsic affinity for coal matrices [18–20].

Activated carbon, a porous carbon material, has garnered significant attention as a potential base material for advanced proppants due to its distinctive properties. Its large specific surface area (usually >1000 m^2/g) and well-developed pore structure provide ample space for gas diffusion and adsorption [3,9], while its lightweight nature (1.3–1.5 g/cm^3) helps slow sedimentation rates and improve suspendability in fracturing fluids [21]. Moreover, the abundant surface active groups (such as hydroxyl and carboxyl groups) on activated carbon enable chemical modification, facilitating the formation of stable composite structures with other materials [22,23]. These characteristics make activated carbon an ideal skeleton for developing proppants that can simultaneously address mechanical strength, conductivity, and gas desorption challenges [24,25].

While quartz sand and ceramsite suffer from limitations like low porosity and water retention, previous attempts to apply carbon-based materials as proppants have been reported. Early studies exploited their high porosity for gas diffusion, but their practical application was hindered by critical drawbacks: insufficient mechanical strength (easily crushed under reservoir pressure), poor wettability control (prone to water locking), and weak compatibility with fracturing fluids [26,27]. These unresolved issues left a gap in developing robust carbon-based proppants suitable for coalbed methane reservoirs—motivating our work to address these limitations through silica–alumina coating modification.

Building on these advantages, this study proposes a novel approach to fabricate high-strength composite proppants by integrating activated carbon with silica–alumina oxides via the sol–gel method. The preparation process involves nitric acid pretreatment to clean the activated carbon surface, remove impurities, and enhance its surface reactivity, coating with silica–alumina sol to form a uniform and dense protective layer, and calcination to strengthen the bonding between the activated carbon skeleton and the coating. By systematically investigating the effects of key parameters such as activated carbon particle size, silica–alumina molar ratio, calcination temperature, and holding time on the proppant's performance [9,24,28], this research aims to optimize the preparation conditions and achieve a proppant with superior mechanical properties, high conductivity, and excellent coalbed adaptability. The comprehensive evaluation of the proppant's microstructure, mechanical strength, corrosion resistance, and gas desorption promotion ability will provide valuable insights into its application potential, ultimately contributing to the technical advancement of efficient coalbed methane extraction [3,29–33].

2. Experimental Design

2.1. Experimental Materials

Activated carbon particles (20–40 mesh, with a specific surface area of 1030 m^2/g and a density of 1.34 g/cm^3, purity 95%) were procured from Aladdin. Nitric acid (30%), sodium silicate, and sodium aluminate (all of analytical grade) were purchased from Sinopharm. Deionized water was prepared in-house. The main raw materials (activated carbon, sodium silicate, sodium aluminate, etc.) used in this study have a laboratory-scale unit cost of approximately USD 2.5–3.0 per kg, slightly higher than quartz sand but comparable to ceramsite, with the potential for a 30–40% cost reduction in large-scale production.

2.2. Experimental Instruments

A Fourier transform infrared spectrometer (FTIR, Nicolet iS50, Thermo Fisher Scientific, Madison, WI, USA) was used to analyze surface functional groups. An automatic specific surface area and porosity analyzer (Autosorb-IQ, Quantachrome, Tallahassee, FL, USA) was utilized to measure the specific surface area and pore size distribution. A universal testing machine (WDW-100, Jinan, China) was used to test the single-particle compressive strength. Scanning electron microscopy (SEM, SU8020, Hitachi, Tokyo, Japan) was employed to observe the surface morphology and pore structure of the proppant. A thermogravimeter (TGA Q500, TA Instruments, New Castle, NSW, USA) was used to assess the thermal stability of the sample under N_2 atmosphere. A high-temperature and high-pressure conductivity meter (FCES-100, Haian Petroleum Scientific Research Instruments, Haian, China) was employed for measuring the conductivity of the proppant packing under simulated coalbed conditions. A gas adsorption and desorption analyzer (WY-98A, Chongqing, China) was used to test the methane desorption rate. Volumetric adsorption apparatus (Micromeritics ASAP 2020, Norcross, GA, USA) and a contact angle goniometer (Model: DSA100, Krüss, Hamburg, Germany) were utilized.

2.3. Proppant Preparation Process

The pretreatment of activated carbon involved immersing the particles in 30% nitric acid solution, then subjecting them to a 60 min treatment in a 60 °C water bath, followed by thorough washing with distilled water until the pH reached neutrality. Finally, they were dried at 110 °C for 6 h to remove surface impurities and activate the pores (Figure 1).

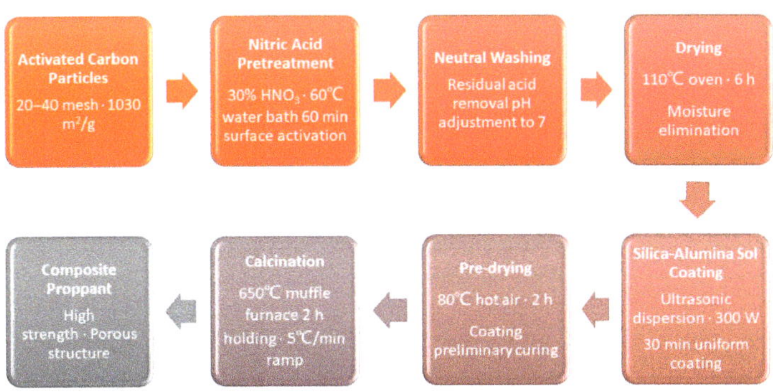

Figure 1. Flow chart of proppant preparation process.

Silica–alumina sol was selected as the coating material for the subsequent process due to its intrinsic properties, which synergistically enhance proppant performance from multiple dimensions: the silica–alumina network it forms (composed of Si-O-Si and Al-O-Si bonds) exhibits excellent chemical stability in the typical acidic coalbed environment (pH 3–5), effectively isolating the activated carbon skeleton from corrosive media and ensuring long-term structural integrity. Meanwhile, it possesses outstanding thermal resistance, capable of withstanding calcination temperatures up to 650 °C in this study without decomposition, forming a dense and firmly bonded coating with the activated carbon skeleton during calcination and preventing cracking or detachment caused by thermal stress. Furthermore, the reactive silanol groups (Si-OH) and aluminate groups in the sol can chemically react with oxygen-containing functional groups (e.g., -OH, -COOH) on the surface of nitric acid-pretreated activated carbon, significantly enhancing coating adhesion and preventing delamination under high closure pressure. Moreover, its colloidal particles are small and uniformly dispersed, resulting in a thin and permeable coating that covers the activated carbon surface without excessively blocking its intrinsic pores. This maintains gas diffusion channels while enhancing mechanical properties, ensuring the proppant retains a high porosity of 33.2% and good permeability.

For the preparation of the composite slurry, sodium silicate and sodium aluminate were dissolved in deionized water according to different molar ratios of silicon to aluminum (3:1, 4:1, 5:1). The solutions were diluted to a concentration of 10% and magnetically stirred for 30 min to form a stable sol (Figure 2).

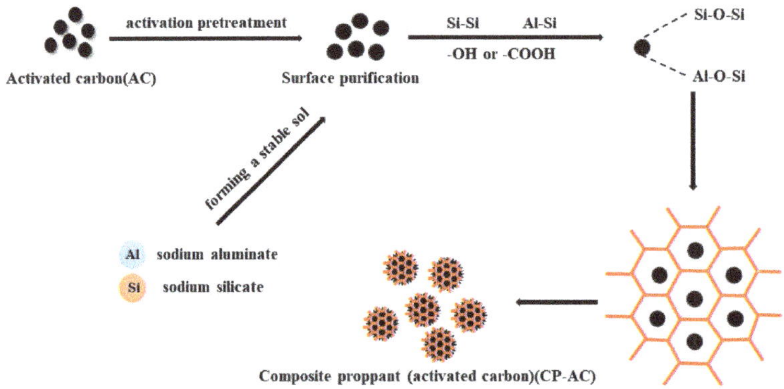

Figure 2. Synthesis mechanism of activated carbon-based composite proppant.

In the coating and forming stage, the pretreated activated carbon was submerged in the prepared sol and ultrasonicated for 30 min (at a power of 300 W) to ensure a uniform coating. Excess slurry was removed by vacuum filtration, and the samples were pre-dried at 80 °C for 2 h.

The calcination treatment entailed placing the pre-dried samples in a muffle furnace, heating them to temperatures of 600 °C, 650 °C, and 700 °C at a rate of 5 °C/min, holding them at these temperatures for 2 h, and then allowing them to cool naturally to room temperature.

2.4. Optimization of Preparation Conditions

Based on the sol–gel reaction mechanism, an L9(3^4) orthogonal experiment was designed, considering factors such as activated carbon particle size (20–30 mesh, 30–40 mesh), silicon–aluminum molar ratio (3:1, 4:1, 5:1), temperature for calcination (600 °C, 650 °C, 700 °C), and holding time (1 h, 2 h, 3 h). The single-particle compressive strength (target > 50 N) and permeability improvement rate (target > 40%) were selected as evaluation indicators to optimize the preparation conditions [34–43].

The permeability improvement rate (η) is calculated as follows:

$$\eta = (K_{CP\text{-}AC} - K_{Quartz\ Sand})/K_{Quartz\ Sand} \times 100\%$$

where

- $K_{CP\text{-}AC}$: Conductivity of CP-AC proppant pack (mD·cm), tested by FCES-100 conductivity meter (Section 2.2) under 30 °C and 50 MPa;
- $K_{Quartz\ Sand}$: Conductivity of quartz sand proppant pack (10.44 mD·cm) under the same testing.

2.5. Characterization Methods

For microstructural analysis, FTIR analysis was carried out with a scanning range of 4000–500 cm^{-1}, and 32 scans were performed to identify surface functional groups and chemical bonding. The Brunauer–Emmett–Teller (BET) method was utilized to quantify the specific surface area and pore size distribution via nitrogen adsorption–desorption isotherms.

Mechanical properties were evaluated using a universal testing machine with a loading rate of 0.5 mm/min to measure the single-particle compressive strength, and the crushing rate was determined at a closure pressure of 50 MPa.

Conductivity was measured using a high-temperature and high-pressure conductivity meter to assess the conductivity of the proppant packing under simulated coalbed conditions (temperature 30 °C, pressure 20–60 MPa).

The gas desorption promotion ability was evaluated using a gas adsorption and desorption analyzer to test the methane desorption rate of a mixed system consisting of a coal sample (200 g) and the proppant (50 g), with the results compared to those of a pure coal sample. Gas adsorption/desorption tests were performed under conditions simulating in situ CBM reservoirs: temperature was controlled at 32 ± 2 °C (matching the average formation temperature of the study coal seam) and pressure was regulated from 0.1 to 6 MPa (covering the typical pressure range of the target coalbed methane reservoir), using 99.99% pure methane as the test gas.

Corrosion resistance was determined by immersing the proppant in a hydrochloric acid solution with a pH of 3, soaking it at 60 °C for 72 h, and then calculating the acid corrosion rate. Acid resistance was evaluated using HCl (pH = 3) because the formation water in the target coal seam is dominated by chloride-type brine (Cl^- concentration ~1200 mg/L, as measured in field samples), making HCl more representative of the in situ acidic environment than H_2SO_4.

To investigate the influence of the flow rate on transport efficiency, comparative experiments were conducted using the composite proppant (CP-AC) and conventional ceramsite under two flow rates (5 L/h and 10 L/h) in a simulated fracture model (length 140 cm, width 2 cm, height 20 cm). The vertical accumulation height of proppants at distances of 20 cm, 40 cm, 60 cm, 80 cm, 100 cm, 120 cm, and 140 cm from the wellbore was measured to evaluate their transport stability. The 140 cm fracture model is a simplified laboratory setup designed to assess proppant transport under controlled conditions (e.g., constant flow velocity, uniform fracture aperture). While it does not fully replicate the complexity of in situ fractures (such as irregular morphology, heterogeneous coal matrix, and dynamic in situ stress), it effectively characterizes key transport behaviors (e.g., proppant migration efficiency and distribution uniformity) that are critical for guiding field applications.

3. Results and Discussion

3.1. Structural Characterization

3.1.1. FTIR Analysis

Figure 3 depicts the FTIR spectrum of the proppant (Figure 3). The broad absorption band spanning from approximately 3400 to 3600 cm^{-1} can be ascribed to the stretching vibration of -OH groups, which are present on both the activated carbon surface and the silica–alumina coating. A prominent peak emerges at around 1000–1050 cm^{-1}, corresponding to the asymmetric stretching vibration of Si-O-Si bonds. The intensity of this peak varies with the silicon–aluminum molar ratio (ranging from 3:1 to 5:1), indicating different contents of silicate in the coating. Moreover, a relatively weak peak is observed at approximately 700–750 cm^{-1}, which is assigned to Al-O-Si bonds, verifying the formation of a cross-linked silica–alumina network within the coating. At a calcination temperature of 650 °C, the peak intensity of Si-O-Si bonds reaches its maximum, implying an optimal condensation of the silica–alumina sol and an enhanced bonding between the coating and the activated carbon skeleton. After nitric acid pretreatment, a distinct small peak emerges in the range of 1650–1700 cm^{-1}, corresponding to the stretching vibration of C=O bonds in carboxyl groups. This observation confirms an increase in surface oxygen-containing functional groups on the activated carbon, which in turn enhances its ability to form chemical bonds with the silica–alumina coating (Table 1).

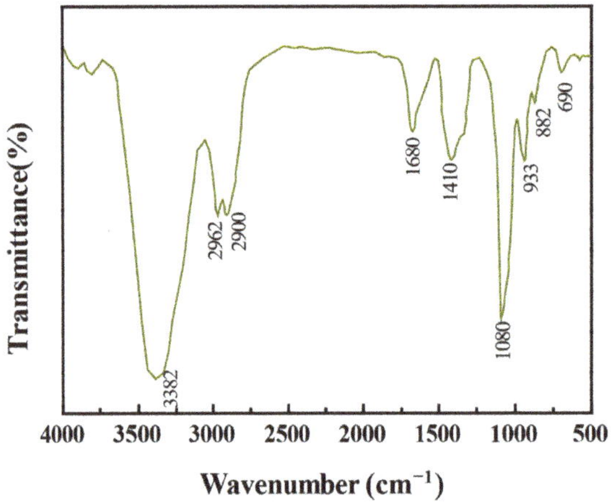

Figure 3. FTIR spectrum of the proppant.

Table 1. Summary of functional groups in the FTIR spectra of the proppant.

Functional Group	Characteristic Wavenumber Range (cm^{-1})	Change in Spectral Intensity
(-OH)	3400–3600	Broad absorption band
(Si-O-Si)	1000–1050	Varies with silicon–aluminum molar ratio (3:1 to 5:1); reaches maximum at 650 °C calcination
(Al-O-Si)	700–750	Relatively weak
(C=O)	1650–1700	Distinct small peak emerges after nitric acid pretreatment (intensity increases)

3.1.2. BET Analysis

The BET results indicated that the specific surface area of the original activated carbon (AC) was 1030 m^2/g. After the pretreatment and coating process (CP-AC), the specific surface area decreased to 590 m^2/g. This was due to the filling of some pores by the coating material. However, the pore volume increased slightly, and the average pore size shifted towards a larger value, which was beneficial for gas flow and proppant–coal interaction.

To better investigate the changes in the internal pore structure of activated carbon and the type of adsorption toward polymers, we employed the N_2 adsorption–desorption method to measure the adsorption isotherms of the two activated carbon samples. As shown in Figure 4, according to the IUPAC classification [30], the adsorption isotherm of the original activated carbon (AC) presents a typical Type I isotherm, with an adsorption hysteresis loop appearing at relative pressures of $0.4 < P/P_0 < 0.8$. This confirms that AC has a rich pore structure and is a microporous adsorbent with a uniform pore size distribution.

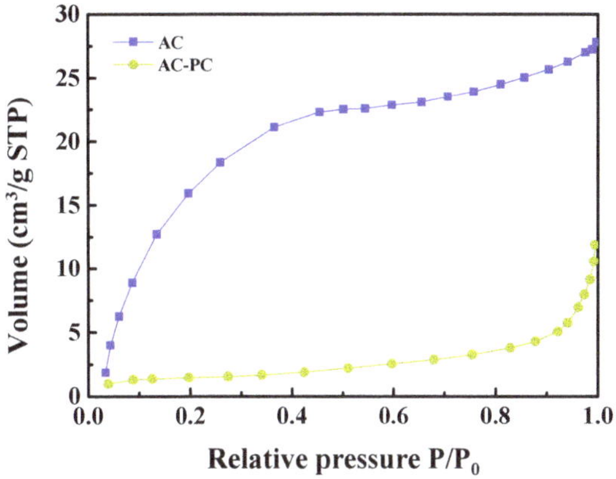

Figure 4. N_2 adsorption–desorption isotherms and schematic diagrams of activated carbon (AC) before and after pretreatment and coating (CP-AC).

The activated carbon after pretreatment and coating (CP-AC) exhibits an adsorption isotherm that differs from AC. Its isotherm shows an enlarged hysteresis loop compared to AC, spanning a wider range of relative pressures ($0.5 < P/P_0 < 0.9$), indicating a mixed adsorption behavior involving both micropores and mesopores. Additionally, the isotherm of CP-AC tends to rise more sharply at higher relative pressures ($P/P_0 > 0.8$), which may be attributed to the increase in average pore size. This phenomenon suggests the presence of more mesopores and larger pores in CP-AC, leading to enhanced capillary condensation and multi-molecular layer adsorption as the relative pressure increases [31], which aligns with the earlier BET results of an increased pore volume and larger average pore size.

3.1.3. SEM Analysis and Particle Size Distribution

Scanning electron microscopy was used to observe the surface morphology and microstructure of the composite proppant (CP-AC). A single SEM image of CP-AC revealed that after nitric acid pretreatment, silica–alumina sol coating, and calcination at 650 °C, the composite proppant exhibited a relatively regular particle morphology, with a uniform and dense silica–alumina coating covering the surface. The coating was tightly bonded to the activated carbon skeleton without obvious cracks or detachment, further confirming the formation of stable chemical bonds between the silica–alumina network and the activated carbon surface, as indicated by FTIR analysis. Under a high magnification, it was observed that the coating retained some micron-scale pores, which were interconnected with the pore structure of the activated carbon itself. These pores contributed significantly to the proppant's porosity of 33.2% and provided channels for gas flow.

Particle size statistical analyses were performed on 100 randomly selected CP-AC particles in the SEM image using image analysis software (Image-Pro Plus 6.0). The results showed that the particle size distribution range of CP-AC was 340–590 μm, with an average particle size of 470 ± 30 μm (Figure 5). This narrow particle size distribution benefited from the selection of 30–40-mesh activated carbon raw materials and the uniform coating process, which helped the proppant form a stable packing structure in fractures and reduced fluctuations in conductivity caused by uneven particle gradation under high pressure. Size distribution analysis was performed on 100 particles randomly selected from three independent SEM images (captured in different regions of the sample) to ensure representativeness. Consistent size distribution patterns (peak at 2–5 μm) were observed across all images, indicating the uniformity of the coating morphology. This multi-region sampling strategy minimizes bias from local inhomogeneity, supporting the reliability of the observed size characteristics.

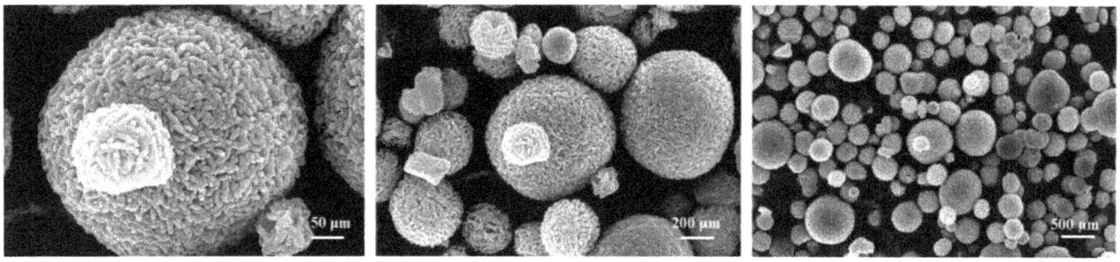

Figure 5. SEM morphology and particle size distribution of composite proppant.

3.1.4. Thermogravimetric Analysis (TGA)

The thermal stability of the composite proppant (CP-AC) and the original activated carbon (AC) was evaluated using a thermogravimeter (TGA Q500, TA Instruments, New Castle, NSW, USA) under an N_2 atmosphere. The samples were heated from room temperature to 800 °C at a heating rate of 10 °C/min, with a gas flow rate of 50 mL/min to ensure complete oxidation.

The TGA curve of AC (Figure 6a) showed a two-stage weight loss process. The first stage (25–180 °C) exhibited a weight loss of approximately 4.95%, which was attributed to the evaporation of adsorbed water and volatile impurities. The second stage (180–680 °C) involved a significant weight loss of 74.59%, corresponding to the oxidation and combustion of the carbon skeleton. In contrast, the TGA curve of CP-AC (Figure 6b) demonstrated enhanced thermal stability. A slight weight loss of 3.85% occurred in the low-temperature range (25–200 °C), indicating reduced moisture adsorption due to the silica–alumina coating. The main weight loss stage shifted to 200–700 °C, with a total weight loss of 45.65%, which was significantly lower than that of AC. This delay in oxidation temperature and reduction in weight loss confirmed that the silica–alumina coating effectively protected the activated carbon skeleton from early oxidation, thereby improving the overall thermal stability of the proppant.

3.1.5. Wettability Analysis

Wettability, a critical parameter reflecting the interaction between proppants and coalbed fluids, directly affects fracturing fluid flowback efficiency and gas migration behavior. In this study, the wettability of activated carbon (AC) and the composite proppant (CP-AC) was evaluated using a contact angle goniometer (Model: DSA100, Krüss, Hamburg, Germany). Static contact angles of deionized water droplets on the sample surfaces were measured at room temperature (25 °C) via the sessile drop method, with five replicate measurements for each sample to ensure accuracy (Figure 7).

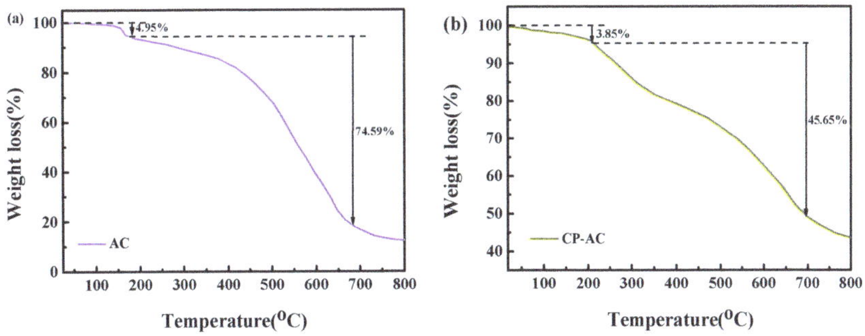

Figure 6. Thermogravimetric analysis curves of activated carbon (**a**) and composite proppant (**b**).

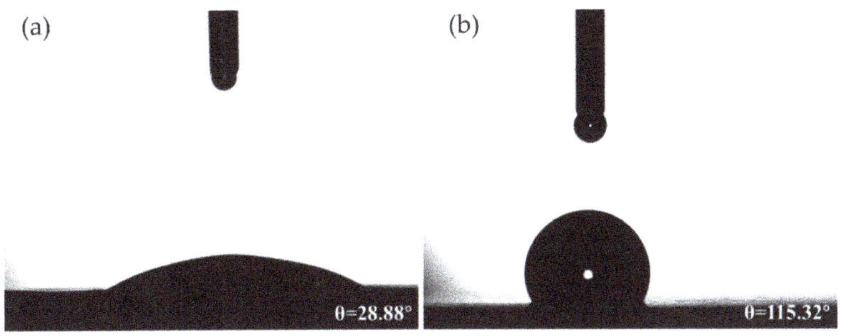

Figure 7. (**a**) Contact angle of activated carbon. (**b**) Contact angle of composite proppant.

The results indicated a significant shift in wettability after the composite modification. The original activated carbon exhibited a contact angle of 28.88°, demonstrating strong hydrophilicity. This characteristic is attributed to the abundant oxygen-containing functional groups (e.g., -OH and -COOH) on the AC surface, which form hydrogen bonds with water molecules and enhance surface affinity for water. In contrast, the composite proppant (CP-AC) showed a substantially increased contact angle of 115.32°, indicating distinct hydrophobicity. This transformation is primarily due to the formation of a nonpolar silica–alumina network (composed of Si-O-Si and Al-O-Si bonds) in the coating, which covers part of the polar functional groups on the activated carbon surface and reduces the interaction energy with water molecules. Additionally, the micron-scale pores retained in the coating induce a "Cassie–Baxter" effect, further reinforcing the hydrophobicity.

The transition in proppant wettability holds practical significance for coalbed methane extraction. The hydrophobic surface of CP-AC can reduce the retention of fracturing fluids on the proppant surface, accelerating fluid flowback and mitigating permeability damage caused by water blocking. Meanwhile, the contact angle of 115.32° enhances the affinity between the proppant and nonpolar methane gas, facilitating gas diffusion and desorption in pore channels. This hydrophobic property enables the composite proppant to adapt better to high-water-content or water-blocking-prone coalbed environments. Notably, contact angle testing serves as a quantitative basis for evaluating such wettability: the measured 115.32° far exceeds the 90° threshold for hydrophobicity, providing objective evidence for the strong water-repellent capability. This quantifiable characteristic, compared to hydrophilic quartz sand (contact angle typically <60°), directly links laboratory measurements to field performance, confirming the proppant's potential to alleviate water blocking and improve gas production efficiency in practical applications.

3.2. Mechanical Properties

3.2.1. Single-Particle Compressive Strength

The single-particle compressive strength of the proppants was significantly affected by the preparation conditions. The optimal conditions for achieving high compressive strength were identified through orthogonal experiments: an activated carbon particle size of 30–40 mesh, a silicon–aluminum molar ratio of 4:1, a calcination temperature of 650 °C, and a holding time of 2 h. Under these conditions, the single-particle compressive strength reached 55.5 N, which was 20% higher than that of proppants prepared under non-optimal conditions (Table 2).

Table 2. L9 (3^4) Orthogonal experiment results for single-particle compressive strength of proppants.

Test No.	Activated Carbon Particle Size (Mesh)	Si/Al Molar Ratio	Calcination Temperature (°C)	Holding Time (h)	Single-Particle Compressive Strength (N)
1	20-30 (A1)	3:1 (B1)	600 (C1)	1 (D1)	46.25 (minimum value)
2	20-30 (A1)	4:1 (B2)	650 (C2)	2 (D2)	48.5
3	20-30 (A1)	5:1 (B3)	700 (C3)	3 (D3)	47.0
4	30-40 (A2)	3:1 (B1)	650 (C2)	3 (D3)	50.0
5	30-40 (A2)	4:1 (B2)	700 (C3)	1 (D1)	52.0
6	30-40 (A2)	5:1 (B3)	600 (C1)	2 (D2)	49.0
7	20-30 (A1)	3:1 (B1)	700 (C3)	2 (D2)	47.5
8	20-30 (A1)	4:1 (B2)	600 (C1)	3 (D3)	48.0
9	30-40 (A2)	5:1 (B3)	650 (C2)	1 (D1)	51.0
Optimal Condition	30-40 (A2)	4:1 (B2)	650 (C2)	2 (D2)	55.5 (maximum value)

The proppants exhibited a crushing rate of 3.2% under 52 MPa closure pressure, which is well below the 10% maximum allowed by API RP 56 (for medium-depth reservoirs) and meets the stricter ISO 13503-2 requirement (<5% for high-performance proppants) [44]. This confirms its compliance with industry standards and superior structural stability compared to conventional proppants.

Figure 8 is a schematic diagram illustrating the mechanical strength enhancement mechanism of the silica–alumina-coated activated carbon proppant (CP-AC). On the left, uncoated activated carbon (AC) undergoes severe fragmentation under applied pressure (indicated by orange "Pressure" arrows)—its inherently porous and relatively loose structure lead to stress concentration, causing extensive crack propagation and subsequent breakage into small fragments. In contrast, on the right, CP-AC is enveloped by a dense silica–alumina coating, and its internal structure is reinforced by cross-linked networks formed via **-Si-O-Si-** and **-Al-O-Si-** chemical bonds. When pressure is exerted, the silica–alumina coating not only fills the original microcracks of AC to reduce stress concentration points but also enables uniform stress distribution through the rigid Si-O-Si/Al-O-Si cross-linked architecture, thus maintaining the structural integrity of CP-AC without breakage. This diagram clearly demonstrates the core mechanism: the silica–alumina coating enhances the mechanical strength of the activated carbon proppant via chemical bonding and structural support.

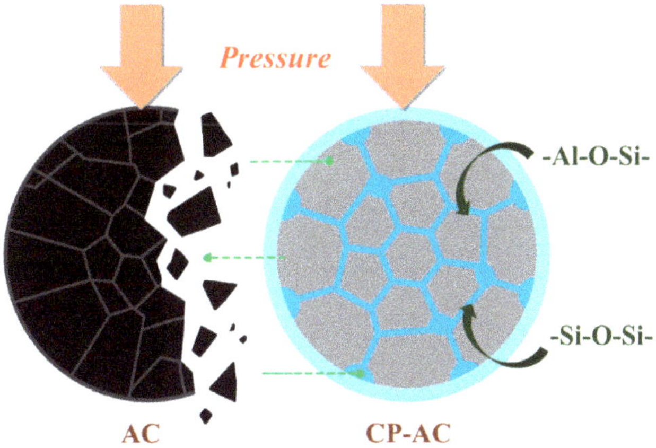

Figure 8. Schematic diagram of mechanical strength enhancement of silica–alumina coated activated carbon proppant.

3.2.2. Crushing Rate

The crushing rate of the proppants under a closure pressure of 50 MPa was also measured. Tests were conducted using 50 g of proppant particles immersed in distilled water (simulating in situ formation water), with pressure maintained for 30 min before sieving through a 40-mesh screen to quantify crushed fragments. The proppants prepared under the optimal conditions had a crushing rate of only 2.3%, much lower than the industry standard of 10%. The enhanced mechanical strength (55.5 N compressive strength, 2.3% crushing rate at 50 MPa) arises from synergistic effects of the silica–alumina coating: its cross-linked network (Si-O-Si/Al-O-Si bonds, confirmed by FTIR) forms a rigid layer to disperse stress; chemical bonding with pretreated activated carbon's oxygen-containing groups strengthens interfacial adhesion; and the dense coating reduces skeleton defects—avoiding brittle failure like quartz sand, thus boosting robustness. This low crushing rate indicated that the proppants had good mechanical stability and could maintain their integrity under high pressure in the coalbed.

3.3. Conductivity

To assess the flow capacity of the proppant pack, conductivity measurements were conducted under simulated coalbed environments, with results presented in Figure 9. Proppants fabricated under the optimized preparation conditions exhibited a conductivity of 15.5 mD·cm, representing a 48.5% improvement over conventional quartz sand. This enhanced conductivity is linked to the activated carbon skeleton's well-developed porous structure and the uniformity of its coating, both of which promote efficient gas migration within fractures. The synergistic enhancement of conductivity stems from the coordinated effects of the activated carbon skeleton and silica–alumina coating: The microporous–mesoporous structure of activated carbon, retained after pretreatment (confirmed by BET analysis), provides a primary channel for gas diffusion. Even after coating, it maintains 33.2% interconnected porosity, reducing flow resistance and increasing gas transport channels compared to dense, non-porous quartz sand. The silica–alumina coating, formed by uniform sol particles, avoids blocking the activated carbon's pores; instead, it preserves micron-scale pores (observed via SEM) that connect with the skeleton's pores, forming a continuous gas flow network—unlike quartz sand, whose irregular particle stacking often causes narrow or blocked channels. Additionally, the coating enhances compressive strength (55.5 N), limiting the crushing rate to 2.3% under 50 MPa and reducing fine particle blockage, which plagues quartz sand due to its low strength and resultant permeability decline. Moreover, the hydrophobic coating (contact angle 115.32°) reduces fracturing fluid retention and avoids water blockage, unlike quartz sand's hydrophilicity, which traps water and impairs conductivity. Together, the activated carbon provides a porous foundation, while the coating ensures structural stability and controlled porosity, overcoming quartz sand's limitations to achieve a 48.5% conductivity improvement.

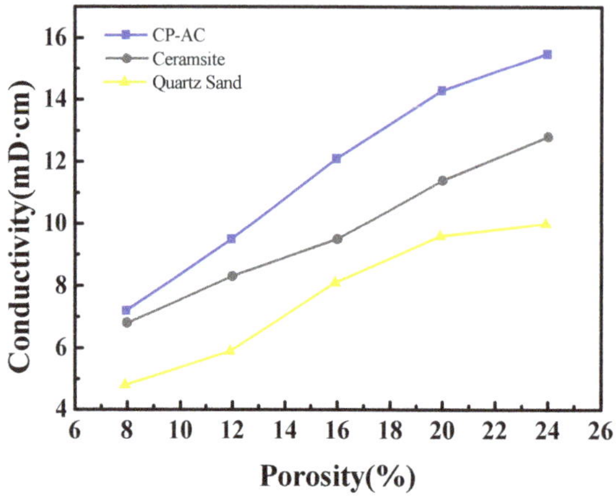

Figure 9. Schematic diagram of conductivity in proppant pack: quartz sand vs. CP-AC and ceramsite.

3.4. Gas Adsorption–Desorption Behavior and Mechanism Analysis

3.4.1. Methane Desorption Promotion Effect

The gas desorption promotion ability of the proppants was evaluated by comparing the methane desorption rate of the CP-AC with that of the AC sample. The results demonstrated that the proppants could enhance the gas desorption rate by 16.2%. The large specific surface area and porous structure of the activated carbon skeleton provided more adsorption sites for gas molecules, promoting the desorption of methane from the coal surface. The results are presented in Figure 10.

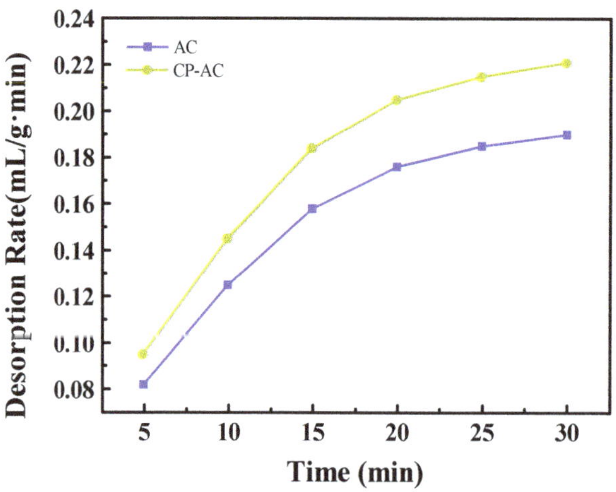

Figure 10. Comparison of methane desorption rates between AC and CP-AC system over time.

To further elucidate the underlying mechanism of gas desorption promotion by the composite proppant, molecular simulations were performed to investigate the interactions between the proppant surface and methane molecules, as well as between the coal matrix and methane molecules. The simulation models were constructed based on the actual structures obtained from experimental characterizations. The activated carbon skeleton surface with oxygen-containing functional groups (carboxyl and hydroxyl groups identified by FTIR) and a representative coal matrix model (composed of polycyclic aromatic hydrocarbons) were built. Methane molecules were introduced into the simulation system, and the interactions were analyzed using the COMPASS force field. The results showed that the interaction energy between methane molecules and the composite proppant surface was lower than that between methane and the coal matrix. Specifically, the van der Waals force between methane and the proppant was weakened by approximately 18.3% compared to that with the coal matrix, while the hydrogen bonding between the proppant's carboxyl groups and water molecules in the coalbed further reduced the adsorption affinity of methane to the coal surface (Table 3). Moreover, the porous structure of the activated carbon skeleton provided channels for methane diffusion, with a calculated diffusion coefficient of 2.3×10^{-9} m^2/s, which was 32.6% higher than that in the pure coal matrix model. The results are presented in Table 4. These molecular simulation results directly confirmed that the composite proppant promotes methane desorption by reducing the adsorption energy barrier and enhancing molecular diffusion, consistent with the experimental observation of a 16.2% increase in gas desorption rate. Figure 11 illustrates the comparison of methane (CH_4) desorption behavior between the silica–alumina-coated activated carbon proppant (CP-AC, left) and uncoated activated carbon (AC, right). The optimized pore structure of the silica–alumina coating in CP-AC facilitates the diffusion and desorption of CH_4 molecules, as schematically shown in the center. In contrast, uncoated AC (with a more irregular and less favorable pore configuration) exhibits less efficient CH_4 desorption. Notably, CP-AC achieves a 16.2% increase in the desorption rate compared to AC, attributed to the coating's tailored pore structure, which enhances the accessibility and release of adsorbed CH_4.

Table 3. Interaction energy analysis of different systems.

System	Van der Waals Energy (kcal/mol)	Hydrogen Bond Energy (kcal/mol)	Total Adsorption Energy (kcal/mol)
Methane–Coal Matrix	-28.6 ± 1.2	-3.2 ± 0.5	-31.8 ± 1.5
Methane-CP-AC	-23.4 ± 0.9	-5.7 ± 0.3	-29.1 ± 1.1

Table 4. Methane diffusion behavior in different matrices.

Parameter	Coal Matrix	CP-AC Porous Structure	Enhancement Ratio
Diffusion Coefficient (m^2/s)	$1.73 \times 10^{-9} \pm 0.08 \times 10^{-9}$	$2.30 \times 10^{-9} \pm 0.11 \times 10^{-9}$	32.9%
Mean Square Displacement (MSD) at 500 ps	65.2 ± 3.1 Å^2	86.6 ± 4.2 Å^2	32.8%

3.4.2. Nitrogen Adsorption Behavior and Competitive Interaction with Methane

To investigate the adsorption behavior of N_2 in coalbed methane, a series of experiments were carried out. Activated carbon (AC) was selected as the adsorbent due to its widespread use in gas separation processes. The experiments were conducted using a volumetric adsorption apparatus (Micromeritics ASAP 2020), which allowed for accurate measurement of gas adsorption under controlled conditions.

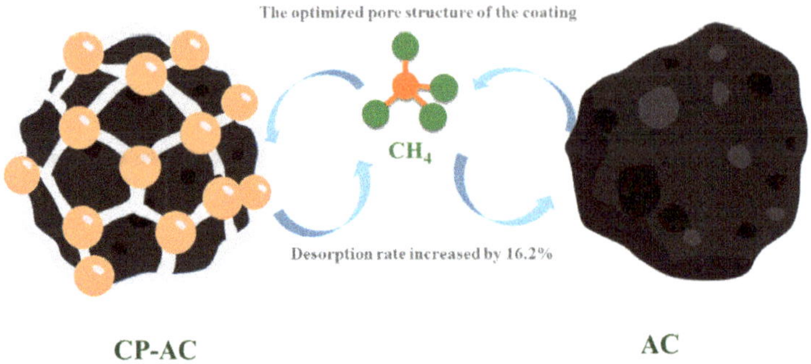

Figure 11. Schematic diagram of the principle of methane adsorption on coal surface.

The AC samples were first pretreated at 300 °C under vacuum for 3 h to remove any pre-adsorbed impurities. Then, pure N_2 gas (purity > 99.99%) was introduced into the adsorption chamber. The temperature was maintained at 298 K (simulating in situ coalbed temperature), and the pressure range was varied from 0 to 100 kPa. At each pressure point, the system was equilibrated for 2 h to ensure that adsorption equilibrium was reached. The amount of N_2 adsorbed was calculated based on the change in pressure and volume in the system, according to the ideal gas law.

In addition to the single-component N_2 adsorption experiments, binary-component adsorption experiments with N_2 and CH_4 (simulating coalbed methane composition) were also performed. The gas mixture with a fixed ratio of N_2 and CH_4 (e.g., 70% N_2 and 30% CH_4 by volume) was introduced into the adsorption chamber, and the adsorption isotherms were measured under the same temperature and pressure conditions as the single-component experiments.

The adsorption of N_2 on activated carbon can be divided into three main stages, as illustrated in Figure 12. At low pressures, N_2 molecules are adsorbed onto the surface of the activated carbon via van der Waals forces, forming a monolayer. The activated carbon has a complex pore structure, including micropores (pore diameter < 2 nm) and mesopores (2–50 nm). As shown in the pore size distribution curves in Figure 12a, before adsorption (represented by the red dashed line), the AC has a certain pore volume distribution in both micropore and mesopore regions. After monolayer adsorption (blue solid line), the pore volume in the mesopore region shows a slight decrease, while the micropore region remains relatively unchanged. This indicates that monolayer adsorption mainly occurs on the surfaces of mesopores and at the entrances of some micropores. The surface area of the activated carbon, especially the outer surface of mesopores, provides active sites for N_2 molecules to attach.

As the pressure increases, N_2 molecules start to penetrate into the micropores of the activated carbon. The small size of N_2 molecules allows them to fit into the narrow micropores. The micropores in activated carbon have a high surface-to-volume ratio, providing strong adsorption forces for N_2. As can be seen from the pore size distribution curves in Figure 12b, after micropore filling, the pore volume in the micropore region decreases significantly, and there is also a further reduction in the mesopore region. This is because some N_2 molecules that were initially adsorbed on the mesopore surface may migrate into the micropores. The filling of micropores is crucial for the overall adsorption capacity of activated carbon for N_2, as micropores contribute a large proportion of the total surface area of the adsorbent.

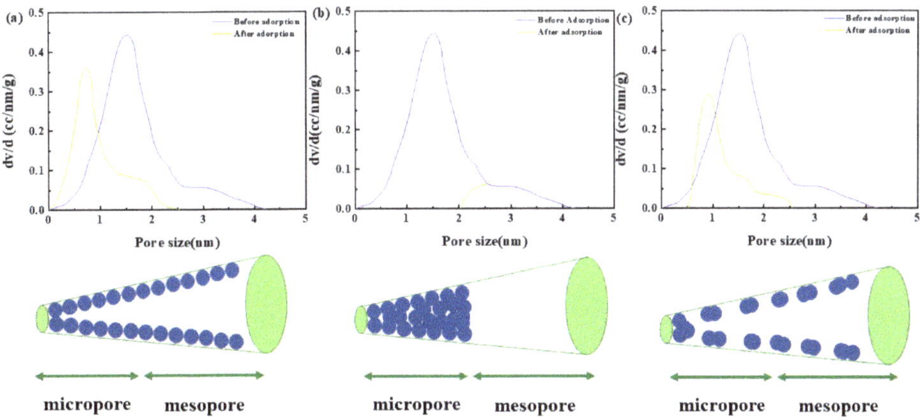

Figure 12. Adsorption models and pore size changes in activated carbon during N_2 adsorption of coalbed methane. (**a**) Monolayer adsorption, (**b**) micropore adsorption, and (**c**) agglomeration adsorption.

At higher pressures, close to the saturation vapor pressure of N_2 at the experimental temperature, multilayer adsorption occurs on the surface of the activated carbon, and capillary condensation takes place in the mesopores and larger pores. In Figure 12c, the pore size distribution curves show a significant decrease in the pore volume of both mesopore and macropore (pore diameter > 50 nm) regions after this stage. Multilayer adsorption is due to the continuous attraction of N_2 molecules to the already-adsorbed N_2 layers on the carbon surface. Capillary condensation occurs when the pressure is high enough for N_2 to form a liquid-like phase in the pores. This is because the curved liquid–vapor interface in the pores reduces the vapor pressure required for condensation (Kelvin effect). The presence of moisture in the coalbed environment can also affect this stage, as water molecules may compete with N_2 for adsorption sites, especially in the polar regions of the activated carbon surface.

In the binary-component adsorption system of N_2 and CH_4, competitive adsorption occurs. CH_4 has a larger molecular size and a higher polarizability compared to N_2. The activated carbon has a certain selectivity for CH_4 over N_2 due to differences in adsorption energy. However, the adsorption capacity for both gases is affected by their partial pressures in the mixture. At low partial pressures of CH_4, N_2 can still be adsorbed to a significant extent. As the partial pressure of CH_4 increases, it competes more effectively with N_2 for adsorption sites, reducing the amount of N_2 adsorbed.

In conclusion, the adsorption of N_2 in coalbed methane on activated carbon is a complex process influenced by factors such as pore structure, gas pressure, temperature, and the presence of other gases. Understanding these mechanisms is crucial for optimizing the separation and purification of coalbed methane.

3.5. Corrosion Resistance

In the simulated acidic coalbed environment (pH = 3–5), the acid corrosion rate of the proppants was less than 2.8% after 72 h soaking, which was significantly lower than that of conventional ceramsite (e.g., 11.3% under the same conditions) and comparable to high-purity quartz sand (e.g., 3.4% under the same conditions). The silica–alumina coating on the surface of the proppants played a critical protective role, effectively preventing the activated carbon skeleton from acid corrosion. This favorable corrosion resistance, combined with its comprehensive performance advantages over traditional proppants, ensures the long-term stability of the proppants in the coalbed. Figures 13 and 14 illustrates the acid resistance performance of uncoated activated carbon (AC, left) and silica–alumina-coated activated carbon proppant (CP-AC, right) after being exposed to acid for 72 h. On the left, uncoated AC shows severe fragmentation—acid droplets penetrate its porous structure, causing extensive cracking and breakage over the 72 h period. In contrast, CP-AC on the right maintains its structural integrity because the silica–alumina coating acts as a protective barrier, preventing the acid from corroding the inner activated carbon skeleton.

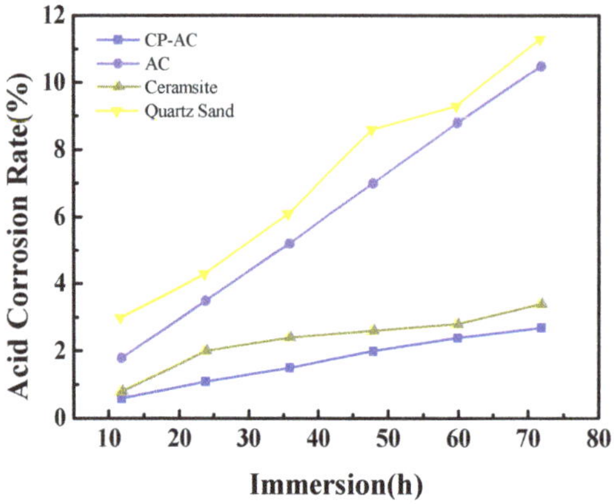

Figure 13. Acid corrosion rate of different proppants vs. immersion time.

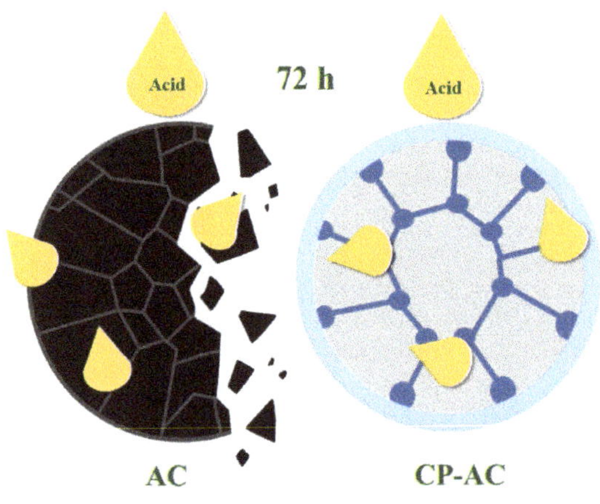

Figure 14. Schematic diagram of proppant corrosion protection in acidic environment.

3.6. Proppant Transport Performance

The results showed significant differences in transport performance between CP-AC and ceramsite under both flow rates (Figure 15). At a flow rate of 5 L/h, CP-AC maintained a higher accumulation height than ceramsite across all measured distances. Near the wellbore (20 cm), CP-AC reached 18.6 cm, slightly higher than ceramsite (16.5 cm). As the distance increased, the height of both proppants decreased, but CP-AC showed slower and more irregular decay. For instance, at 60 cm, CP-AC retained 16.5 cm, while ceramsite dropped sharply to 9.8 cm (a 26.9% reduction compared to CP-AC's 4.0% drop). At the fracture tip (140 cm), CP-AC still maintained a height of 7.8 cm, which was five times that of ceramsite (1.6 cm), demonstrating its superior ability to reach distal fractures.

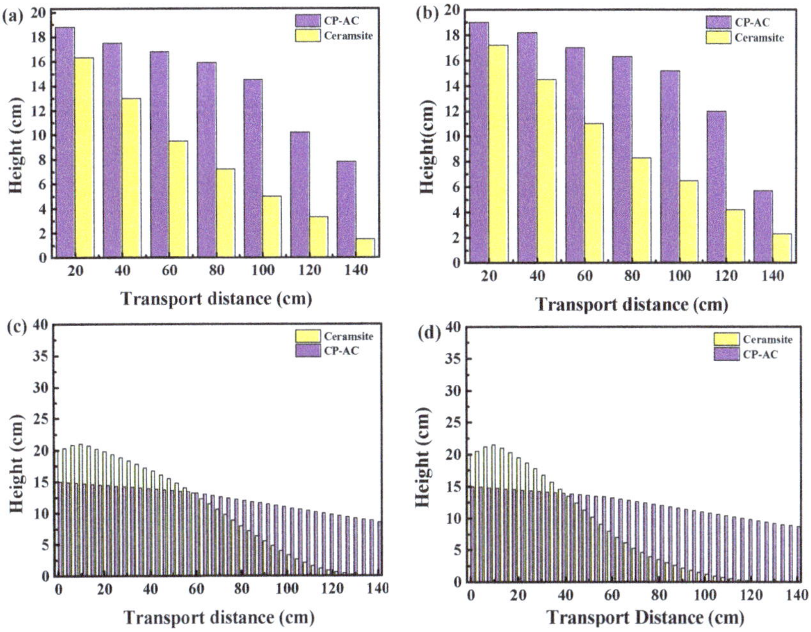

Figure 15. (**a**,**b**) Bar chart of distribution of CP-AC and ceramsite under flow rates of 5 L/h and 10 L/h. (**c**,**d**) Trend chart of migration and distribution of CP-AC and ceramsite under flow rates of 5 L/h and 10 L/h.

At an increased flow rate of 10 L/h, both proppants showed an improved transport efficiency due to an enhanced fluid carrying capacity, but CP-AC remained advantageous. Near the wellbore (20 cm), CP-AC reached 19.0 cm, marginally higher than ceramsite (17.2 cm). At 100 cm, CP-AC's height (15.2 cm) was 2.3 times that of ceramsite (6.5 cm). A notable observation was the irregular decay of CP-AC: its height decreased moderately from 100 cm (15.2 cm) to 120 cm (12.0 cm) but dropped more sharply to 5.7 cm at 140 cm, simulating the natural weakening of flow at fracture tips. Even so, CP-AC's tip height (5.7 cm) was still 2.5 times that of ceramsite (2.3 cm), confirming its superior retention under high flow conditions.

The superior transport performance of CP-AC is attributed to its low density (1.45 g/cm^3) compared to ceramsite (3.1 g/cm^3), which reduces the settlement tendency, and its porous structure, which enhances interaction with fracturing fluids to form a "viscous buffer layer" that slows settlement. The irregular decay trends reflect real-world fracture heterogeneities, such as varying flow velocities and local turbulence, further validating CP-AC's adaptability to complex coalbed environments.

4. Conclusions

In this study, a high-performance coalbed fracturing proppant with an activated carbon (AC) skeleton was successfully synthesized through a systematic process involving nitric acid pretreatment, silica–alumina sol coating, and calcination. Orthogonal experimental optimization confirmed that the optimal preparation conditions—30–40-mesh AC particles, a Si/Al molar ratio of 4:1, and calcination at 650 °C for 2 h—yielded a proppant with exceptional comprehensive properties. Mechanically, the proppant exhibited a single-particle compressive strength of 55.5 N and a crushing rate of only 2.3% under 50 MPa closure pressure, outperforming traditional quartz sand and meeting the demands of high-geostress coalbed environments. Its permeability improvement rate, at 48.5% higher than quartz sand, combined with a porosity of 33.2%, ensured efficient gas flow through fractures, supported by a well-developed porous structure and stable silica–alumina coating. In terms of coalbed adaptability, the proppant demonstrated strong acid corrosion resistance (<2.8% corrosion rate in pH 3–5 environments after 72 h) and a 16.2% enhancement in methane desorption efficiency. Molecular simulations and experimental data confirmed that this desorption promotion stemmed from reduced methane–surface interaction energy and improved diffusion dynamics. Additionally, its low density (1.45 g/cm^3) enhanced transportability in fracturing fluids, with distal fracture accumulation five times higher than ceramsite, while its hydrophobic surface (contact angle 115.32°) facilitated fracturing fluid flowback.

This novel proppant integrates high mechanical strength, superior conductivity, gas desorption promotion, and environmental stability, addressing key limitations of conventional proppants in coalbed methane extraction. However, it is important to acknowledge certain limitations: the proppant is currently suitable for medium–shallow coalbeds (≤2000 m depth, ≤50 MPa closure pressure, single acidic environments) but not for deep seams (>2000 m, >50 MPa) or multi-ion extreme conditions. It remains at the lab scale (<100 g per batch) without pilot validation, and assumptions like using regular simulated fractures may differ from actual irregular coal seams. Uncertainties include an untested long-term performance (only 72 h tests) and unconfirmed long-term compatibility with coal matrices. Future work should focus on scaling up production to reduce costs and validating the performance under field conditions with complex geological variations, further solidifying its potential as a transformative material for efficient CBM recovery.

Author Contributions: Conceptualization, K.W.; methodology, K.W.; formal analysis, Q.G. and G.L.; investigation, K.W. and C.G.; writing—original draft preparation, K.W., C.G., Q.G. and G.L.; writing—review and editing, X.Z., P.Z. and C.C.; funding acquisition, K.W. All authors have read and agreed to the published version of the manuscript.

Funding: This study was funded by Chongqing's Talent Program—Innovative Leading Talent Project (Project Number: CQYC20220304204), a Key Project of the National Natural Science Foundation of China (Joint Fund for Regional Innovation and Development) (Grant No. U24A2090) and key Project of Tiandi Science and Technology Co., Ltd. Innovation and Entrepreneurship Special Fund (2022-2-TD-ZD009).

Data Availability Statement: The original contributions presented in this study are included in the article material. Further inquiries can be directed to the corresponding author.

Conflicts of Interest: Author Kai Wang, Chenye Guo and Gen Li were employed by the Southwest Branch of China National Coal Group Corp. Author Qisen Gong, Xiaoyue Zhuo, Peng Zhuo and Chaoxian Chen were employed by the Chongqing Energy Investment Group Technology Co., Ltd. The authors declare that the research was conducted in the absence of any commercial or financial relationships that could be construed as a potential conflict of interest.

References

1. Chen, Q.; Huang, Z.; Huang, H.; Chen, Q.; Ling, X.; Xin, F.; Kong, X. An Experimental Study on the Impact of the Particle Size and Proportion of Composite Proppant on the Conductivity of Propped Fractures in Coalbed Methane Reservoirs following Pulverized Coal Fines Infiltration. *Processes* **2023**, *11*, 2205. [CrossRef]
2. Huang, J.; Chen, J.; Lv, L.; Yu, Z.; Yao, W.; Cheng, H.; Niu, W.; Wang, J.; Zhang, J.; Qi, H. Design and Verification of a Wearable Micro-Capacitance Test System for POC Biosensing. *IEEE Trans. Instrum. Meas.* **2025**, *74*, 2007411.
3. Liu, Y.; Zhou, L.; Wan, X.; Tang, Y.; Liu, Q.; Li, W.; Liao, J. Synthesis and Characterization of a Temperature-Sensitive Microcapsule Gelling Agent for High-Temperature Acid Release. *ACS Omega* **2024**, *9*, 20849–20858. [CrossRef] [PubMed]
4. Chen, Z.; Liu, W. Proppant embedding-based optimal design method for fracturing coalbed methane wells in Qingshui basin. *Sci. Rep.* **2025**, *15*, 8508. [CrossRef] [PubMed]

5. Hao, H.; Yao, E.; Pan, L.; Chen, R.; Wang, Y.; Xiao, H. Exploring heterogeneous drivers and barriers in MaaS bundle subscriptions based on the willingness to shift to MaaS in one-trip scenarios. *Transp. Res. Part A Policy Pract.* **2025**, *199*, 104525. [CrossRef]
6. Kurnia, J.C.; Xu, P.; Sasmito, A.P. A novel concept of enhanced gas recovery strategy from ventilation air methane in underground coal mines—A computational investigation. *J. Nat. Gas Sci. Eng.* **2016**, *35*, 661–672. [CrossRef]
7. Dai, S.F.; Xie, P.P.; French, D.; Ward, C.R.; Graham, I.T.; Yan, X.Y.; Guo, W.M. The occurrence of buddingtonite in super-high-organic-sulphur coals from the Yishan Coalfield, Guangxi, southern China. *Int. J. Coal Geol.* **2018**, *195*, 347–361. [CrossRef]
8. Wu, K.; Zou, J.; Jiao, Y. Focal Mechanism of Strong Ground Seismicity Induced by Deep Coal Mining. *Rock Mech. Rock Eng.* **2023**, *56*, 779–795. [CrossRef]
9. Zeng, Q.S.; Wang, Z.M.; Liu, L.Q.; Ye, J.P.; Brian, J. Modeling CH_4 Displacement by CO_2 in Deformed Coalbeds during Enhanced Coalbed Methane Recovery. *Energy Fuels* **2018**, *32*, 1942–1955. [CrossRef]
10. Dai, S.F.; Xie, P.P.; Ward, C.R.; Yan, X.Y.; Guo, W.M.; French, D.; Ian, T. Graham. Anomalies of rare metals in Lopingian super-high-organic-sulfur coals from the Yishan Coalfield, Guangxi, China. *Ore Geol. Rev.* **2017**, *88*, 235–250. [CrossRef]
11. Guo, X.; Yan, Q.; Wang, A. Evaluation of Relative Permeability in Coalbed Methane Reservoirs Based on Production Data: A Case Study in Qinshui Basin, China. *Nat. Resour. Res.* **2019**, *28*, 187–198. [CrossRef]
12. Office, J.E. Acknowledgement to Reviewers of Journal of Marine Science and Engineering in 2017. *J. Mar. Sci. Eng.* **2018**, *6*, 8. [CrossRef]
13. Colin, R.W.; Ken, R.; David, F. Dalway John (Dal) Swaine: 1920–2013. *Int. J. Coal Geol.* **2013**, *113*, 60–62.
14. Li, J.; Wen, M.; Yang, J.; Liu, Y.; Jiang, Z.; Chen, J. Synthesis and analysis of magnetic nanoparticles within foam matrix for foam drainage gas production. *Geoenergy Sci. Eng.* **2024**, *238*, 212887. [CrossRef]
15. Fu, X.Y.; Zhao, C.P.; Lun, Z.M.; Wang, H.T.; Wang, M.; Zhang, D.F. Influences of controlled microwave field radiation on pore structure, surface chemistry and adsorption capability of gas-bearing shales. *Mar. Pet. Geol.* **2021**, *130*, 105134. [CrossRef]
16. Zhou, Y.; Xia, H.; Yu, D.; Cheng, J.; Li, J. Outlier detection method based on high-density iteration. *Inf. Sci.* **2024**, *662*, 120286. [CrossRef]
17. Gu, F.; Chalaturnyk, R.J. Sensitivity Study of Coalbed Methane Production with Reservoir and Geomechanic Coupling Simulation. *J. Can. Pet. Technol.* **2005**, *44*, 23–32. [CrossRef]
18. Hu, S.Y.; Chen, Y.B.; Hao, Y.X.; Chen, Z.Y.; Feng, G.R.; Li, G.F.; Guan, S.W.; Zhang, X.T.; Li, S.Y. Experimental study of the effects of fine retention on fracturing proppant permeability in coalbed methane reservoirs. *J. Nat. Gas Sci. Eng.* **2020**, *83*, 1875–5100. [CrossRef]
19. Shen, S.C.; Fang, Z.M.; Li, X.C.; Jiang, Q.; Shen, H.M.; Shi, L. Enhanced Coalbed Methane Recovery from Lignite Using CO_2 and N_2. *Energy Fuels* **2025**, *39*, 10394–10409. [CrossRef]
20. Liu, H.; Li, Z.; Wang, H. Numerical Simulation Investigation of N_2 Injection for Enhanced Coalbed Methane Recovery. *Arab. J. Sci. Eng.* **2025**, *50*, 4661–4672. [CrossRef]
21. Guo, J.H.; Zhang, Z.S.; Guo, G.S.; Xia, H.; Zhu, L.Q.; Zhang, C.M.; Tang, X.; Zhou, X.Q.; Zhang, Y.N.; Wang, C. Evaluation of Coalbed Methane Content by Using Kernel Extreme Learning Machine and Geophysical Logging Data. *Geofluids* **2022**, *2022*, 3424367. [CrossRef]
22. Yin, T.T.; Zhang, J.J.; Liu, D.M.; Cai, Y.D.; Dong, Z.T.; Zhang, L. Application of molecular simulation in coalbed methane reservoirs: A systematic review. *Unconv. Resour.* **2022**, *2*, 124–132. [CrossRef]
23. Azarpour, A.; Mohammadzadeh, O.; Rezaei, N.; Zendehboudi, S. Current status and future prospects of renewable and sustainable energy in North America: Progress and challenges. *Energy Convers. Manag.* **2022**, *269*, 115945. [CrossRef]
24. Flores, R.M. Coalbed methane: From hazard to resource. *Int. J. Coal Geol.* **1998**, *35*, 3–16. [CrossRef]
25. Liu, G.; Tang, R.; Mu, C.; Liu, X.; Zhang, J. Physical Properties of High-Rank Coal Reservoirs and the Impact on Coalbed Methane Production. *Processes* **2024**, *12*, 1754. [CrossRef]
26. Alzanam, A.A.; Ishtiaq, U.; Muhsan, A.S.; Mohamed, N.M. A Multiwalled Carbon Nanotube-Based Polyurethane Nanocomposite-Coated Sand/Proppant for Improved Mechanical Strength and Flowback Control in Hydraulic Fracturing Applications. *ACS Omega* **2021**, *6*, 20768–20778. [CrossRef] [PubMed]
27. Alzanam, A.A.; Muhsan, A.S.; Abdulelah, H.; Mohamed, D.; Iglauer, S. Novel Nanocomposite Coated Proppants for Enhanced Scale Inhibition Lifetime in Tight Shale Formations. *Energy Fuel* **2022**, *36*, 14136–14147. [CrossRef]
28. Zhang, C.; Hu, Q.; Liu, C.; Jia, H.; Sang, G.; Wu, D.; Li, K.; Wang, Q. The Gas Production Characteristics of No. 3 Coal Seam Coalbed Methane Well in the Zhengbei Block and the Optimization of Favorable Development Areas. *Processes* **2024**, *12*, 2018. [CrossRef]
29. Zhang, B.F.; Chen, J.; Sha, J.D.; Zhang, S.; Zeng, J.; Chen, P.; Yao, D.X.; Liu, W.Z.; Wang, X.M.; Zhang, P.S.; et al. Geochemistry of coal thermally-altered by igneous intrusion: A case study from the Pansan Coal Mine of Huainan Coalfield, Anhui, Eastern China. *J. Geochem. Explor.* **2020**, *213*, 106532. [CrossRef]

30. Fu, X.X.; Lun, Z.M.; Zhao, C.P.; Zhou, X.; Wang, H.T.; Zhou, X.T.; Xu, Y.; Zhang, H.; Zhang, D.F. Influences of controlled microwave field irradiation on physicochemical property and methane adsorption and desorption capability of coals: Implications for coalbed methane (CBM) production. *Fuel* **2021**, *301*, 121022. [CrossRef]
31. Tasque, J.E.; Vega, I.N.; Marco, S.; Raffo, P.A.; D'ACcorso, N.B. Ultra-light weight proppant: Synthesis, characterization, andperformance of new proppants. *J. Nat. Gas Sci. Eng.* **2021**, *85*, 103717. [CrossRef]
32. Wang, J.; Guo, X. Adsorption isotherm models: Classification, physical meaning, applicationandsolving method. *Chemosphere* **2020**, *258*, 127279. [CrossRef]
33. Burhan, M.; Shahzad, M.K.; Ng, K.C. Energy distribution function based universal adsorption isotherm model for all types of isotherm. *Int. J. Low-Carbon Technol.* **2018**, *13*, 292–297. [CrossRef]
34. Yang, D.; Cui, G.; Elsworth, D.; Wang, C.; Cheng, W.; Yang, C.; Chen, T. Damage-Based Microscale Model Explains Observed Two-Stage Shale Gas Depletion Profiles at Reservoir Scale. *Energy Fuels* **2025**, *39*, 13488–13504. [CrossRef]
35. Cheng, W.; Cui, G.; Elsworth, D.; Tan, Y.; Pan, Z.; Guo, Y.; Zhang, Y. Constrained study of nanoindentation-based upscaling of mechanical properties of shales. *Acta Geotech.* **2025**, *20*, 4515–4533. [CrossRef]
36. Cheng, W.; Cui, G.; Tan, Y.; Elsworth, D.; Wang, C.; Yang, C.; Chen, T.; Jiang, C. A multi-layer nanocased model to explain the U-shaped evolution of shale gas permeability at constant confining pressure. *Fuel* **2024**, *359*, 130478. [CrossRef]
37. Lu, W.; Wang, J.; Wang, T.; Zhang, K.; Jiang, X.; Zhao, H. Visual style prompt learning using diffusion models for blind face restoration. *Pattern Recognit.* **2025**, *161*, 111312. [CrossRef]
38. Cheng, W.; Guo, Y.; Cui, G.; Elsworth, D.; Tan, Y.; Pan, Z. Impact of micro-scale characteristics of shale reservoirs on gas depletion behavior: A microscale discrete model. *Adv. Geo-Energy Res.* **2025**, *15*, 143–157. [CrossRef]
39. Cui, G.; Tan, Y.; Chen, T.; Feng, X.; Elsworth, D.; Pan, Z.; Wang, C. Multidomain two-phase flow model to study the impacts of hydraulic fracturing on shale gas production. *Energy Fuels* **2020**, *34*, 4273–4288. [CrossRef]
40. Cui, G.; Feng, X.; Pan, Z.; Chen, T.; Liu, J.; Elsworth, D.; Tan, Y.; Wang, C. Impact of shale matrix mechanical interactions on gas transport during production. *J. Pet. Sci. Eng.* **2020**, *184*, 106524. [CrossRef]
41. Cui, G.; Liu, J.; Wei, M.; Shi, R.; Elsworth, D. Why shale permeability changes under variable effective stresses: New insights. *Fuel* **2018**, *213*, 55–71. [CrossRef]
42. Zhang, J.; Qi, H.; Wu, J.; Mao, X.; Zhang, H.; Amin, N.; Xu, F.; Dong, C.; Wang, C.; Wang, P.; et al. Disposable peptidoglycan-specific biosensor for noninvasive real-time detection of broad-spectrum Gram-positive bacteria in exhaled breath condensates. *Anal. Chem.* **2024**, *96*, 9817–9825. [CrossRef] [PubMed]
43. Sun, L.; Shi, W.; Tian, X.; Li, J.; Zhao, B.; Wang, S.; Tan, J. A plane stress measurement method for CFRP material based on array LCR waves. *NDT E Int.* **2025**, *151*, 103318. [CrossRef]
44. *ISO 13503-2:2024*; Oil and Gas Industries Including Lower Carbon Energy—Completion Fluids and Materials. International Organization for Standardization: Geneva, Switzerland, 2024.

energies

Article

Evaluation of Development Performance and Adjustment Strategies for High Water-Cut Reservoirs Based on Flow Diagnostics: Application in the QHD Oilfield

Yifan He [1,2], Yishan Guo [2,3], Li Wu [1], Liangliang Jiang [4], Shouliang Wang [1,*], Shangshu Ning [1,*] and Zhihong Kang [1]

[1] School of Energy, China University of Geosciences (Beijing), Beijing 100083, China; heyifan0@126.com (Y.H.); wuli_edu@163.com (L.W.)
[2] Tianjin Branch, CNOOC (China) Co., Ltd., Tianjin 300459, China
[3] School of Ocean Sciences, China University of Geosciences (Beijing), Beijing 100083, China
[4] Department of Chemical and Petroleum Engineering, University of Calgary, Calgary, AB T2N 1N4, Canada
* Correspondence: wangshuoliang@cugb.edu.cn (S.W.); 17662508246@163.com (S.N.)

Abstract: Offshore reservoirs in the high water-cut stage present significant development challenges, including declining production, complex remaining oil distribution, and the inadequacy of conventional evaluation methods to capture intricate flow dynamics. To overcome these limitations, this study introduces a novel approach based on flow diagnostics for performance evaluation and potential adjustment. The method integrates key metrics such as time-of-flight (TOF) and the dynamic Lorenz coefficient, supported by reservoir engineering principles and numerical simulation, to construct a multi-parameter evaluation system. This system, which also incorporates injection–production communication volume and inter-well fluid allocation factors, precisely quantifies and visualizes waterflood displacement processes and sweep efficiency. Applied to the QHD32 oilfield, this framework was used to establish specific thresholds for operational adjustments. These include criteria for infill drilling (water-flooded ratio < 45%, remaining oil thickness > 6 m, TOF > 200 days), conformance control (TOF < 50 days, dynamic Lorenz coefficient > 0.5), and artificial lift optimization (remaining oil thickness ratio > 2/3, TOF > 200 days). Field validation confirmed the efficacy of this approach: an additional cumulative oil production of 165,600 m^3 was achieved from infill drilling in the C29 well group, while displacement adjustments in the B03 well group increased oil production by 2.2–3.8 tons/day, demonstrating a significant enhancement in waterflooding performance. This research provides a theoretical foundation and a technical pathway for the refined development of offshore heavy oil reservoirs at the ultra-high water-cut stage, offering a robust framework for the sustainable management of analogous reservoirs worldwide.

Keywords: high water-cut reservoirs; flow diagnostics; time of flight; Lorenz coefficient; hierarchical adjustment

1. Introduction

The effective development of mature oilfields, particularly in offshore environments, is a critical challenge for global energy security [1,2]. As these fields enter the high water-cut stage (Figure 1), they are characterized by increasingly complex geological conditions, highly discrete remaining oil distribution, and significant interlayer heterogeneity, which collectively diminish recovery efficiency [3,4]. For instance, challenges such as poor well pattern adaptability in the Shengtuo oilfield [5] and complex fluid dynamics in the faulted P oilfield [6] exemplify the limitations of conventional development strategies. With China's crude oil import dependency exceeding 70% and mature fields constituting the bulk of its domestic reserves, optimizing production from these assets is a strategic imperative for ensuring national energy security [7–9].

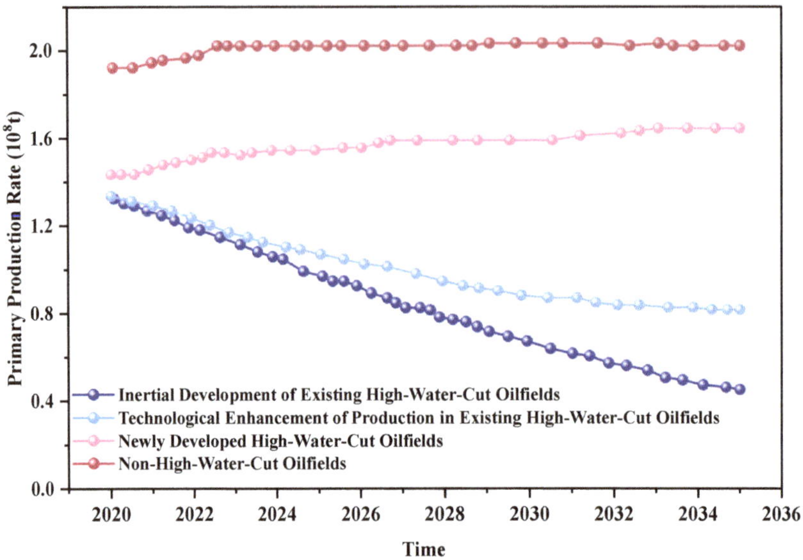

Figure 1. Predicted production composition of high water-cut oilfields in China [8].

To address the challenges inherent in high water-cut reservoirs—strong heterogeneity, severe water channeling, and dispersed remaining oil—a precise evaluation of fluid flow dynamics is essential for optimizing recovery. Historically, reservoir performance evaluation has relied on traditional reservoir engineering techniques, such as decline curve analysis and well testing. Methods like the Arps decline curve [10] provide production forecasts but their accuracy is often compromised in heterogeneous reservoirs where complex fluid and rock properties are not fully accounted for. While modified approaches have been proposed to accommodate data limitations [11], and well testing can yield physical parameters, these methods often fail to capture the dynamic, layer-specific performance in complex settings [12]. More advanced techniques that incorporate water-cut curves [13] or vector flow fields from numerical simulations [14] offer improved characterization of channeling paths and sweep efficiency. However, they can be computationally intensive and may not fully resolve the dynamic heterogeneity of the flow field.

Streamline simulation has emerged as a powerful tool for visualizing fluid flow and integrating dynamic data [15]. It offers significant computational speed advantages over conventional numerical methods and provides excellent flow visualization, particularly for thermal recovery processes [16]. Similarly, tracer monitoring integrated with numerical simulation can clarify sand body connectivity and injection–production correspondence [17]. Nevertheless, the accuracy of all simulation-based approaches is fundamentally dependent on the quality of the underlying geological model. Uncertainties in reservoir characterization and the idealizations inherent in numerical models can lead to significant prediction biases, especially in reservoirs with complex fracture networks.

The limitations of both traditional empirical methods and computationally demanding simulation techniques highlight the need for an alternative evaluation framework that can efficiently integrate dynamic data and precisely capture complex flow field characteristics. This study proposes such a framework, founded on the principles of flow diagnostics. Our approach utilizes key metrics derived from streamline or numerical models—principally the time-of-flight (TOF) and a dynamic Lorenz coefficient—to quantify displacement heterogeneity. TOF measures the travel time of fluid particles along streamlines, while the Lorenz coefficient quantifies the degree of flow uniformity. These metrics are integrated into a multi-parameter evaluation system that also includes injection–production communication volume and inter-well fluid allocation factors. This system enables a precise and visualized assessment of the waterflood process, overcoming the experiential dependency and computational latency of conventional methods in the ultra-high water-cut stage. By applying this methodology to the QHD32-6 oilfield, we establish quantitative criteria for targeted interventions such as infill drilling, conformance control, and artificial lift optimization. This research aims to provide a robust theoretical and technical pathway for the refined development of offshore heavy oil reservoirs, offering a valuable reference for the sustainable management of similar assets, as shown in Table 1.

Table 1. Comparison of three evaluation methods.

Method	Evaluation Parameters	Features	Advantages	Application
Traditional Reservoir Engineering	Waterflood Characteristic Curves, Production Decline Curves	Traditional curve-fitting methods are computationally efficient but possess low diagnostic accuracy, with their application being constrained by empirical assumptions.	Wide applicability; relies on empirical and engineering judgment.	Daqing Oilfield, Qinjiatun Oilfield Bohai Oilfield [18–20]
Reservoir Numerical Simulation	Saturation, Streamlines, Flow Velocity, etc.	Reservoir numerical simulation is computationally intensive and time-consuming, making it difficult to satisfy real-time diagnostic needs. However, it excels at meticulously characterizing the internal physical dynamics and remaining oil distribution within the reservoir, though its application scale is generally restricted to localized studies.	High accuracy; capable of evaluating multi-parameter changes during the waterflooding process; offers 3D visualization.	Gangxi Oilfield, Jilin Oilfield [21]
Flow Diagnostics (The proposed method)	TOF, Dynamic Lorenz Coefficient, Streamlines, Oil Saturation, Pressure Field, Waterflood Characteristic Curve, Production Decline Curve	The flow diagnostics method is computationally efficient, provides targeted diagnostics, and its application is both flexible and scalable.	Rapid evaluation of waterflood performance for different well groups at various times. Facilitates the screening of candidate interventions and remedial actions.	QHD Oilfield

2. Methods

This study proposes and applies a flow diagnostics-based methodology to address the complex development challenges in the high water-cut QHD32-6 oilfield. The core of this approach is a multi-parameter evaluation system that integrates advanced diagnostic metrics—specifically time-of-flight (TOF) and a dynamic Lorenz coefficient—with established reservoir engineering techniques and numerical simulation outputs. The comprehensive system incorporates data from streamline fields, oil saturation and pressure distributions, waterflooding characteristic curves, and production decline analyses to achieve a precise, quantitative characterization of the waterflood displacement process. Based on this evaluation, a hierarchical adjustment strategy is formulated, encompassing infill drilling, conformance control, and artificial lift enhancement. The methodology's effectiveness is validated through field application and optimization in the QHD32-6 oilfield, with the overall research workflow detailed in Figure 2.

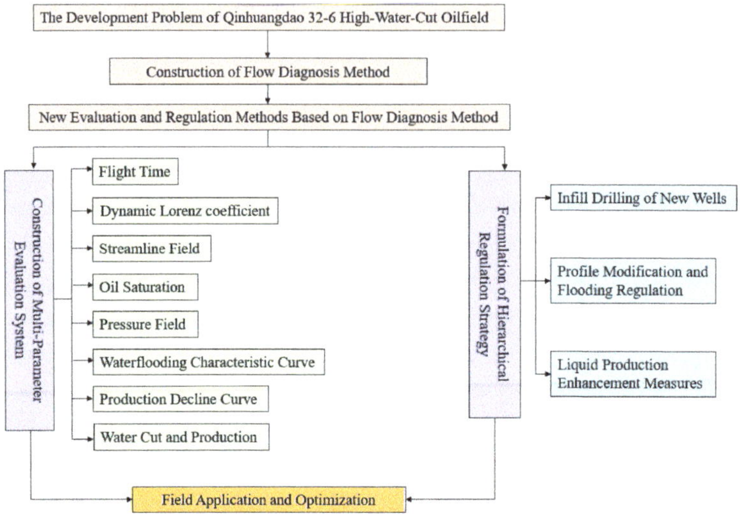

Figure 2. Research workflow diagram.

2.1. Flow Diagnostics Method

To tackle issues in ultra-high water-cut reservoirs such as high water cut, declining oil production, and complex remaining oil distribution, flow diagnostics introduce time of flight (TOF) and dynamic Lorenz coefficient for precise waterflooding performance evaluation and adjustment optimization, providing scientific basis for development. The flow diagnostics model by Shahvali et al. [22] is based on incompressible fluid steady-state flow theory. For single-phase flow, the continuity and motion equations are:

$$\nabla \cdot \vec{v} = \tilde{q}, \qquad \vec{v} = -\frac{K}{\mu}\nabla p \tag{1}$$

where ∇ is the divergence; $\vec{v}$ is velocity (m/s); $\tilde{q}$ is source/sink strength (positive for sinks, negative for sources); K is permeability (μm^2); μ is fluid viscosity (Pa·s); p is pressure (Pa). For waterflooding oilfields involving oil–water or oil–gas–water multiphase flow with significant mobility differences, the multiphase flow equation is:

$$\nabla \cdot \vec{v} = \tilde{q}_t, \qquad \vec{v} = -K\lambda_t \nabla p \tag{2}$$

where $\tilde{q}_t$ is total multiphase source/sink strength; λ_t is total mobility, calculated from relative permeability curves given saturation distribution. Equation (2) unknowns are pressure and velocity, termed the pressure equation. With boundary conditions, pressure distribution is solved via linear algebra, yielding fluid velocity. If multiphase unsteady flow simulations are available, more accurate pressure values can be used directly for velocity computation.

(1) TOF Equation

Pressure or velocity fields alone cannot fully describe actual fluid migration; visualization aids like streamline methods are needed to track fluids effectively, characterizing fluid migration paths from injectors to producers in reservoirs. Streamlines are curves tangent to fluid particle velocity vectors at a given time, non-intersecting, reflecting flow paths. Let x(r) be a streamline at a moment, with r as arc length; by definition:

$$\frac{d\vec{x}}{dr} \times \vec{v}\left(\vec{x}, \vec{t}\right) = 0 \tag{3}$$

For a streamline at a given moment, the rate of change of x with respect to the arc length r is equal to the component of the velocity vector in the x direction divided by the magnitude of the velocity vector.

This can also be written as:

$$\frac{d\vec{x}}{dr} = \frac{\vec{v}(\hat{t})}{\left|\vec{v}(\hat{t})\right|} \tag{4}$$

To further describe displacement intensity within different injection sweep ranges, the concept of TOF is adopted instead of arc length. The mathematical expression for TOF is:

$$\tau(r) = \int_0^r \frac{\phi\left(\vec{x}(s)\right)}{\left|\vec{v}\left(\vec{x}(s)\right)\right|} \tag{5}$$

where τ is the time (s) for a particle to travel a distance along the streamline; defined as interstitial velocity; ϕ is the distance coordinate along the streamline.

Computing TOF via the above requires solving the velocity field, tracing streamlines $\vec{v}\left(\vec{x}(s)\right)$, and numerical integration along them. This study establishes a TOF equation based on finite integration, directly solved to avoid cumbersome streamline tracing and integration. Differentiating Equation (5) using calculus and directional derivatives yields:

$$\frac{d\tau}{dr} = \frac{\phi}{\left|\vec{v}\right|} = \frac{\vec{v}}{\left|\vec{v}\right|} \cdot \nabla\tau \tag{6}$$

Considering both injectors and producers, forward and backward TOF equations are defined.

The forward TOF equation is:

$$\vec{v} \cdot \nabla\tau_f = \phi, \qquad \tau_f\Big|_{inflow} = 0 \tag{7}$$

This describes the time for a particle to travel from the nearest source or inflow boundary along a streamline to a point. $\tau_f\Big|_{inflow}$ is the time at the nearest source (starting point). The backward TOF equation is:

$$-\vec{v} \cdot \nabla\tau_b = \phi, \qquad \tau_b|_{outflow} = 0 \tag{8}$$

$\tau_b|_{outflow}$ is the time to the nearest sink or outflow. The sum of forward and backward TOF reflects the time for a particle from source along the streamline to sink. TOF values are determined by static properties like porosity and permeability, and dynamic attributes like injection–production intensity and fluid mobility. Compared to pressure and velocity fields,

TOF comprehensively characterizes seepage resistance and dominant flow directions in water-injection reservoirs, evaluating injection sweep range, displacement intensity, and waterflooding efficiency. High-TOF regions often indicate poor properties or imperfect well patterns, potentially areas difficult for injected fluids to sweep.

(2) Tracer Concentration Equation

The time-of-flight (TOF) method only reflects fluid migration velocity, making it challenging to characterize injected fluid distribution and inter-well connectivity in reservoirs. Tracer simulation adds compatible tracers at injectors to track movement trajectories, assessing fluid allocation and inter-well connectivity. To delineate well control regions, tracers label well positions, establishing tracer equations along fixed streamlines, neglecting adsorption and diffusion. As development time increases, tracer sweep expands, precisely characterizing fluid distribution and dynamic properties, providing key basis for reservoir management. The tracer concentration equation mirrors conventional mass transfer equations:

$$\frac{\partial(\phi C)}{\partial t} + \nabla\cdot\left(\vec{v}C\right) = \tilde{q}C \tag{9}$$

where C is tracer concentration (mg/L); ∇ is the gradient operator; $\vec{v}$ is velocity (m/s); $\tilde{q}$ is source/sink strength (positive for sinks, negative for sources).

Consider a simple tracer simulation experiment: set tracer concentration at one injector to 1, others to 0; neglecting diffusion, tracer concentration reflects regions swept by injected fluid. At steady state, tracer concentration distribution is solved by:

$$\vec{v}\cdot\nabla C = 0 \tag{10}$$

where ∇C is the tracer concentration gradient from injection to production. Similar to TOF equations, injector and producer tracer equations are defined.

The injector tracer equation is:

$$\vec{v}\cdot\nabla C_i^k = 0, \qquad C_i^k\Big|_{inflow} = 1 \tag{11}$$

In Equation (11), ∇C_i^k represents concentration gradient for a specific injector or inflow boundary point, describing the extent of influence from that source or boundary on a reservoir point.

The producer tracer equation is:

$$-\vec{v}\cdot\nabla C_p^k = 0, \qquad C_p^k\Big|_{outflow} = 1 \tag{12}$$

In Equation (12), $C_p^k\Big|_{outflow}$ represents a producer or outflow boundary point, describing the extent of influence from that sink or boundary on a reservoir point. After solving tracer concentration equations for injection and production wells, concentrations C label injection–production pairs, enabling delineation of pairs, computation of control volumes, fluid allocation coefficients, etc.

Table 2 provides a more detailed explanation of the aforementioned equations.

Table 2. Key Equations in Flow Diagnostics: Functional Objectives and Intuitive Explanations.

Brief Description	Functional Objective	Role in Decision-Making in Reservoir Development
Continuity equation for single-phase flow	Ensures mass conservation in steady-state incompressible flow, deriving baseline velocity fields.	Supports initial sweep efficiency evaluation for well pattern screening
Multiphase pressure equation	Solves pressure and velocity in oil–water systems, accounting for mobility contrasts from relative permeability.	Identifies flow paths; aids conformance control (TOF < 50 days).
Time-of-flight (TOF) along streamline	Computes particle travel time based on interstitial velocity and streamline distance.	Quantifies intensity; flags unswept areas for infill (TOF > 200 days).
Differentiated TOF equation	Derives partial differential form for direct numerical solution without streamline tracing.	Accelerates tortuosity assessment for flow field planning.
Forward TOF equation	Measures transit time from nearest injector to grid block along streamline.	Defines sweep volume; guides modification for channeling.
Backward TOF equation	Measures transit time from grid block to nearest producer.	Defines drainage; supports lift optimization (TOF > 200 days).
Tracer concentration equation	Models advective transport of tracers along velocity fields, neglecting diffusion.	Tracks allocation; quantifies heterogeneity for oil targeting.
Steady-state tracer equation	Solves equilibrium concentration distribution from injectors.	Maps swept regions; informs connectivity adjustments.
Injector tracer equation	Computes concentration gradients from specific injectors.	Quantifies control volumes; balances injection allocation.
Producer tracer equation	Computes concentration gradients to specific producers.	Quantifies drainage; enables communication volume for shutoff.

2.2. Quantitative Diagnostic Evaluation System for Water Injection Development Performance

Conventional waterflood evaluation systems exhibit significant limitations. First, traditional reservoir engineering methods, which rely on metrics such as water cut and production decline rates, effectively capture the performance of individual wells or well groups but fail to quantitatively delineate injection–production correspondence or subsurface fluid migration pathways. Second, while numerical simulation can visualize saturation and velocity fields, it often lacks the sensitivity required to quantify changes in sweep efficiency or displacement intensity, particularly in the ultra-high water-cut stage.

To overcome these deficiencies, this study establishes a quantitative diagnostic evaluation system that serves as a robust supplement to existing methods. This system is founded on principles derived from time-of-flight (TOF) calculations and tracer analysis and is structured around four core indicators:

(1) Time-of-Flight (TOF): Quantitatively describes the injection sweep range, defines displacement fronts, and assesses displacement intensity.
(2) Dynamic Lorenz coefficient: Characterizes the uniformity of displacement and the heterogeneity of flow distribution within an injection–production pattern.
(3) Injection–Production Communication Volume: Measures the degree of inter-well connectivity and the effective sweep efficiency of the injected fluids.
(4) Inter-well Fluid Allocation Factor: Assesses the dynamic correspondence and fluid transfer architecture between injectors and producers.

The calculation and application of each indicator are detailed in the subsequent sections.

(a) Time-of-Flight (TOF)

Time-of-flight is computed by solving the TOF equation, yielding three distinct metrics: forward TOF, backward TOF, and total TOF. Forward TOF represents the transit time from the nearest injection source to a specific grid block, thereby characterizing the injection sweep volume. Conversely, backward TOF measures the time from a grid block to the nearest production sink, defining the drainage volume. The sum of these two, the total TOF, reflects the overall hydraulic resistance, or tortuosity, of the flow path between an injector–producer pair.

The spatial distribution of TOF effectively delineates high-velocity flow paths from stagnant, unswept regions. Areas with low τ values indicate rapid fluid transit and high displacement intensity, often corresponding to preferential flow channels. In contrast, regions with high τ values signify areas that remain poorly swept under the current operational strategy, thereby identifying them as priority targets for subsequent interventions, such as infill drilling or conformance control.

(b) Dynamic Heterogeneity

Reservoir heterogeneity, particularly in permeability, is a primary factor governing fluid flow and connectivity. Static measures of heterogeneity, such as the Lorenz coefficient or Dykstra–Parsons coefficient, have been widely used but often exhibit a weak correlation with ultimate recovery. This is because static parameters like permeability and porosity do not account for the dynamic interactions between geological heterogeneity, the evolving flow field, and the specific well pattern. Consequently, assessments based solely on static data may be unreliable for predicting dynamic reservoir performance.

As a dynamic indicator, TOF provides a more insightful characterization of multiphase flow by inherently capturing the interplay between reservoir heterogeneity and the flow field under defined operational and boundary conditions. This offers a superior description compared to relying on pressure or velocity fields alone. Static heterogeneity governs the distribution of storage capacity and intrinsic fluid mobility, whereas dynamic heterogeneity reflects variations in the effective flow path lengths and communication structures that emerge during production. By integrating TOF-based analysis with static heterogeneity evaluation, we construct a dynamic heterogeneity evaluation system. This provides a more precise and physically grounded basis for analyzing the reservoir flow field and optimizing development strategies.

(1) F-Φ Diagnostic Curve

Assume the reservoir consists of N independent flow tubes, each with volume V_i and flow rate q_i; total TOF is the fluid transit time or residence time in the tube, $\tau_i = V_i/q_i$. Sort N flow tubes by ascending residence time, $\tau_1 \leq \tau_2 \leq \ldots \leq \tau_N$. Assuming piston-like displacement in each tube, define normalized storage capacity Φ_i and flow capacity F_i as:

$$\text{storage capacity}: \ \Phi_i = \sum\nolimits_{j=1}^{i} V_j / \sum\nolimits_{j=1}^{N} V_j \tag{13}$$

$$\text{flow capacity}: \ F_i = \sum\nolimits_{j=1}^{i} q_i / \sum\nolimits_{j=1}^{N} q_j \tag{14}$$

Thus, Φ_i is the ratio of flow tube volumes with breakthrough at τ_i to total volume, and F_i is the corresponding fractional flow, i.e., produced injected fluid ratio to total. The F-Φ relationship curve resembles the fractional flow curve f-Sw, reflecting displacement uniformity. Figure 3 shows typical F-Φ curves, with the light blue solid line for uniform displacement and dotted line for nonuniform.

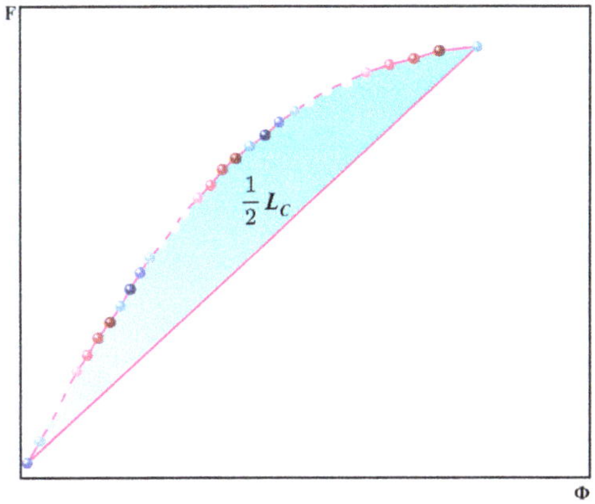

Figure 3. Typical *F-Φ* curves.

(2) Dynamic Lorenz Coefficient

From storage and flow capacities, the dynamic Lorenz coefficient is further defined as:

$$L_c = 2\int_0^1 (F(\Phi) - \Phi)d\Phi \tag{15}$$

When considering remaining oil distribution, it is modified to:

$$L_c = 2\int_0^1 (F(\Phi) - \Phi)S_o d\Phi \tag{16}$$

The dynamic Lorenz coefficient quantitatively indicates reservoir displacement heterogeneity, as the envelope area between the *F-Φ* curve and the 45-degree line. Higher *Lc* indicates more uneven displacement; for fully uniform displacement, *Lc* = 0. By computing dynamic Lorenz coefficients for the reservoir, layers, and well groups, waterflooding efficiency can be measured hierarchically, identifying remaining oil potential tapping targets.

(3) Injection–Production Communication Volume

From tracer equations, tracer concentration distributions for each well are solved, labeling reservoir regions to determine injection–production control ranges.

$$\text{Injection Well Sweep Volume } V_i = \sum_k V_k^i = \sum_k c_k^i V_k^{GC} \tag{17}$$

$$\text{Production Well Drainage Volume } V_j = \sum_k V_k^j = \sum_k c_k^j V_k^{GC} \tag{18}$$

$$\text{Injection–Production Communication Volume } V_{ij} = \sum_k V_k^{ij} = \sum_k c_k^i c_k^j V_k^{GC} \tag{19}$$

where c_k^i is injector i tracer concentration distribution; c_k^j is producer j tracer concentration distribution; k is grid index; V_k^{GC} is grid pore volume.

The injection–production communication volume calculated from tracer concentration equations reflects the ultimate achievable communication state under current well patterns and production conditions. Combined with TOF between injection–production pairs, it effectively guides well pattern optimization and provides scientific basis for assessing remaining oil potential tapping.

(4) Inter-Well Fluid Allocation Factor

Similar to communication volume, tracer concentration distributions also label inter-well fluid allocation.

Injection allocation factor:

$$a^{i}_{nm}[w^{n}_{k}] = q^{i}[w^{n}_{k}]c^{p}_{m}[w^{n}_{k}] \tag{20}$$

Production allocation factor:

$$a^{p}_{nm}[w^{m}_{k}] = q^{p}[w^{m}_{k}]c^{i}_{n}[w^{m}_{k}] \tag{21}$$

where c^{i}_{n} is the injector n tracer concentration (mg/L); c^{p}_{m} is the producer m tracer concentration (mg/L); $w^{n(m)}_{k}$ is the perforation grids for well n or m; q is the injection–production rate.

The selection of the four core indicators—time-of-flight (TOF), dynamic Lorenz coefficient, injection–production communication volume, and inter-well fluid allocation factor—is based on their collective ability to precisely quantify waterflood displacement processes, flow field heterogeneity, and dynamic inter-well relationships. Each metric provides a unique diagnostic perspective: TOF characterizes hydraulic resistance by identifying high- and low-permeability pathways; the dynamic Lorenz coefficient quantifies flow field non-uniformity; communication volume assesses inter-well connectivity; and the allocation factor details the fluid distribution between injector–producer pairs. Together, these indicators form a comprehensive evaluation system that integrates dynamic and static data across multiple scales to precisely analyze waterflood performance in high water-cut reservoirs.

The comparative strengths of these advanced diagnostics versus traditional methods are illustrated in Figure 4. The radar charts evaluate various metrics against key performance criteria. Figure 4a demonstrates the complementary nature of the flow diagnostic indicators. For instance, TOF excels at evaluating displacement intensity and mapping oil–water movement, while the dynamic Lorenz coefficient and oil saturation provide superior quantification of dynamic heterogeneity and material balance. The streamline field is unparalleled in characterizing displacement direction. This visualization highlights how the indicators collectively provide a holistic assessment. In contrast, Figure 4b shows the limitations of conventional tools. While metrics like waterflooding characteristic curves and pressure fields can effectively describe oil–water movement, they, along with production decline curves, are less effective at quantifying material basis, characterizing displacement direction, and assessing dynamic heterogeneity. Although simpler to implement, their diagnostic precision is significantly lower. This analysis validates the superiority of the proposed flow diagnostics system, emphasizing its ability to provide a more accurate basis for identifying remaining oil potential and guiding targeted development adjustments.

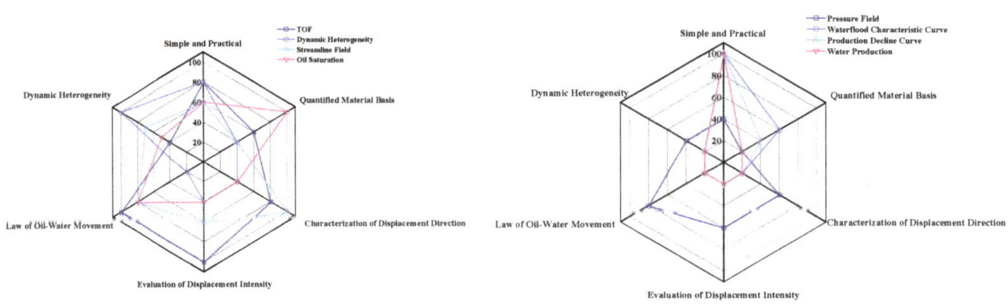

(a) (b)

Figure 4. Evaluation system based on flow diagnostics. (**a**) Dynamic Indicator Evaluation System; (**b**) Static Indicator Evaluation System.

These visualizations, complemented by field-validated thresholds (e.g., dynamic Lorenz >0.5 for conformance control), demonstrate the method's enhanced sensitivity over traditional tools, as evidenced in the QHD32-6 adjustments detailed in Section 3.

2.3. Hierarchical Adjustment Strategies for Water Injection Development

Building upon the diagnostic evaluation, a hierarchical adjustment strategy was developed to provide a systematic management framework, as depicted in the workflow in Figure 5. This decision-making process begins with a primary screening for material basis, where a region is considered a candidate for intervention only if its water-cut ratio is below 45% and remaining oil thickness exceeds 6 m. For qualified regions, the strategy then bifurcates based on a TOF threshold of 200 days. Areas identified as "Strongly Water-flooded" (TOF < 200 days) are further assessed for dynamic heterogeneity; if high (Lorenz coefficient > 0.5), profile modification and water plugging are prescribed to mitigate channeling, otherwise, no immediate action is needed. Conversely, for "Weakly Water-flooded" or "Immovable" regions (TOF > 200 days), which hold significant remaining oil potential, interventions are similarly guided by heterogeneity. Low-heterogeneity zones are targeted for artificial lift enhancement or infill drilling, while high-heterogeneity zones require conformance control or well pattern adjustments to improve sweep efficiency and access unswept reserves. This structured approach ensures that specific interventions are precisely matched to the diagnosed reservoir conditions.

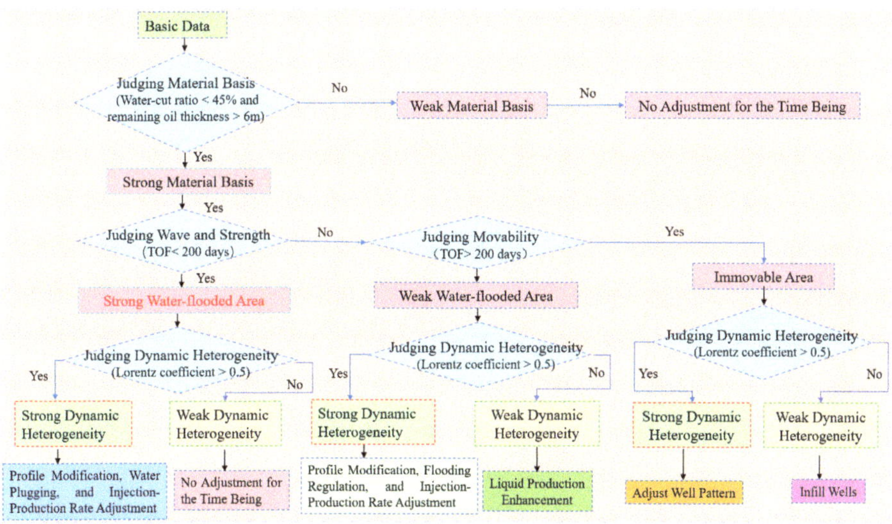

Figure 5. Adjustment strategies for different levels of channeling paths.

2.4. Limitations and Uncertainty Considerations

The flow diagnostics method relies on inputs like permeability (k) and porosity (φ), which affect total mobility in Equation (2) and TOF computations (Equations (5)–(8)). Uncertainties from core/log data ($\pm$10–20% for k in heterogeneous reservoirs) may alter TOF, potentially misidentifying sweep thresholds (e.g., TOF <50 days for channeling). Similarly, Lorenz coefficient (Equation (15)) sensitivities to φ variations could overstate heterogeneity. Tracer Equations (9)–(12) assume steady-state, limiting transient accuracy.

Mitigation in this study used QHD32-6 history-matched simulations to calibrate inputs, aligning diagnostics with field trends. Future work could include $\pm$15% perturbation analyses for threshold robustness. This qualitative approach complements the method's visualization strengths and field validations.

3. Block Application and Practice

3.1. Block Overview

The QHD32-6 oilfield, the subject of this study, is a fluvial heavy oil reservoir located in the central Bohai Sea (Figure 6, blue dashed circle). Situated on a low-amplitude anticline structure over a paleo-uplift, the reservoir is characterized by medium porosity (32%), high permeability (2300 mD), and a crude oil viscosity range of 22–260 mPa·s. The field is subject to an active edge-bottom water drive and features a loose sandstone structure prone to sand production. Currently, it is in an ultra-high water-cut, mature development stage, with an overall water cut of 94% and a cumulative recovery factor of 18.4%. Although early operational adjustments effectively improved waterflooding performance, the current well pattern has led to a complex remaining oil distribution. This complexity, compounded by inaccurate performance characterization, has exacerbated the effects of reservoir heterogeneity and resulted in inefficient drainage of remaining reserves.

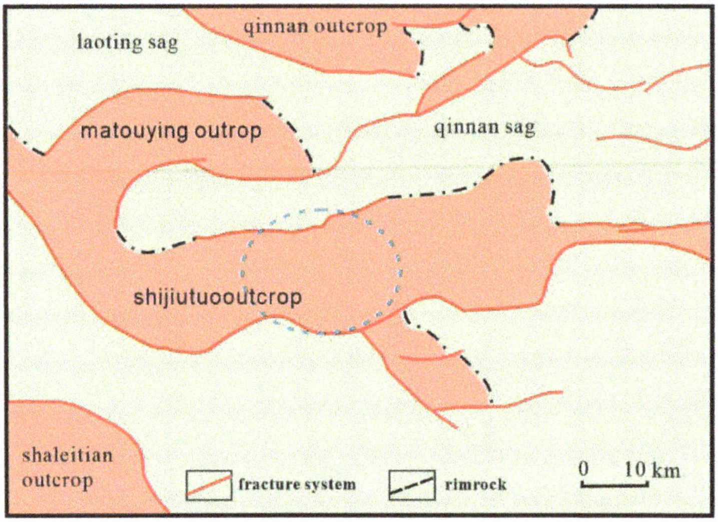

Figure 6. Regional location map.

3.2. Water Injection Development Performance Evaluation in Target Block

An evaluation of the B14 well group using flow diagnostics illustrates the system's ability to identify preferential flow paths. As shown in Figure 7, streamline analysis reveals that injected water from the B14 well flows towards eight surrounding producers, with significantly stronger flow intensity directed at wells A10 and A22. This qualitative observation is quantitatively confirmed by the dynamic Lorenz coefficients (Figure 8), which were calculated to be 0.6679 for A10 and 0.7840 for A22. These high coefficient values are indicative of severe channeling and strong planar heterogeneity, where injected fluid preferentially sweeps along high-permeability pathways. This phenomenon leads to poor sweep efficiency in other parts of the pattern and contributes to an accelerated production decline.

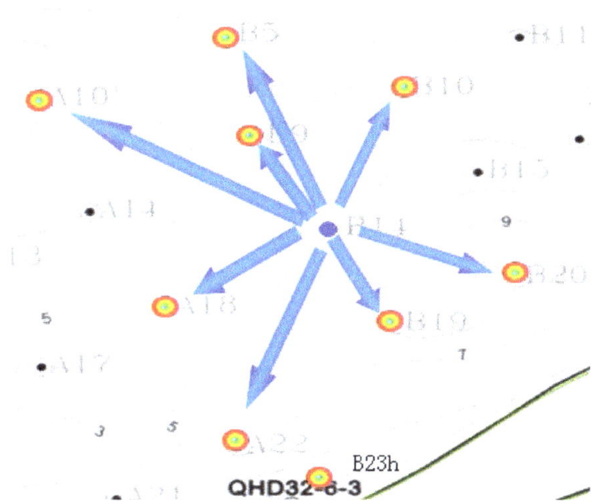

Figure 7. Waterflooding direction and intensity in B14 well group.

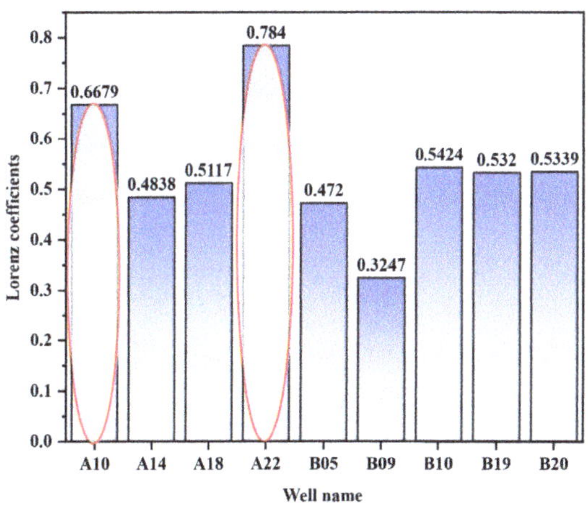

Figure 8. Dynamic heterogeneity in B14 well group before conformance control.

A comparative analysis of two horizontal infill wells, H08H and H09H, drilled in January 2014, further highlights the utility of flow diagnostics in explaining production disparities. Despite their similar locations and initial remaining oil saturations, their five-year cumulative production profiles diverged significantly (Figure 9). Well H09H exhibited robust production, reaching 9.2 m^3, whereas H08H produced only 2.5 m^3 over the same period—a difference of 268%. The diagnostic evaluation provides a clear explanation for this performance variance. The time-of-flight (TOF) distribution map (Figure 10) shows that H09H was drilled in a low-resistance area (blue, TOF $\approx$ 75 days), indicating favorable hydraulic connectivity. In contrast, H08H is situated in a high-resistance region (red, TOF $\approx$ 300 days), characteristic of poor reservoir connectivity. This disparity in local flow dynamics is the primary cause of the observed production difference. Further validation is provided by the water saturation and velocity fields (Figure 11), which show that the area surrounding the underperforming H08H well is characterized by high water saturation and low fluid velocity. This indicates that the region contains largely bypassed oil that the infill well is unable to effectively mobilize due to the poor local flow properties.

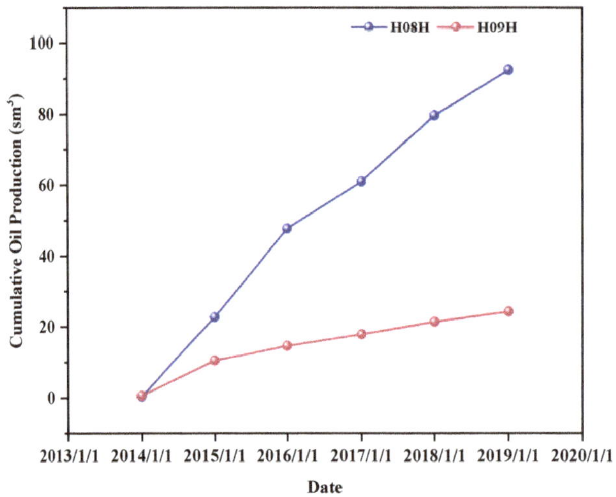

Figure 9. Cumulative production comparison of two infill wells.

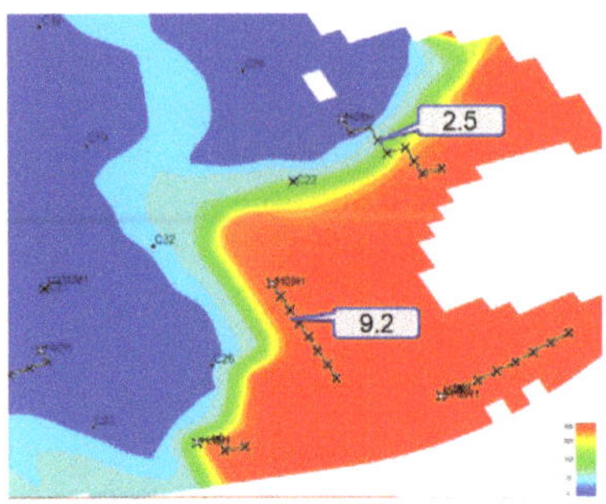

Figure 10. Time-of-Flight Distribution Map Around Wells in Layers 1–3 of the Southern Block.

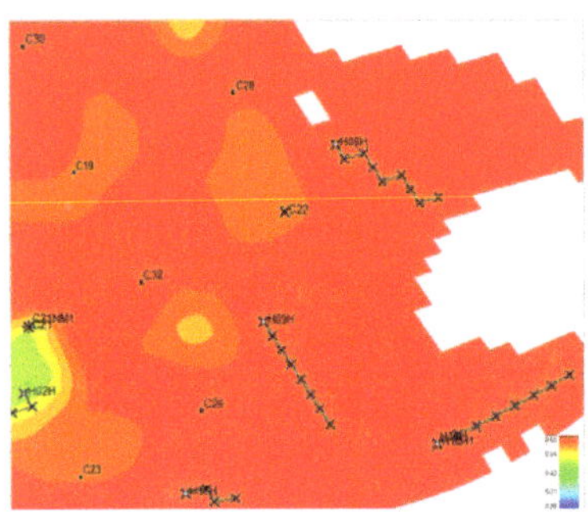

Figure 11. Oil Saturation Distribution Map Around Wells in Layers 1–3 of the Southern Block.

Table 1 underscores the proposed method's advantages in diagnostic precision and scalability, with empirical support from field applications in Section 3, such as the B14 well group's heterogeneity evaluation (Figure 8) and production gains in the C29 (165,600 m^3) and B03 (2.2–3.8 tons/day increase) groups.

3.3. *Adjustment Thresholds*

3.3.1. Infill Drilling Thresholds

To establish quantitative thresholds for infill drilling, a statistical analysis was conducted on 105 infill wells drilled in the study area prior to 2017 (Figure 12). This analysis correlated pre-infill reservoir parameters—waterflooded ratio, remaining oil thickness, and TOF—with post-infill cumulative oil production. The analysis demonstrated a clear inverse correlation between the pre-infill waterflooded ratio and production (Figure 13); wells in areas with a ratio below 45% yielded significantly higher production. Furthermore, a positive linear relationship was observed between remaining oil thickness and cumulative production (Figure 14), with a notable increase in performance when the thickness exceeded 6 m. Finally, production exhibited an exponential relationship with pre-infill TOF (Figure 15), confirming that infill wells are most effective in high-resistance, poorly connected regions (TOF > 200 days).

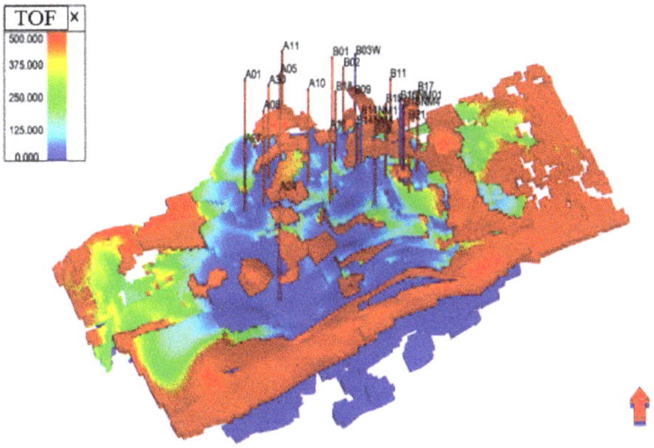

Figure 12. Southern block time-of-flight map.

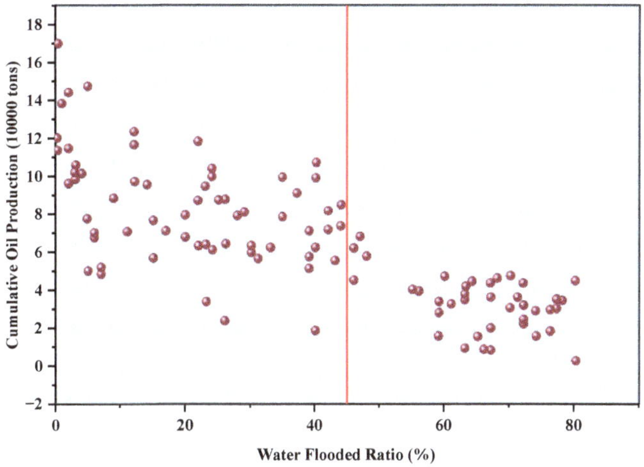

Figure 13. Correlation between cumulative oil and waterflooded ratio.

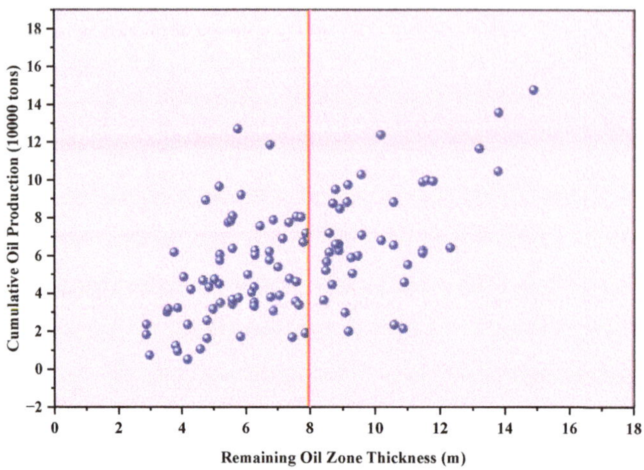

Figure 14. Correlation between cumulative oil and remaining oil thickness.

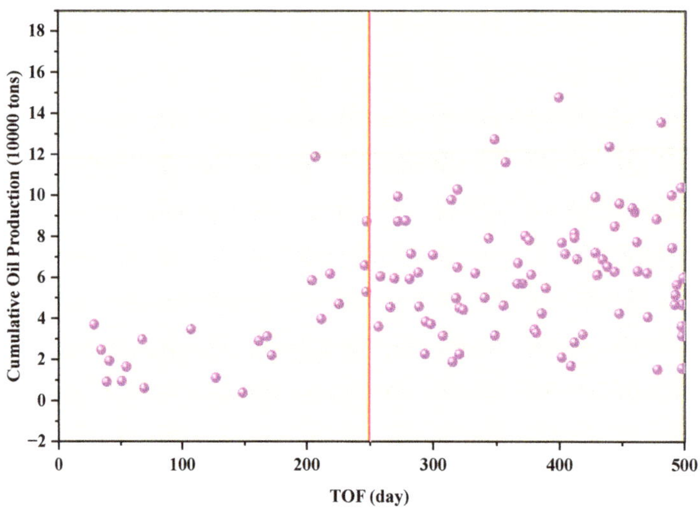

Figure 15. Correlation between cumulative oil and time of flight.

The direct impact of infill drilling on reservoir dynamics is illustrated in Figures 16 and 17. The pre-infill TOF distribution (Figure 16) was characterized by extensive high-TOF regions (>500 days), reflecting inefficient waterflooding. Post-infill, the TOF in these areas was significantly reduced, indicating that the new wells successfully optimized the flow field and improved sweep efficiency(the red dashed circles). Similarly, the pre-infill oil saturation map (Figure 17, the black dashed circles) showed an uneven distribution, which became more uniform following infill drilling as the new wells mobilized previously bypassed oil.

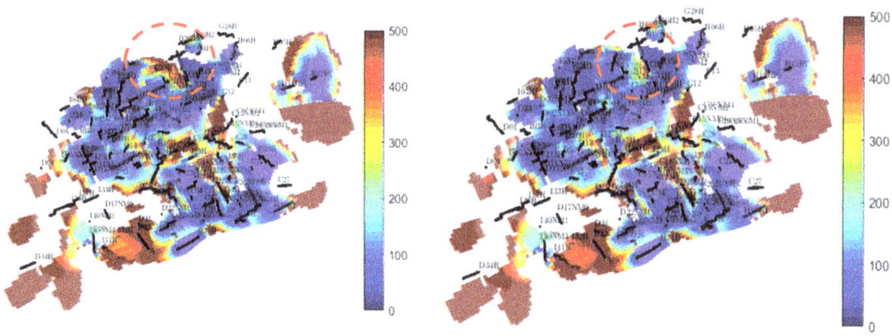

Figure 16. Southern block layers 1–3 initial well positions pre- and post-infill time of flight.

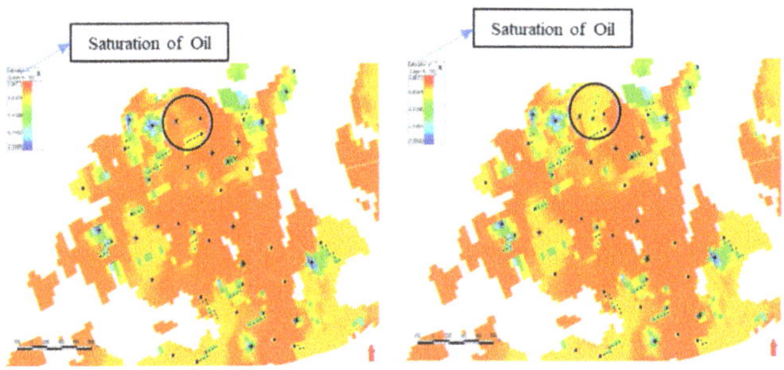

Figure 17. Southern block layers 1~3 well positions pre- and post-infill oil saturation distribution.

Synthesizing these findings, this study proposes a set of comprehensive screening criteria for identifying optimal infill drilling locations. The established thresholds are: a waterflooded ratio <45%, a remaining oil thickness >6 m, and a time-of-flight >200 days. Adherence to these criteria ensures that infill wells are strategically placed in areas with high remaining oil potential and inefficient existing sweep patterns. Field applications have confirmed that wells meeting these standards consistently deliver significant production gains, providing a scientific basis for refined field management.

3.3.2. Conformance Control and Displacement Adjustment Thresholds

To define quantitative criteria for conformance control, dynamic data from 25 well groups subjected to tracer injection were analyzed. This investigation revealed a direct correlation between the dynamic Lorenz coefficient and the incidence of channeling. As depicted in Figure 18, a sharp increase in the number of channeling wells occurs when the Lorenz coefficient surpasses 0.5, establishing this value as a robust indicator of severe inter-well heterogeneity and the formation of preferential flow paths.

Furthermore, the efficacy of conformance treatments was found to be contingent upon the degree of reservoir heterogeneity. A comparative analysis of the D08 and C03 well groups exemplifies this principle (Table 3). The D08 group, which presented with a high initial Lorenz coefficient, exhibited a marked improvement in sweep uniformity post-treatment; this signifies the successful diversion of injected fluid and the mobilization of previously bypassed oil (Figure 19). In contrast, the C03 group, characterized by low heterogeneity and a TOF below 50 days, showed a negligible response to the intervention, with minimal change in oil saturation (Figure 20). This disparity underscores that conformance control is most impactful in highly heterogeneous settings and offers limited benefit in relatively homogeneous reservoirs lacking well-defined thief zones.

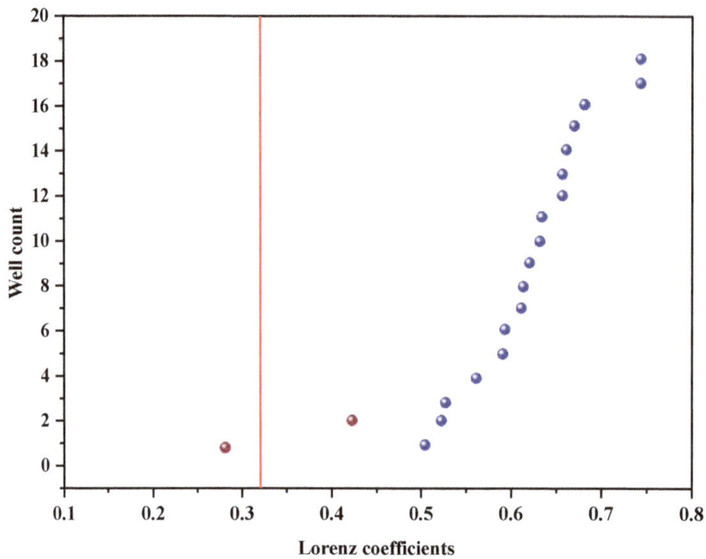

Figure 18. Lorenz coefficients for different wells.

Table 3. Conformance control and displacement adjustment results for tracer well numbers.

Intervention Well Number	Treatment Well Number	Distance to Water Well (m)	Treatment Duration (d)	Treatment Rate (m/d)	TOF	Maximum Lorenz Coefficient	Average Lorenz Coefficient	Lorenz Coefficient Surge Factor
D08Well Group (NmI3Segment) /2020.9.2	D07	437	10	43.78	14	0.6621	0.62	1.068
	D13	311	62	5.02	58			
	D27M	346	66	5.24	78			
	D28H	257	41	6.26	30			
	C04	352.1	105	3.35	139			
C03Well Group (NmISegment) /2019.7.18	H03H	400	32	12.5	32	0.6692	0.71	0.943
	C02	370	38	9.7	33			
	H19H	250	47	5.3	54			
	B26HS	270	37	7.3	26			

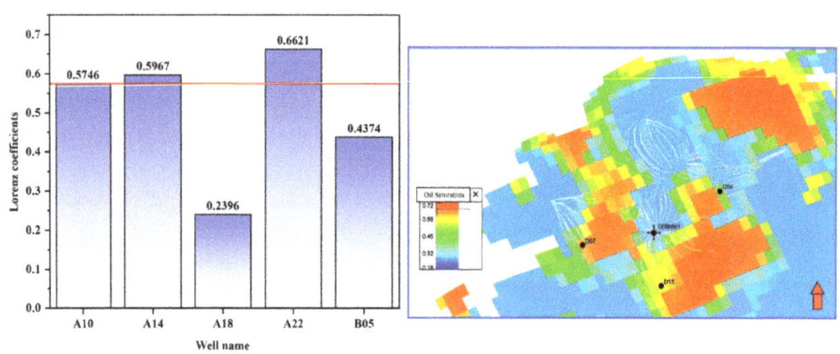

Figure 19. Lorenz Coefficient and Oil Saturation Distribution Map for the D08 Well Group.

Consequently, based on this empirical evidence, the following thresholds were established for identifying candidate well groups for conformance control: a time-of-flight (TOF) of less than 50 days combined with a dynamic Lorenz coefficient greater than 0.5. Applying interventions to well groups that meet these criteria enables the targeted sealing of dominant flow paths, which in turn redirects injection fluids into unswept, lower-permeability zones to substantially improve overall sweep efficiency.

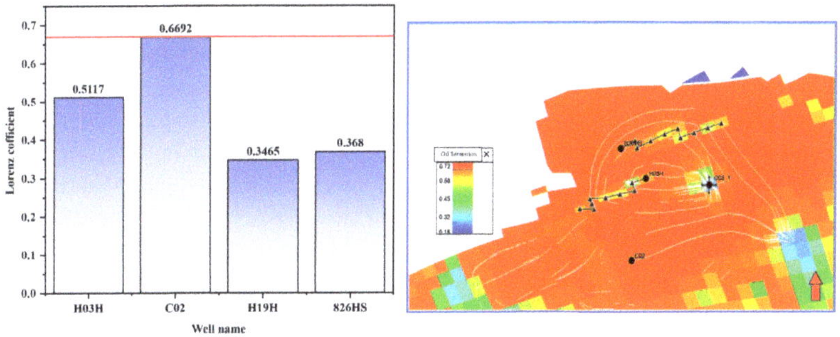

Figure 20. Lorenz Coefficient and Oil Saturation Distribution Map for the C03 Well Group.

3.3.3. Liquid Lifting Thresholds

To define criteria for optimizing production through liquid lifting, historical data from measures implemented before 2023 were analyzed (Table 4). The analysis showed a strong positive correlation between post-treatment production gains and two key parameters: the remaining oil thickness ratio and the TOF. A significant "oil increase without water increase" effect was observed in wells where the TOF was greater than 200 days, yielding a success rate of 87%—a substantial improvement over the sub-40% rate achieved with traditional methods. Moreover, the most substantial production gains occurred when the remaining oil thickness ratio exceeded two-thirds. By combining these parameters, a refined screening methodology was developed that increased the success rate for liquid lifting to 78% across 23 interventions.

Table 4. Statistics on producer liquid lifting measures.

Production Well	Water Zone Thickness (m)	Remaining Oil Zone Thickness (m)	Reservoir Thickness Ratio (%)	Pre-Intervention Production Rate (m^3)	Pre-Intervention Water Cut (%)	Post-Intervention Production Rate (m^3)	Post-Intervention Water Cut (%)	TOF
A01	20	86.96	20	90	95	83	75	264
A05	63	75.90	14	93	45	93	31	273
A10	43	66.15	20	94	40	95	20	273
B09	78	89.66	20	95	50	95	30	322
B15	76	91.57	23	76	40	83	17	304
B23H	10	83.33	36	71	75	70	39	350
G29H	20	100.00	15	56	45	68	30	270
G34H	8	72.73	22	95	38	95	16	292
C01	55	66.27	21.8	89	52.7	92	30.8	275
C24	37	75.51	7.6	88	27.0	94	19.4	261
C30	41	71.93	9.8	91	30.7	92	21.0	214
D07	57	76.00	49.3	86	77.5	87	28.2	329
D28H	3	50.00	29.7	70	124.4	78	94.7	253
H02H	3	100.00	14.4	96	39.6	94	25.2	287
H03H	12	100.00	47.6	70	70.0	84	22.4	323
H04H	10	100.00	22.1	95	44.8	96	22.6	343
H09H	7	100.00	9.6	94	27.3	95	17.7	211
H10H	7	100.00	26.4	75	54.6	81	28.2	283
H11H	14	100.00	18.1	94	42.2	94	24.0	268
H17H	5	100.00	30.1	76	53.8	84	23.7	240
I09H	8	66.67	27.4	83	110.9	77	83.5	343
I09H1	18	81.82	20.3	97	38.5	97	18.2	237
I28H	13	59.09	21.8	86	42.4	89	20.6	247
I33H	14	66.67	24.2	95	58.9	92	34.7	277
I36H	4	66.67	25.2	86	56.0	89	30.8	252
B26HS	4	50.00	12.3	93	30.7	95	18.4	257

Accordingly, the final thresholds for implementing liquid lifting measures were established as: a remaining oil thickness ratio >2/3 and a time-of-flight (TOF) >200 days.

3.4. Adjustment Strategies and Schemes

The practical application of this diagnostic framework is demonstrated through two case examples. First, an analysis of the southern block identified a large, poorly swept area adjacent to the C29 well, even after liquid lifting measures were applied (Figure 21). This region, characterized by high remaining oil potential, was prioritized as a target for infill drilling, leading to the planned deployment of a new horizontal well to enhance local recovery (Figures 22 and 23 show the flooded thickness and oil reservoir thickness of Layers 1–3 in the Southern Block, Within the area enclosed by the black coil is the hard-to-produce portion of this region, where significant remaining oil is present.).

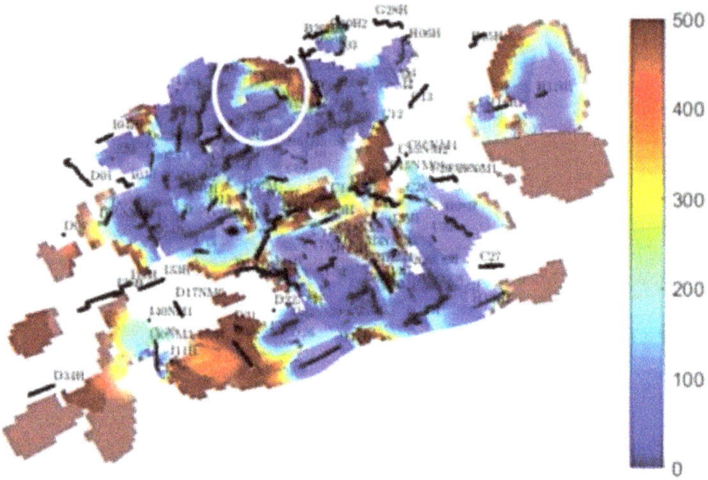

Figure 21. Southern block layers 1–3 pre-infill flow diagnostics results.

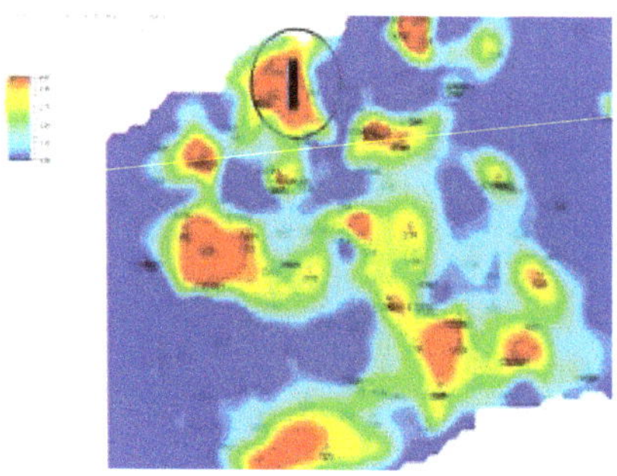

Figure 22. Waterflooded Thickness Map for Layers 1–3 of the Southern Block.

As a further illustration, a detailed diagnostic evaluation was performed on a well group including producers F21, F22, F25, and F26. The analysis revealed high remaining oil saturation between wells F21, F26, and F05 (Figure 24), which coincided with a zone of low permeability (Figure 25). While static permeability data alone might suggest this oil is immobile, the dynamic diagnostics revealed a different mechanism. The dynamic Lorenz coefficient for well F26 was high (>0.5), indicating that active channeling was bypassing the oil within the lower-permeability matrix (Figure 26). A cross-sectional view confirmed the presence of a thick, insufficiently swept remaining oil column between wells F25 and F22 (Figure 27). This integrated analysis, which combined static and dynamic indicators, correctly identified the primary production issue as channeling rather than merely poor reservoir quality (Figure 28). Consequently, a conformance control treatment was recommended to optimize the flow field and mobilize the substantial remaining oil in this area.

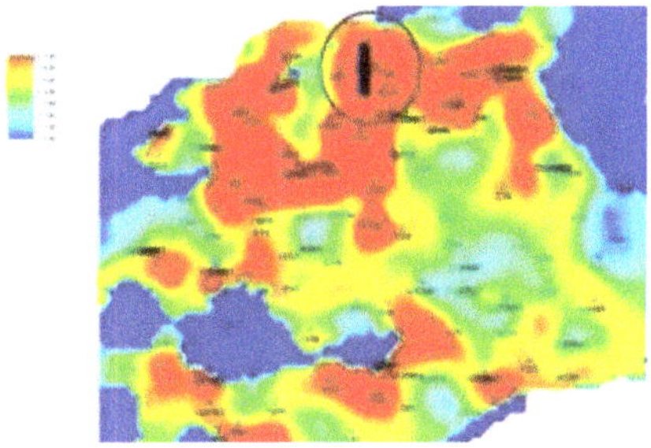

Figure 23. Oil Layer Thickness Map for Layers 1–3 of the Southern Block.

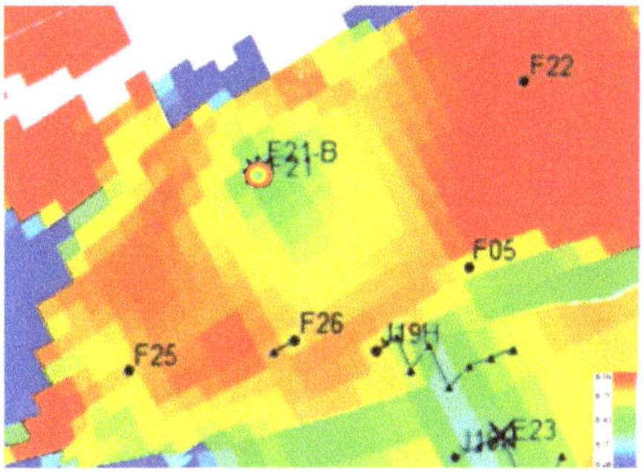

Figure 24. Near-well remaining oil.

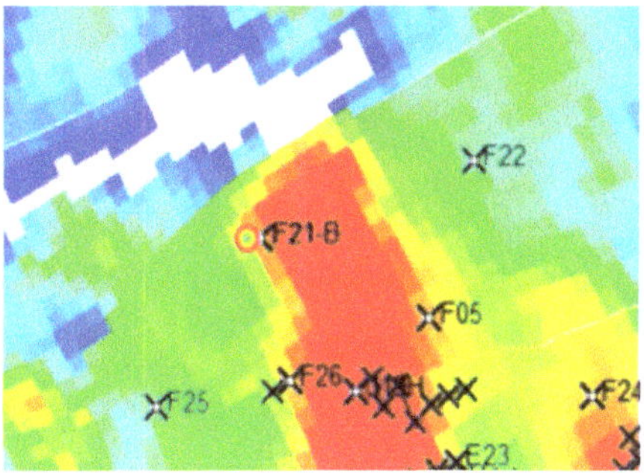

Figure 25. Near-well permeability.

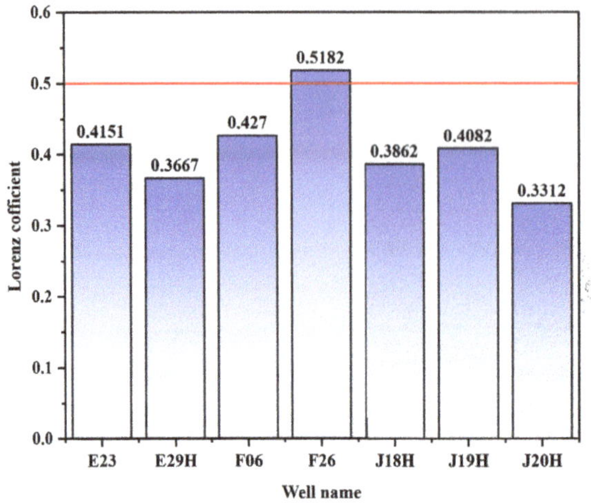

Figure 26. Dynamic Lorenz Coefficient Map.

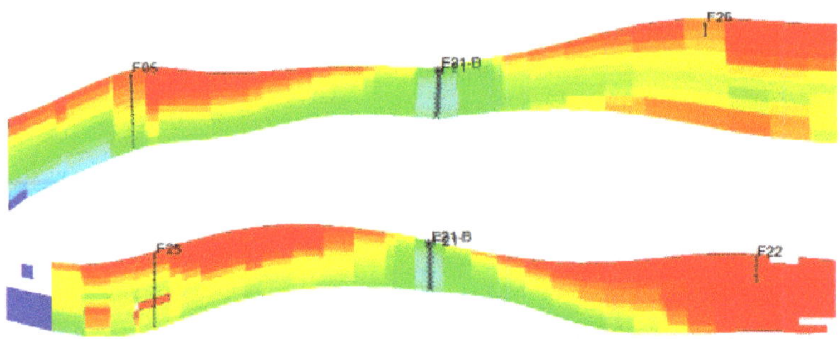

Figure 27. Remaining oil profile.

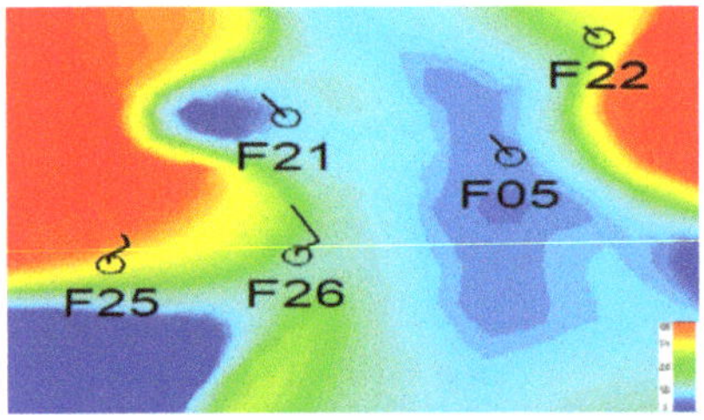

Figure 28. Time-of-Flight Distribution Map of Well F21 and Its Surrounding Wells.

Based on measure effect assessment and numerical simulation, propose 31 well-time measures: 13 liquid lifting, 8 conformance and displacement, 2 layered injection, 6 water shutoff, 2 cyclic injection groups. See Figure 29.

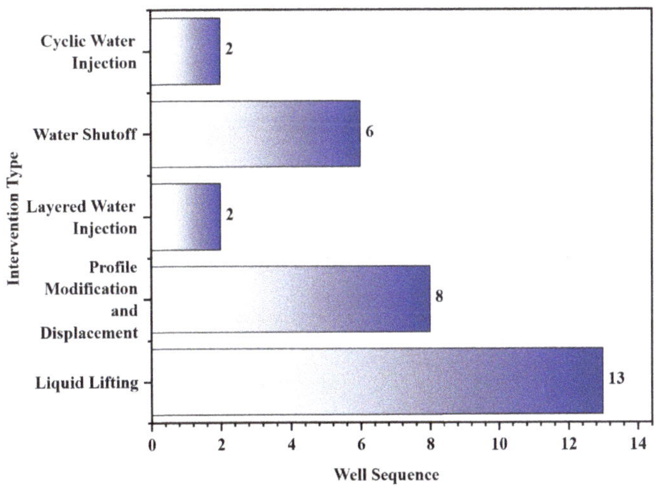

Figure 29. Block measure adjustment suggestions.

3.5. Adjustment Effects

Field results confirmed the efficacy of the proposed adjustment strategies. In a pilot test, a synergistic operation was implemented, coupling water injection in well D18H1 (Nm03 layer) with coordinated liquid lifting in the adjacent well H07H. This combined intervention successfully mobilized previously bypassed inter-well oil, resulting in a sustained incremental production of approximately 7 tons/day. Subsequent diagnostic analysis confirmed the physical mechanism behind this success: the flow field had been significantly optimized (Figures 30 and 31). The waterflooded ratio in the target area became more uniform, falling below the 45% threshold, while the TOF increased to over 200 days. This indicates that the injected fluid was successfully diverted, enhancing sweep into lower-permeability zones and mitigating the risk of channeling.

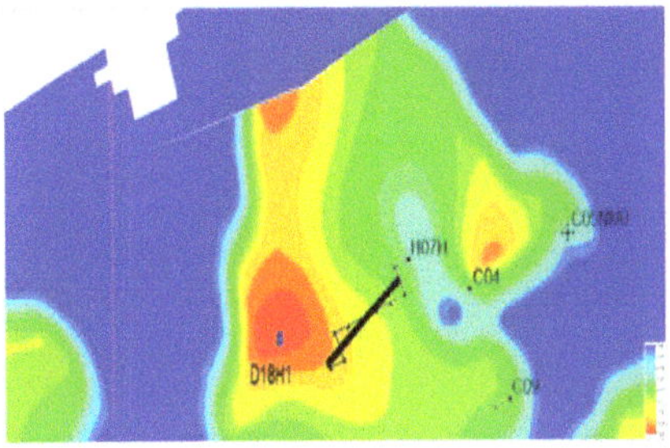

Figure 30. Waterflooded Ratio Map.

The diagnostic thresholds for conformance control were validated in the A31NM1 well group. The area exhibited a TOF of less than 50 days and a dynamic Lorenz coefficient of 0.6438 (well above the 0.5 threshold), confirming the presence of strong heterogeneity and dominant flow paths suitable for treatment (Figures 32 and 33). The subsequent success of the conformance treatment in the analogous G31H1 well demonstrated the validity of these criteria. Post-treatment, the well showed a significant and sustained increase in oil production (Figure 34) and a corresponding stabilization of the water cut at a lower level (Figure 35), indicating that the treatment successfully redirected injected fluid to previously unswept oil zones.

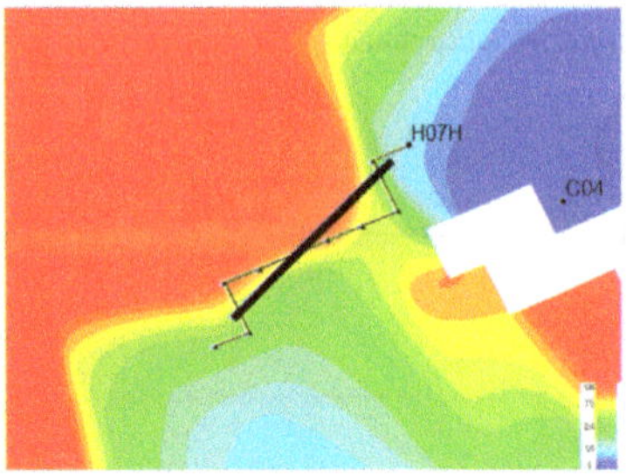

Figure 31. Time-of-Flight Distribution Map.

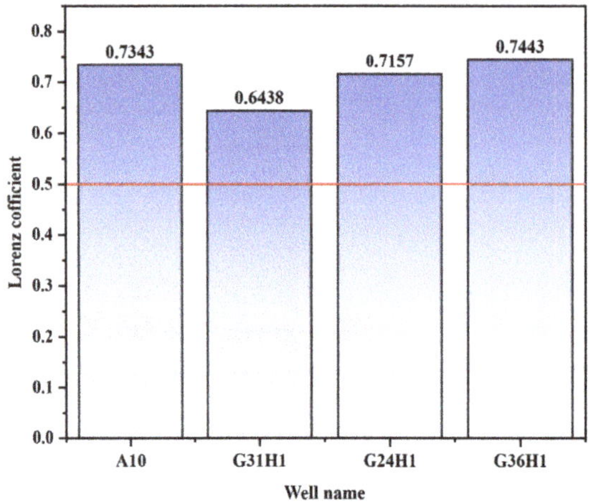

Figure 32. A31NM1 well group Lorenz coefficient map.

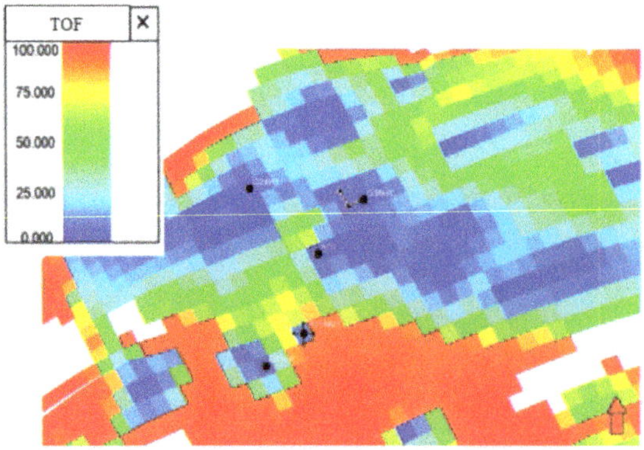

Figure 33. A31NM1 well group time-of-flight map.

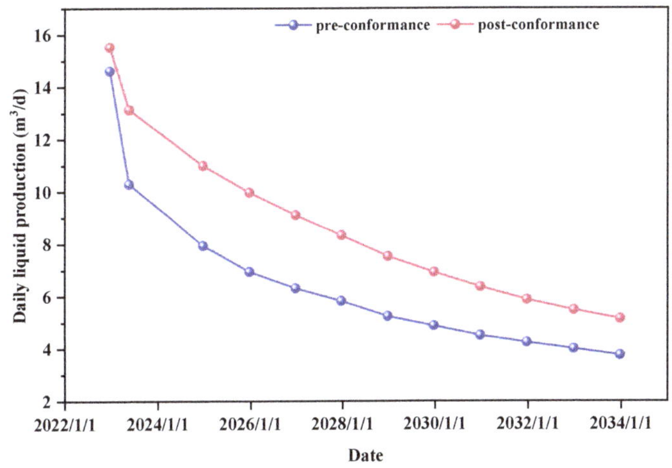

Figure 34. G31H1 pre- and post-conformance daily liquid production.

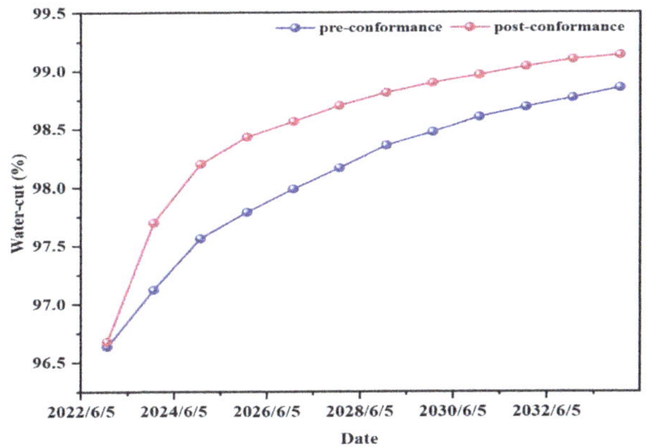

Figure 35. G31H1 pre- and post-conformance daily water cut.

Figures 34 and 35 show G31H1 pre- and post-conformance daily oil and water cut changes. Figure 34 post-conformance significant oil rise, upward trend, sealing dominant paths diverted to low-permeability, utilizing prior unswept oil. Figure 35 post water cut decrease, stable curve, effective channeling suppression. Validates dynamic Lorenz >0.5 and TOF <50 days thresholds optimize uniformity, enhance efficiency.

A31NM1 and G31H1 via conformance achieved marked oil increase, effective water cut reduction, reflecting sealing optimized distribution. Dynamic Lorenz >0.5 and TOF <50 days thresholds engineering-validated.

Collectively, these field results validate the diagnostic thresholds and adjustment strategies proposed in this study. The application of this methodology has enabled the successful implementation of the world's smallest offshore infill spacing (100–150 m) and has significantly improved the field's performance. As shown in Figure 36, the average oil production rate in the QHD oilfield increased from 0.95 m^3/d to 1.6 m^3/d. The ultimate recovery factor is now projected to reach 26.63%, demonstrating that stable, efficient production can be achieved and sustained even in the ultra-high water-cut stage of a complex heavy oil reservoir.

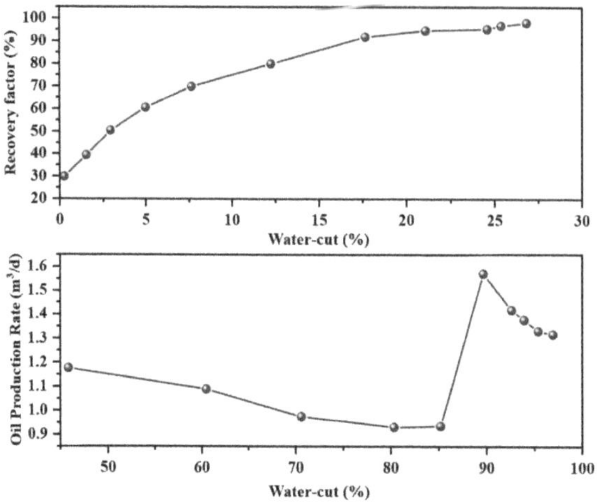

Figure 36. QHD water cut, oil production rate, and recovery degree effects.

Based on QHD water injection adjustment strategies, 15 field well-time injection–production adjustments were carried out: 8 lifting, 4 conformance, and 3 shutoff, with a success rate of 93.3%. Field implementation is carried out as shown in Table 5:

Table 5. Field adjustment measure implementation table.

Intervention Recommendations	Implementation Well Group	Implementation Time	Implementation Effectiveness	
			Monthly Incremental Oil Production (10,000 m^3)	Cumulative Incremental Oil Production (up to October 2024)
Liquid Lifting	G34H	22 January 2024	0.0	0.0153
	G44H	2024 August 22	0.0606	0.1324
	G46H	27 August 2024	0.0409	0.0741
	G37H	19 September 2024	0.0164	0.0164
	G47H	13 August 2024	0.0192	0.1088
	D13	28 May 2024	0.0228	0.1175
	I16H	17 March 2024	0.0032	0.0119
	D28H	3 June 2024	0.0398	0.2757
Profile Modification and Displacement	C03	20 August 2024	0.1104	0.6973
	C05	22 August 2024	0.0737	0.5523
	C06	20 August 2024	0.1161	0.7248
	C29	20 August 2024	0.1100	0.6993
Water Shutoff	B17	1 March 2024	0.012	0.0956
	B21	7 March 2024	0.0094	0.0412
	I18H	30 March 2024	0.0032	0.0119

4. Conclusions

This study established and applied a new flow diagnostics methodology to enhance oil recovery in the challenging QHD32-6 heavy oilfield. We developed a multi-parameter evaluation system, using time-of-flight (TOF) and the dynamic Lorenz coefficient, to precisely map the distribution of remaining oil. This system provided quantitative, data-driven thresholds to guide targeted interventions, including infill drilling and conformance control. Field results directly confirmed the method's effectiveness, leading to significant production gains and validating its practical value.

The specific quantitative thresholds derived in this work are, however, tailored to the unique geological and fluid properties of the QHD32-6 field. Consequently, while the diagnostic framework itself is robust, applying these exact numerical criteria to other reservoirs will require field-specific recalibration. The scalability of the methodology across geologically diverse formations warrants further investigation.

In conclusion, this work delivers a validated, practical framework for the refined management of mature heavy oilfields. Its application in QHD32-6 demonstrates a targeted approach crucial for unlocking remaining oil in complex, high water-cut reservoirs. Future work will focus on adapting and testing this methodology in a wider range of fields to establish its broader utility for improving oil recovery.

Author Contributions: Conceptualization, Y.H.; Methodology, L.W.; Software, Y.G.; Validation, Y.H. and Y.G.; Formal analysis, Y.H.; Resources, S.W.; Data curation, Y.H.; Writing—original draft, Y.H.; Writing—review and editing, L.W., L.J., S.W., S.N. and Z.K.; Visualization, S.N.; Supervision, L.W., L.J. and Z.K.; Project administration, S.W. All authors have read and agreed to the published version of the manuscript.

Funding: This research was funded by the National Natural Science Foundation of China (Grant Nos. 52374051 and U24B2037), and the project "Key Technologies Research on Deep Potential Tapping in Ultra-High Water Cut Stage of Bohai Water-Drive Heavy Oil Fields" (Grant No. KJZH-2024-2204).

Data Availability Statement: The original contributions presented in this study are included in the article. Further inquiries can be directed to the corresponding author.

Conflicts of Interest: Authors Yifan He and Yishan Guo were employed by the company CNOOC (China) Co., Ltd., Tianjin Branch. The remaining authors declare that the research was conducted in the absence of any commercial or financial relationships that could be construed as a potential conflict of interest. The funder provided financial support solely and did not participate in the scientific research process (e.g., study design, data collection/analysis, manuscript preparation, or publication decisions).

References

1. Yuan, S.; Han, H.; Wang, H.; Luo, J.; Wang, Q.; Lei, Z.; Xi, C.; Li, J. Research progress and potential of new enhanced oil recovery methods in oilfield development. *Pet. Explor. Dev.* **2024**, *51*, 963–980. [CrossRef]
2. Feng, G.; Zhou, Y.; Yao, W.; Liu, L.; Feng, Z.; Yi, Y.; Feng, Y. Countermeasures to Decrease Water Cut and Increase Oil Recovery from High Water Cut, Narrow-Channel Reservoirs in Bohai Sea. *Geofluids* **2021**, *2021*, 6671098. [CrossRef]
3. Temizel, C.; Canbaz, C.H.; Tran, M.; Abdelfatah, E.; Jia, B.; Putra, D.; Irani, M.; Alkouh, A. A comprehensive review heavy oil reservoirs, latest techniques, discoveries, technologies and applications in the oil and gas industry. In Proceedings of the SPE International Heavy Oil Conference and Exhibition, Kuwait City, Kuwait, 10–12 December 2018.
4. Liu, J.; Liu, S.; Gao, P.; Zhang, Q.; Liu, Y.; Chen, X.; Zhong, L.; Lin, Y. Efficient development strategies for heavy oil reservoirs with a focus on rational utilization of water resources. *Sep. Purif. Technol.* **2025**, *362*, 131943. [CrossRef]
5. Yu, H.; Wang, Y.; Zhang, L.; Zhang, Q.; Guo, Z.; Wang, B.; Sun, T. Remaining oil distribution characteristics in an oil reservoir with ultra-high water-cut. *Energy Geosci.* **2024**, *5*, 100116. [CrossRef]
6. Ma, K.; Li, A.; Guo, S.; Pang, J.; Xue, Y.; Zhou, Z. Techniques for improving the water-flooding of oil fields during the high water-cut stage. *Oil Gas Sci. Technol. Rev. d'IFP Energ. Nouv.* **2019**, *74*, 69. [CrossRef]
7. Shi, H.; Liu, Y.; Jiang, L.; Zheng, J.; Gan, L. Integrated Solutions for the Redevelopment Plan of Offshore Mature-Oilfield. In Proceedings of the International Petroleum Technology Conference, Virtual, 23 March–2 April 2021.
8. Liu, H. Challenges and development direction for deep development of high water-cut mature oilfields. *Pet. Geol. Oilfield Deve. Daqing* **2024**, *43*, 15–24. (In Chinese)
9. Arps, J.J. Analysis of decline curves. *Trans. AIME* **1945**, *160*, 228–247. [CrossRef]

10. Miao, F.; Zhang, H.; Zhang, Y. New method for evaluating waterflooding effect by decline rate index. *Fault-Block Oil Gas Field* **2015**, *22*, 353–355. (In Chinese)
11. Miguel, P.R.; Joanna, Z.; Stefano, Z.; Alejandro, S.; Eduardo, T. Fluid dynamics analysis and performance of polymer flow regulators for polymer flooding in multilayered reservoirs. *J. Pet. Sci. Eng.* **2022**, *208*, 109680. [CrossRef]
12. Wang, Q.; Ma, X.; Bi, Y.; He, H.; Gu, X.; Liu, T.; Zhao, L. Analysis of Channeling-Path Phenomena in a Complex Fault-Block Reservoir with Low Recovery Factor and High Water-Cut Ratio. *Fluid Dyn. Amp; Mater. Process.* **2022**, *19*, 361–377.
13. Zhao, P.; He, S.; Cai, M.; Tao, Z.; Zhao, M.; Wu, X.; Ni, T.; Guo, Q.; Wei, P.; Wang, Q.; et al. Streamline simulation based vector flow field characterization and reconstruction method for high water cut reservoir. In Proceedings of the International Petroleum Technology Conference, Shanghai, China, 26–28 August 2020.
14. Guan, C.; Zhang, J.; Li, X. Quantitative Research on Development Status of Heterogeneous Reservoirs in Offshore Oilfields. In Proceedings of the International Petroleum Technology Conference, Riyadh, Saudi Arabia, 21–23 February 2022.
15. Li, L.; Yuan, X.; Dai, D.; Yan, C.; Ping, F.; Wang, W. Evaluation of Water Injection Development Effect in Block B of the Bohai Sea. *Chem. Eng. Manag.* **2024**, *7*, 152–156. (In Chinese)
16. Ning, L.; Yang, L.; Zheng, X.; Zhang, J.; Liu, Y.; Ma, J. Evaluation on injection-production connectivity of low-permeability reservoirs based on tracer monitoring and numerical simulation. *Xinjiang Pet. Geol.* **2021**, *42*, 735.
17. Kulkarni, K.N.; Datta-Gupta, A.; Vasco, D. A Streamline Approach for Integrating Transient Pressure Data into High-Resolution Reservoir Models. *SPE J.* **2001**, *6*, 273–282. [CrossRef]
18. Wen, H.; Sun, N.; Liu, Y. Index system evaluating water flooding development effect of oilfield at ultra-high water cut stage. *J. Pet. Explor. Prod. Technol.* **2017**, *7*, 111–123. [CrossRef]
19. Chu, Q.; Chen, X.; Shi, P. Developed effects analyses and quantitative evaluating method for QinJiaTun Complex fault-block oilfield. *Pet. Geol. Oilfield Dev. Daqing* **2016**, *35*, 58–63. (In Chinese)
20. Li, Z.; Guo, Q.; Zhuang, T.; He, S. Evaluation method of producer development effect in high water-cut reservoirs. *Pet. Geol. Recovery Effic.* **2022**, *29*, 85–93. (In Chinese)
21. Wang, C.; Sun, Z.; Sun, Q.; Zhang, L.; Zhang, X. Comprehensive evaluation of waterflooding front in low-permeability reservoir. *Energy Sci. Eng.* **2021**, *9*, 1394–1408. [CrossRef]
22. Shahvali, M.; Mallison, B.; Wei, K.; Gross, H. An Alternative to Streamlines for Flow Diagnostics on Structured and Unstructured Grids. *SPE J.* **2012**, *17.3*, 768–778. [CrossRef]

energies

MDPI

Review

Review of Formation Mechanisms, Localization Methods, and Enhanced Oil Recovery Technologies for Residual Oil in Terrigenous Reservoirs

Inzir Raupov, Mikhail Rogachev and Egor Shevaldin *

Department of Oil and Gas Fields Development and Operation, Empress Catherine II Saint Petersburg Mining University, 2, 21st Line, 199106 St. Petersburg, Russia; inzirrr@yandex.ru (I.R.); rogatchev@mail.ru (M.R.)
* Correspondence: sss_sheva2@mail.ru

Abstract: Residual oil (RO) in terrigenous reservoirs formed after waterflooding can exceed 60% of the original oil in place; approximately 70% is trapped at the macro-scale in barriers and lenses, whereas about 30% remains at the micro-scale as film and capillary-held oil. This review aims to synthesize current knowledge of RO formation mechanisms, localization methods and chemical recovery technologies. It analyzes laboratory, numerical and field studies published from 1970 to 2025. The physical and technological factors governing RO distribution are systematized, and the effects of heterogeneities of various types, imperfections in pressure-maintenance (waterflood) systems and contrasts in oil–water properties are demonstrated. Instrumental monitoring techniques—vertical seismic profiling (VSP), well logging (WL), hydrodynamic well testing (WT) and geochemical well testing (GWT)—are discussed alongside indirect analytical approaches such as retrospective production-data analysis and neural-network forecasting. Industrial experience from more than 30,000 selective permeability-reduction operations, which have yielded over 50 Mt of additional oil, is consolidated. The advantages of gel systems of different chemistries are evaluated, and the prospects of employing waste products from agro-industrial, metallurgical and petroleum sectors as reagents are considered. The findings indicate that integrating multi-level neural-network techniques with instrumental monitoring and adaptive selection of chemical formulations is crucial for maximizing RO recovery.

Keywords: residual oil (RO); terrigenous reservoirs; reservoir heterogeneity; localization of reserves; polymer gel systems; physico-chemical enhanced oil recovery (EOR); optical well logging; neural-network forecasting

1. Introduction

The development of oil fields inevitably leads to the formation of residual recoverable reserves (RRR). According to [1], the design oil recovery factor (ORF) amounts to 0.375 frac. units; consequently, up to 62.5% of unrecovered reserves remain in the subsurface after the end of production, which, on a national scale, amounts to billions of tons. The low ORF observed in Russian oil fields is a result of insufficient implementation of enhanced oil recovery methods, the application of which requires additional capital investment [2,3].

The largest portion of Russia's oil reserves is concentrated in the West Siberian, Volga-Ural, and Timan-Pechora petroleum provinces (PPs), with the majority of their unique fields currently at the third and fourth stages of development. The final stage of development is characterized by a sharp increase in water cut, a consistently high decline in oil production rates, the decommissioning of a large number of wells, a drop in reservoir pressure, and an increase in the volume of hard-to-recover reserves [4,5].

The above-mentioned petroleum provinces are predominantly represented by terrigenous reservoirs composed of sandstones and siltstones. The oil- and gas-bearing formations are characterized by complex geological structure, heterogeneity in filtration and storage properties, and wide areal distribution [6–8]. Most fields are developed using waterflooding systems to involve a larger portion of reserves in production and to increase the ultimate oil recovery factor [9]. A common phenomenon during waterflooding is the uncontrolled increase in water cut due to water breakthrough along more permeable zones, which leads to the formation of undrained areas. These zones are of particular interest and require localization using modern approaches in order to implement the most effective geological and technical measures (GTMs) [10–12].

This topic has been addressed in the works of leading authors within the global scientific and technical community, such as Surguchev M.L. [13,14], Zheltov Yu.V. [14], Gazizov A.A. [15], Gazizov A.Sh. [16], Devlikamov V.V. [17], Kabirov M.M. [17], Khabibullin Z.A. [17], Craig F.F. [18], Willhite G.P. [19], Rose S.C. [20], Buckwalter J.F. [20], Woodhall R.J. [20], Lake L.W. [21], and others.

2. Classification and Formation of Residual Oil

Within the framework of this review, terrigenous reservoirs of continental origin are considered, where the processes of residual oil formation are most pronounced due to the high lithological and structural heterogeneity. There are various forms and quantitative distributions of residual reserves in oil reservoirs. As noted by the author, the causes may include a range of natural and technological factors.

The general concept of residual oil formation can be represented by three main categories of causes: geological, physico-chemical, and technological. Geological causes include reservoir heterogeneity in terms of filtration and storage properties, the presence of impermeable clay barriers, lack of lithological continuity, and lithological-facies heterogeneity. Physico-chemical causes involve capillary phenomena, interfacial tension at the oil–water boundary, hydrocarbon (HC) adsorption on the surface of rock grains, and end effects in reservoir pores [22]. Technological causes include the characteristics of the selected waterflooding system [20], the efficiency of the applied injection pattern, the distance between injection and production wells, the injection rate, and the resulting pressure gradient [23].

Based on the aforementioned causes, several types of residual recoverable reserves (RRR) are identified:

1. Dispersed oil, which includes capillary-trapped and film oil [14,24];
2. Oil remaining in interlayer low-permeability zones not affected by waterflooding;
3. Oil located in lenses isolated from the productive reservoir by impermeable barriers or baffles and not penetrated by wells;
4. Oil migrating into dead-end zones of homogeneous reservoirs.

According to the data presented in [25], the first type accounts for approximately 30%, the second—27%, the third—24%, and the fourth—19% of the total RRR volume.

Thus, after the primary development phase of the field is completed, two levels of residual recoverable reserves are formed. Residual oil at the microscale is represented by dispersed oil, which in turn consists of capillary-trapped oil (oil globules in water) and film oil (films wetting the surface of the solid rock). Macroscale residual oil, represented by lenses and pillars, accounts for approximately 70% of the total RRR volume, which largely determines the priority of its extraction among the various types of residual reserves.

2.1. Formation of RRR at the Microscale

The presence of dispersed oil in the reservoir is determined by two main factors: rock wettability and the geometry of the reservoir pore space [26,27]. Wettability is a property defined by the interaction between two immiscible phases—water and oil—with a solid surface and a liquid. Wettability is characterized by the contact angle, which indicates the hydrophilic or hydrophobic nature of the rock. A hydrophilic rock is coated with a water phase; in this case, the surface is said to be "wetted." Water is present as a hydrophilic film on the grain surfaces and as bound water located in dead-end pores. Hydrophobic rocks, on the contrary, are coated with oil; however, such rocks are unlikely to exist in practice, as it is generally accepted that all reservoir rocks formed in water-rich environments or were submerged shortly after sediment deposition. Most reservoirs are composed of rocks with mixed wettability.

The authors of [28], using experimental methods with X-ray computed tomography (CT scanner) and a core flooding system, established that residual oil is distributed within the capillary zones of the pore space: in hydrophilic reservoirs it remains in film form, whereas in hydrophobic ones it is retained in isolated droplets. In addition to the results of the study, it should be noted that, despite the high methodological rigor and the use of a modern experimental system (including X-ray computed tomography and controlled water saturation under specified pressure), the paper does not indicate compliance with any industry standards (such as [29–32]) that regulate the conditions of filtration experiments. Parameters such as core dimensions (diameter ~2.5 cm, length 3–5 cm), filtration rate (0.01 mL/min), and the viscosity of the working fluids (white oil with 2 mPa·s and NaI solution) were selected based on laboratory practice and are typical for well-controlled core-flooding studies. However, the lack of standardized reference points may complicate direct comparison of the results with other works, particularly when assessing residual oil saturation. As shown in specialized guidelines, the core size and filtration rate can significantly affect the degree of edge capillary effects and the structure of the displacement front, which, in turn, influences the quantitative value of residual oil. Therefore, when interpreting such results, it is advisable to consider applicable standards and recommendations for the unification of experimental conditions, which is especially important for comparative analysis or attempts to upscale the findings to reservoir conditions.

The geometry of pores also directly influences the formation of dispersed oil. Reservoir porosity is the result of sediment deposition and diagenesis. Depending on the pore size, the degree of pore connectivity, and the sorting of grains, the amount of dispersed oil varies. According to [33], experimental studies have shown that the geometry of the pore space affects the distribution of residual oil by increasing its retention in medium and small pores. In addition, the presence of clays can form dead-end zones and retain capillary-trapped oil. Experimental studies using SEM and XRD methods, conducted on low-permeability sandstone samples from the Daqing Oilfield (China), demonstrated that residual oil is retained in capillary traps formed by clay minerals, which leads to a reduction in permeability after water injection [34].

The distribution of dispersed oil in the reservoir is governed by three types of forces: capillary, viscous, and gravitational. In this case, capillary forces—which are also referred to as surface forces—are the dominant ones, as their magnitude can reach several MPa depending on the wettability of the rock and the size of the pore channels. Hydrodynamic resistance forces are negligible, as they are proportional to the filtration velocity. Gravitational forces exert a constant effect ranging from 0.1 to 10 kPa, which is caused by the density difference between the saturating fluids. Microscopic analysis of pore structure in [35] showed that in reservoirs with high heterogeneity, residual oil is distributed in the form of films on grain surfaces and droplets in pore throats, indicating the significant role of capillary forces in oil displacement processes.

In fact, dispersed oil is regarded as immobile due to the high-pressure gradients generated by capillary forces. It should be noted that such oil is formed as a result of water flooding, which means it can be considered to be 'at the threshold of mobility' [13,14].

According to the study [36], after polymer flooding in sandstone reservoirs with varying permeability, residual oil is distributed as follows: In high-permeability reservoirs (>1100 mD), 50–51% of the oil remains in a bound state, primarily in the form of adsorbed films on particle surfaces, while 37% corresponds to free oil. In medium-permeability reservoirs (470–490 mD), the distribution of free and bound oil is approximately equal (49% and 32–49%, respectively). In low-permeability reservoirs (<180 mD), bound oil dominates (47–49%), and free oil accounts for only 20–38%, which is attributed to the low mobility of oil in small pore channels. Thus, the permeability of the pore space significantly affects both the form and volume of residual oil in the reservoir.

2.2. Formation of RRR at the Macroscale

The formation of residual oil at the macroscale is caused by the incomplete displacement of oil from the reservoir, which is associated with a number of factors. Formally, the resulting oil-saturated pillars can be classified into categories according to their formation mechanism [18].

1. Oil-saturated pillars formed as a result of high heterogeneity of the reservoir in terms of geological and physical properties;
2. Oil-saturated pillars formed due to the manifestation of the viscoplastic properties of oil;
3. Oil-saturated pillars formed as a result of an imperfect well pattern or other technological factors.

The size and shape of such RRR formations in the subsurface may vary widely depending on the properties of the reservoir and the saturating fluids.

When designing a reservoir pressure maintenance (RPM) system, which is used in the development of most existing fields, an important stage is the assessment of the degree, nature, and direction of heterogeneity distribution within the production target [19], since this directly affects the completeness of oil displacement. Reservoir heterogeneity can take various forms, including:

1. The presence of multiple hydrocarbon-bearing interlayers composed of rocks with significantly different properties, such as sandstones interbedded with carbonates;
2. Variations in the continuity, interconnection, and areal distribution of porous and permeable layers within the reservoir;
3. The presence of a natural fracture system in the formation, resulting from regional tectonic stresses in the rock;
4. Directional development of permeability caused by depositional conditions or diagenetic alterations;
5. The presence of faults that affect the connectivity between parts of the reservoir and adjacent zones, either acting as barriers or enabling unrestricted fluid flow along the fault plane;
6. Interbedding of permeable and porous reservoir layers with nearly or completely impermeable layers of anhydrite, shale, or other minerals.

All of the above-mentioned types of heterogeneity, in one way or another, contribute to incomplete displacement, as high-permeability filtration channels are formed and mainly high-productivity zones are drained, while less permeable areas remain undeveloped. Another case leading to the formation of undrained zones containing RRR is the presence of crossflows between two reservoirs with different filtration and storage properties. In such situations, the displacing agent follows the path of least resistance, leaving large portions of the reservoir undrained. A significant difference in the thermobaric conditions of fluid occurrence, particularly reservoir pressure, leads to uneven depletion of layers within a single production target. The mechanisms of oil-saturated pillar formation due to heterogeneity are discussed in detail in the following studies [37,38].

Another major reason for the formation of oil-saturated pillars in the reservoir is the physico-chemical difference between the properties of oil and the injected water. To analyze this cause of stagnant zone formation, it is useful to consider the reservoir in terms of a 2D projection [39,40], that is, as a thin, isotropic horizontal plane. Studying oil and water displacement under such conditions allows for the exclusion of several factors, such as reservoir heterogeneity in thickness, geological and physical properties, and filtration–storage characteristics. In the general case, only the interaction of the two fluids from the standpoint of their physical and chemical behavior remains.

An example of a model representing this type of interaction is the so-called "Hele-Shaw cell." The cell consists of two superimposed plates made of plexiglass, with a gap between them. This gap is considered to be infinitesimally small compared to the thickness of the plates [41]. This assumption allows the flow within the cell to be treated as two-dimensional, which simplifies the analysis of the effect and eliminates other influencing factors. The flow observed in the Hele-Shaw cell is referred to as Hele-Shaw flow and is described by the following Equation (1) [42]:

$$\vec{u} = -\frac{b^2}{12 \cdot \mu} \nabla p, \quad (1)$$

where:
$\vec{u}$—flow velocity (averaged over the gap thickness);
b—gap between the plates;
μ—fluid viscosity;
p—pressure.

Since the fluid is considered incompressible, the following condition also applies (2):

$$\nabla \cdot \vec{u} = 0 \ \nabla^2 p = 0, \quad (2)$$

That is, the pressure satisfies the Laplace equation.

Oil displacement by water in a Hele-Shaw cell proceeds with the formation of breakthrough fingers [43], which, from a geometrical point of view, are referred to as fractals. The primary cause of fractal formation is the difference in viscosity between the displacing and displaced fluids, as well as the wettability and interfacial tension at the water–oil boundary [44,45].

In study [46], the Hele-Shaw cell is used to investigate the influence of the mobility ratio between oil and the displacing fluid on the front behavior and the level of residual oil. The authors show that in the case of high oil viscosity and low water viscosity, pronounced fingering develops, leading to uneven sweep and increased residual oil saturation. Conversely, increasing the viscosity of the displacing fluid (by adding polymer) stabilizes the displacement front. Visualization in the Hele-Shaw cell convincingly demonstrated that the difference in viscosities—and consequently in phase mobilities—is the key factor responsible for the formation of residual oil zones under conditions where other influencing factors are absent.

Similar conclusions are presented in study [47], where a Hele-Shaw cell was used to analyze the effect of the displacing fluid's viscosity on the oil displacement behavior. The authors demonstrated that increasing viscosity by adding cellulose nanoparticles leads to a more stable displacement front that maintains a uniform shape during propagation. Such displacement occurs evenly, without the development of unstable "fingers" or breakthrough channels. The article emphasizes that this effect is achieved only under a favorable mobility ratio between oil and the displacing phase, which is directly related to the viscosity properties of the system. This confirms the key role of viscosity and relative phase mobility in shaping the flow front and determining the volume of residual oil in the Hele-Shaw model.

In article [48], the Hele-Shaw cell is used to study the dynamics of fingering during the displacement of glycerin by water under conditions of complete mutual miscibility of the phases. The authors model displacement in a sealed cell and identify several characteristic flow regimes, including an initial slow front, rapid compression following breakthrough, and uniform expansion. The study emphasizes that even in the absence of porosity, the difference in viscosity between the displacing phase (water) and the displaced phase (glycerin) determines the shape and stability of the displacement front. Visualization made it possible to establish that increasing the viscosity contrast accelerates the transition from a diffusion-dominated to a convection-dominated regime and enhances front instability, increasing the volume of the residual, more viscous fluid.

Study [49] presents a review of micromodels, including Hele-Shaw cells, for analyzing wettability alteration and mobility changes in oil–water systems when using nanofluids and surfactants. Although the main focus of the article is on wettability modification, the authors emphasize that Hele-Shaw cells enable direct visualization of phase redistribution in response to changes in relative mobility and contact angles. Within the scope of this review, it is highlighted that even with fixed pore channel geometry, differences in viscosity and interfacial tension give rise to localized zones of residual oil. The Hele-Shaw cell is regarded as an important tool for isolating capillary and viscous effects in their pure form.

The pressure maintenance systems currently applied at oil fields are intended to ensure the most effective flooding of the entire reservoir. The systems described by the authors in [20] as the main and most widespread are most efficient in homogeneous reservoirs without any natural or artificial disturbances—an idealized scenario that rarely exists in reality. The ratio of production to injection wells depends on the expected volume of injected displacing agent and oil production. The distance between wells can be defined as a function of the expected oil recovery factor increment, oil production rate, reduction in operating costs, and minimization of drilling expenditures for additional wells [50]. The ORF increment, in turn, depends on the geological and physical characteristics and heterogeneity of the reservoir, including porosity, permeability, anisotropy of rock properties, and technological performance of the wells. In general, designing the most efficient development system plays a critical role in the rational exploitation of the field [51].

The main factors contributing to the formation of RRR as a result of development system design include the following:

1. Insufficient water injection—when the injection rate lags behind oil production, it leads to pressure decline, resulting in a reduced pressure gradient. According to Poiseuille's law [52], water begins to preferentially flow through the most permeable zones toward the nearest wells, bypassing distant areas and leaving them unswept. Prolonged exposure to such conditions results in the formation of stable streamlines from injection to production wells, leading to the development of so-called "dead zones."
2. High-intensity reservoir flooding—when the injection rate significantly exceeds the oil production rate. On the one hand, this causes water to push oil through high-permeability channels, forming dominant filtration pathways. On the other hand, it

may lead to the formation of auto-hydraulic fracturing (auto-HF) channels—high-permeability paths from the RPM well—into which the injected water escapes. It is important to note the difference in RRR formation in this case compared to point 1: here, the pressure is sufficient to reach remote zones of the reservoir, but water bypasses large volumes due to high filtration velocity. In other words, the reservoir's filtration capacity was not properly accounted for during RPM system design.

3. The presence of multiple productive horizons within a reservoir, differing in filtration and storage properties (FSP), penetrated by a common wellbore filter. When an impermeable barrier exists between them, water injection occurs predominantly into the most permeable intervals, leaving other layers almost or entirely unaffected. If there is some degree of hydrodynamic communication between the layers (i.e., crossflows), along with lateral variability in reservoir properties, situations may arise in which the injected water enters one formation and then flows into another, more permeable one, forming filtration channels and bypassing other portions of the reserves. For example, analysis of low-permeability formations in China has shown that after 20 years of waterflooding, approximately one-third of the layers in the section had never been swept by water; it is precisely in these layers that residual oil is concentrated [53]. The cause was the absence of an effective layer-by-layer RPM system during the early years of development. A similar issue is observed at mature fields: due to wear and damage of casing strings in old wells, it becomes difficult to implement proper zonal injection across the formations, which complicates the alignment of the water front.
4. The distance between injection and production wells, as well as the density of the well pattern, is another common cause of RRR formation. Large spacing between injection and production wells can lead to low sweep efficiency—the water front may not reach the production wells or may arrive in a weakened state. However, this relationship is most relevant for heterogeneous reservoirs with variable filtration and storage properties (FSP). It has been demonstrated repeatedly—mainly in fields in the western part of Texas, developed in carbonate reservoirs—that in isotropic formations with a high degree of continuity, reducing the spacing between wells did not lead to a significant increase in development efficiency and had little effect on the ultimate oil recovery factor. The optimal well pattern density depends on several factors, including permeability; for example, low-permeability reservoirs require denser well spacing. In practice, most fields started with sparse well patterns, gradually increasing density. At the Romashkinskoye field, well spacing was reduced in 1968, and the development system was switched from a linear to a block-pattern waterflood. In 1976, the grid was further densified by implementing a 3:1 well ratio. It should be noted that well pattern design is a complex task influenced by numerous factors [54].
5. Formation of zones with reduced pressure gradient between wells, where oil remains in an almost immobile state. During waterflooding, a specific pressure distribution is established in the inter-well space. The farther a given point is from the production well, the lower the filtration velocity in that zone, as it is directly proportional to the pressure gradient. High pressure gradients are established near the wellbore, while with increasing distance, the gradients decrease accordingly.
6. Another cause of RRR formation is design errors in the waterflooding system with respect to geological structures within the reservoir. For example, when there is strong hydrodynamic communication between the outer and near-boundary zones and an active aquifer, an improperly designed RPM system may lead to water bypassing into the near-boundary zone. This, in turn, promotes the formation of zones with unswept oil in the central part of the reservoir [55].

From the standpoint of mobility, residual oil can be divided into movable and immovable parts. At present, the scientific literature devoted to the classification of residual oil does not provide a definitive answer to whether it should be considered movable or immovable. Hypothetically, any oil can be regarded as movable if there are technologies capable of altering the physicochemical conditions within the reservoir. In practice, the movable portion of residual oil is defined as that which retains hydrodynamic connectivity with the main drainage zones and can be recovered by modifying pressure gradients, mobility ratio, wettability, or interfacial tension. The immovable oil, on the other hand, is defined as the oil that is difficult to affect due to technological limitations; it is typically localized within capillaries and dead-end pores and exists in the form of thin films as a result of dominant adhesive and capillary retention forces. Furthermore, part of the oil existing on the macroscopic scale can also be considered immovable if it is hydrodynamically isolated from the drainage zones or if pressure gradients are absent, precluding its movement.

The analysis results are presented in the form of a conceptual diagram (Figure 1) illustrating the structure, hierarchy, and main factors governing the formation of residual recoverable oil reserves (RRO) in terrigenous reservoirs. The diagram demonstrates the relationship between microscopic (≈30%) and macroscopic (≈70%) forms of residual oil, as well as the physical and technological processes controlling their distribution.

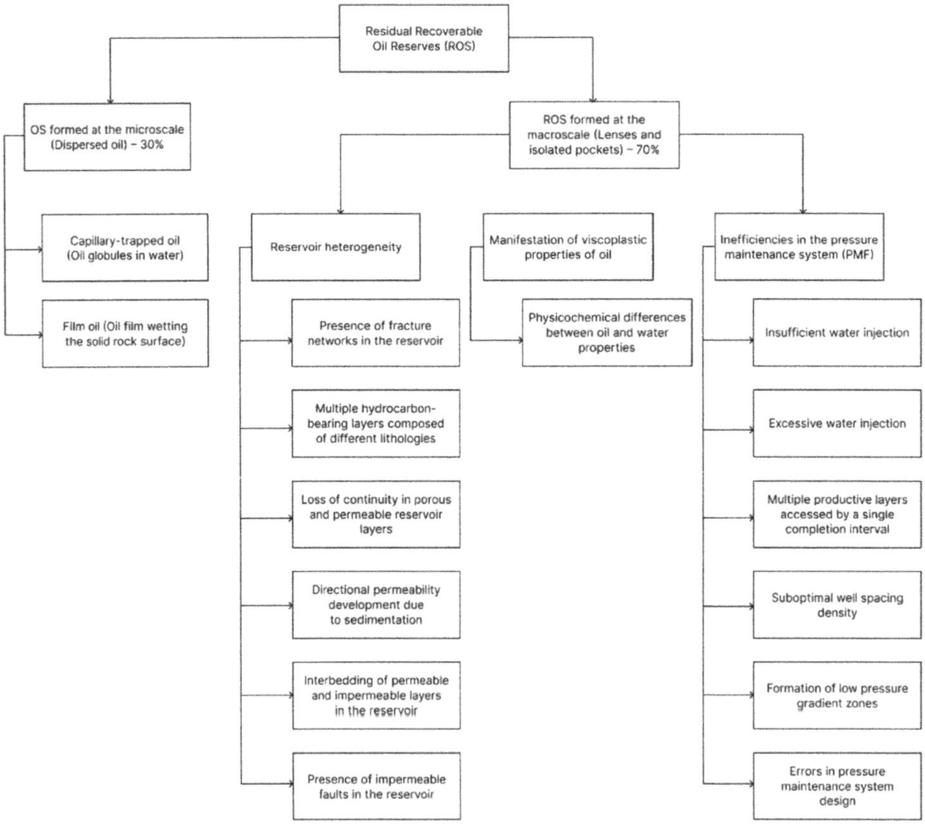

Figure 1. Generalized block diagram of RRR classification [compiled by the authors].

Microscopic forms (capillary-trapped and film oil) are formed within the pore space under the influence of capillary and adhesive forces, which depend on wettability, pore geometry, and the surface characteristics of mineral grains.

Macroscopic forms (lensed and isolated accumulations) are determined by reservoir heterogeneity, gravitational redistribution of fluids, and imperfections in the pressure-maintenance system (PMS). Their formation is associated with the presence of fracturing, interbedding of rocks with differing permeabilities, and uneven water injection within the PMS.

3. Methods for Localizing Residual Recoverable Reserves

The RRR types classified above represent a source of additional oil production. Accordingly, the identification and targeted development of such reserves can significantly increase the ultimate oil recovery factor (ORF) at oil fields by utilizing existing infrastructure. Various methods are used for localization, which are conventionally divided into two groups: direct (instrument-based) methods and indirect methods.

Instrument-based methods include a number of investigations, among which are:

1. Seismic technologies, such as Vertical Seismic Profiling. This is a high-resolution seismic exploration method in which geophones are placed along the wellbore, while the seismic source is located on the surface—or vice versa. VSP provides higher accuracy compared to conventional 3D surface seismic surveys. It is used both in single-well mode for detailed examination of a local zone and in multi-offset mode (e.g., walkaway VSP, where the seismic source is moved along a profile) [56,57].

This method allows for predicting hydrocarbon saturation near the wellbore and for identifying poorly drained zones where oil accumulates in the form of isolated pockets. Its high vertical resolution enables detailed evaluation of reservoir intervals adjacent to the well, while the recording of multicomponent P- and S-waves allows differentiation of fluid types and qualitative assessment of their migration and volume. The method is widely used at modern oil fields; however, it should be noted that each VSP session covers only the vicinity of a specific well. Therefore, assessing the entire field requires

2. Well Logging. This group of instrument-based methods includes both open-hole logging, i.e., conducted during drilling, and cased-hole logging. The fundamental difference lies in the objectives of these investigations. Open-hole logging is aimed at recording the initial parameters of the wellbore. Cased-hole logging—commonly referred to in modern literature as production geophysical investigations —is considered a separate category and focuses on recording parameters during field development.

In the context of residual recoverable reserve localization, production geophysical investigations include pulsed neutron logging and carbon–oxygen (C/O) logging. The operating principle of neutron-based methods is based on the generation of neutrons by a source placed in the wellbore and the subsequent detection of thermal or epithermal neutrons, or gamma rays. The emitted neutron undergoes changes as a result of interactions with the surrounding formation material. Neutron logging responses primarily depend on the hydrogen and chlorine content in the vicinity of the well. C/O logging is one of the few methods that allows for the evaluation of reservoir fluid saturation independently of formation water salinity [58]. C/O logging directly measures the ratio of carbon to oxygen concentrations in the rock: a high C/O ratio indicates the presence of oil (carbon), whereas when oil is replaced by water, the oxygen content increases and the ratio decreases [59].

These methods are routinely used by major oil companies for reservoir development monitoring. The ability to determine the current oil saturation of the reservoir around the wellbore without prolonged well shutdown, along with the possibility of repeated measurements to track dynamic changes, is a clear advantage. However, the sensitivity of these methods decreases in low-porosity formations, in the presence of thin productive layers, and—specifically for neutron logging—in cases of low formation water salinity, since low-salinity water produces signals similar to those of oil. The investigation radius of these methods typically spans several tens of centimeters from the wellbore.

Well thermometry, noise logging, flowmetry, moisture logging, and densitometry are also noteworthy methods widely used for development monitoring and assessment of current oil saturation. These methods help answer the question of which fluids, and in what quantities, are entering the wellbore from each penetrated reservoir interval. Accordingly, they allow detection of oil–water contact (OWC) movement, behind-casing flow (BCF), and interlayer crossflows, which result in the formation of stagnant oil zones [60].

3. Hydrodynamic well testing is another group of instrument-based investigation method. Special attention should be given to the modern impulse-coded hydrodynamic listening (ICHL) method and tracer studies. ICHL is an advanced version of inter-well hydrodynamic listening. Instead of continuous excitation, it uses a series of short pulses delivered into one or several excitation wells. The reception of the coded signal in the observation well eliminates the need to shut down surrounding wells, making the method highly resistant to various types of interference. The method allows for the identification of stagnant oil zones based on the intensity of the received signal.

Tracer studies involve injecting a labeled fluid—chemically inert with respect to reservoir oil and rock—into an injection well and detecting it in a production well after a certain time interval. This method makes it possible to evaluate inter-well hydrodynamic connectivity, filtration heterogeneity, and the filtration and storage properties (FSP) of the reservoir. When combined with other methods, tracer studies provide a comprehensive picture of RRR formation in the reservoir [61]. The main advantage of this method lies in its simplicity of application, which allows its use at any stage of field development [62,63].

4. Geochemical well investigations, when applied to the task of localization, represent a group of methods based on the detection of anomalies and deviations from natural background values during oil field development. Methods that are actively used in practice include sampling of formation water and gas, determination of the component composition of formation gas, determination of the physico-chemical properties of formation fluids, optical studies (in the UV–visible–NIR spectral range), infrared (IR) spectral analysis, and luminescent-bituminological analysis of cuttings and core samples. Unlike the previous groups of methods, this group is less labor-intensive and does not require the suspension of production well operations.

Special attention should be given to optical studies of oil, as numerous studies have demonstrated that the optical properties of oil are highly sensitive to changes in its chemical composition during field development. The method is based on measuring the optical density of oil across different spectral ranges and subsequently correlating these values with the amount of residual oil in the reservoir. In other words, the analysis evaluates how strongly oil absorbs light at specific wavelengths, and this data is used as an indicator of whether the oil sample originates from a "depleted" or "unaffected" part of the reservoir.

Optical studies are based on the Bouguer–Lambert–Beer law (3) [64]:

$$K_{ac} = \frac{D}{0.4343 \cdot C \cdot l}, \tag{3}$$

where:

K_{ac}—absorption coefficient,
D—optical density of the analyzed solution,
C—concentration of the solution,
l—path length (thickness of the solution layer).

The physical basis of the method lies in the observation that the composition and optical properties of oil change as the reservoir is depleted. During displacement processes, lighter oil fractions are extracted first, while heavier oil—rich in asphaltenes and resins—remains in stagnant zones. Such heavy oil has higher optical density and a greater absorption coefficient.

The advantages of the method include its relative simplicity and the speed of optical density analysis in laboratory conditions. However, it should be noted that during transportation, samples undergo chemical changes due to the adsorption of natural surfactants and the evaporation of light oil fractions. Another benefit of the method is its complementarity with other techniques—optical studies can fill gaps where other methods, such as well logging, are limited or infeasible. Moreover, the method's high sensitivity to resin and asphaltene content—which is directly associated with the accumulation of RRR—offers strong potential for its further development in localization technologies. It is important to emphasize that the method characterizes the oil entering a specific well. If the well receives a mixture of oil from multiple zones, the optical response becomes averaged, making it difficult to attribute the signal to specific intervals. This requires either selective sampling from wells that penetrate only a specific zone or the combined use of optical data with interval-resolved well logging results. It is also evident that the presence of mechanical impurities, additional water inflow, and other indirect factors can distort the final results.

In study [11], a method for localizing residual oil reserves was proposed based on measuring the light absorption coefficient of wellhead samples. The method was tested on 50 wells of the Minnibayevo section of the Romashkinskoye field and demonstrated comparability with the results of geological modeling. The authors note the need for further validation of the method when adapting it to various geological conditions, and they emphasize the influence of laboratory factors as well as the importance of automating measurements for field-scale application.

Indirect methods include approaches that involve the analysis and processing of large datasets obtained from sensors and instruments installed at the wellhead or downhole. These instruments may be stationary or handheld/mobile operator tools. This group includes sensors from thermobarometric systems (TMS), downhole thermometers, pressure gauges, the "SUDOS Automat 2" device, optical sensors, various types of flowmeters, and other wellbore instrumentation.

1. Methods based on neural network algorithms represent a rapidly developing set of approaches for localizing residual recoverable reserves. Neural network algorithms apply machine learning (ML) techniques to analyze large volumes of heterogeneous field data and generate distribution forecasts. The main objective of using neural networks is to identify hidden nonlinear relationships that are not detectable through manual analysis. A key advantage of these methods lies in their ability to automatically account for massive datasets, including development history, well logging, hydrodynamic studies, and geochemical investigations. It is important to emphasize that neural network models are most effective at fields with a large, accumulated

data base and a sufficient number of examples for training. For example, in a large producing field where RRR has been assessed in several pilot zones, a neural network can be trained on those data and then applied to the rest of the area [65,66]. The application of neural network algorithms is especially justified in multilayered and highly heterogeneous systems, where traditional methods struggle to generalize information—neural networks can independently uncover complex dependencies. They are also well suited for real-time forecast updates as new data become available: the model can be retrained using fresh inputs. In small or poorly studied fields, neural network methods are less reliable due to the scarcity of training data; in such cases, they are more appropriate when combined with physical models—as part of history matching or for hypothesis validation. Thus, the application of neural network algorithms can be viewed as an integration of data from all types of investigations into a unified base for further analysis and pattern discovery. The authors further propose to explore the combined use of ML and optical oil analysis for RRR localization.

One of the most comprehensive and technologically advanced examples of integrating machine learning methods into residual oil exploration is presented in study [67]. The authors developed a unique architecture for forecasting residual oil saturation based on convolutional neural networks (CNNs), in which spatially normalized tensors are used as input data. These tensors reflect the dynamics of cumulative oil production, water injection, porosity, permeability, oil saturation, and results from geochemical investigations. Particular attention is given to chemical indicators—such as the ratio of light to heavy n-alkanes in associated gas—which allow for model prediction validation and the identification of unswept zones that are inaccessible to conventional hydrodynamic simulators. The methodology incorporates a stochastic analog of a hydrodynamic model based on the Kalman filter, which ensures high adaptability when working with real field data. Implementation of this technology enabled not only the accurate localization of residual reserves but also the automation of geological and technical measure (GTM) selection at mature fields.

2. Retrospective analysis is a method for identifying undeveloped or poorly developed reservoir zones based on the analysis of field development over the entire production history. Essentially, it is a search for missed oil-bearing intervals using development data. Within this approach, all available information is collected: initial reserves, inflow profiles for each reservoir object, well logging results, monthly production reports (MPR), workover histories, well reassignments to other horizons, and so on. Then, using a decision-making algorithm, potential zones containing RRR are identified, along with applicable geological and technical measures for their recovery.

Based on a proprietary RRR classification algorithm developed by the authors and implemented within the retrospective analysis of data from 2605 wells at one of the Romashkinskoye field sites, large-scale localization of previously unswept intervals was performed. More than 96% of the identified RRR were classified as Category 1—zones with high residual oil saturation that had not been previously affected during field development. The main localization zones included inter-injection-well rows, central rows in three-row patterns, areas near formation junctions, pinch-out zones, and oil–water contact regions. The study proposed GTM selection algorithms tailored to each RRR category. Their effectiveness was confirmed during field trials—in particular, additional perforation in wells No. 16 and No. 6 led to actual oil inflow. The authors emphasize that the method is still being refined, including optimization of production forecasting and adaptation of the algorithm to other areas of the field, taking into account geological characteristics and well stock condition [10].

3. Voronoi diagrams, or analysis of specific production reserves by well. This method is based on dividing the field area into so-called Voronoi polygons—filtration zones in the form of polygons that are closer to a given well than to any other. Each such polygon is considered the "zone of responsibility" of the corresponding well, from which it is expected to extract oil. The initial oil reserves within each polygon are then compared with the actual cumulative production. The difference is considered the residual reserve associated with the well. The method is simple and visually intuitive; however, it relies on a simplified model of streamlines, assuming that the entire selected area is hydraulically connected to a single well—an assumption that often does not reflect reality. The method requires highly reliable input data, as errors in the initial reserves map will directly affect the accuracy of residual reserve estimates. The approach is compatible with retrospective analysis—in fact, the Voronoi diagram method is a formalized version of well-by-well retrospective analysis. It also integrates well with categorical classification, allowing polygons to be classified as fully depleted or underdeveloped.

From the standpoint of applicability ranges, the research methods span a scale from 10^{-2} m to several kilometers. Thus, static core analysis methods allow for refinement at a scale of 10^{-2}–10^{-1} m, depending on the core sample size. It should be noted that the data obtained vary significantly with the selected core size: researchers from TatNIPIneft have reported permeability variations by an order of magnitude when transitioning from small to full-size core samples. Routine well-logging methods provide data with a resolution on the order of 10^{-1}–10^{0} m, establishing a link between laboratory investigations and reservoir-scale measurements. Hydrodynamic well testing (HWT) methods interpret filtration processes over a range from 10 to 10^{3} m, enabling analysis of inter-well connectivity and reservoir permeability. Field geochemical methods are applicable within the range of 10^{0}–10 m, providing additional information on flow directions and fluid composition. Tracer studies, which combine geochemical and hydrodynamic approaches, offer comparable resolution—from 1 to 10^{3} m—and make it possible to identify active flow zones and stagnant regions. Finally, seismic profiling covers the range of 10^{0}–10^{4} m, providing insights into the large-scale structure and heterogeneity of the reservoir.

The accuracy and applicability ranges of localization methods belonging to the class of indirect approaches are entirely dependent on the results obtained through instrumental measurements. Their reliability decreases as the analysis extends beyond the applicability range of the instrumental techniques. However, within the limits of field-scale investigations, indirect approaches make it possible to refine data in areas where instrumental interpretation was ambiguous.

Summarizing the above, it should be noted that there is no universal method for the localization and evaluation of residual recoverable reserves; no single method is perfect or self-sufficient. In practice, the highest efficiency is achieved through the integrated application of multiple approaches. A consolidated classification of RRR localization and evaluation methods is presented in Figure 2.

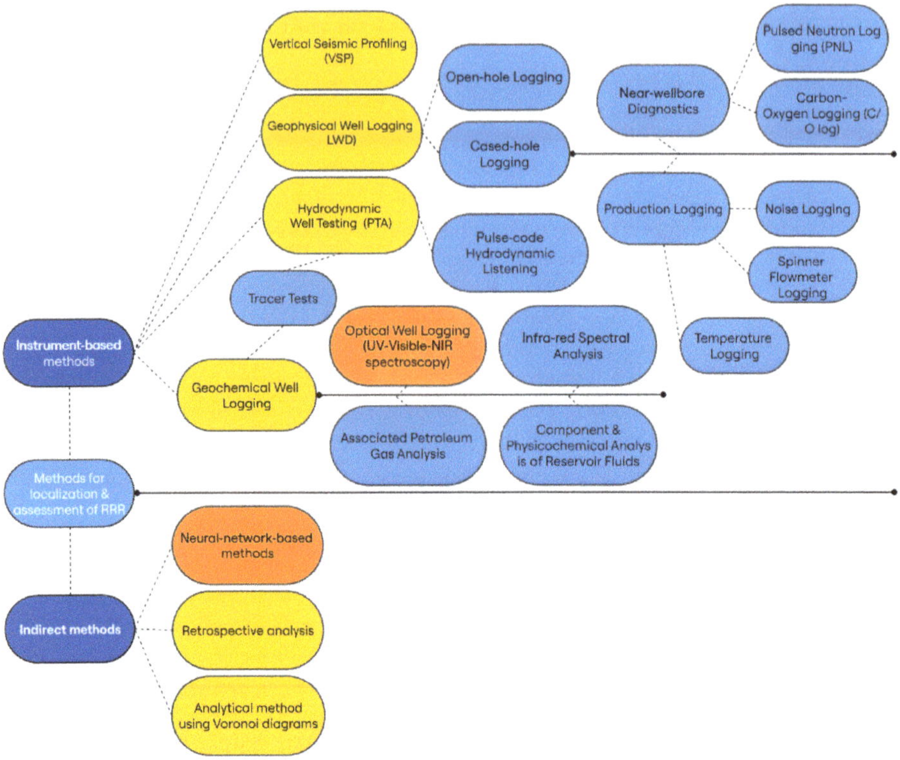

Figure 2. Block diagram of the classification of RRR localization and evaluation methods [compiled by the authors].

4. Chemical Methods for Enhanced Oil Recovery at the Macroscale

As previously noted, a significant portion of oil reserves remains in the reservoir after the primary development phase. In practice, regulation of the development process through methods such as increased injection pressure, cyclic injection, intensified fluid withdrawal, isolation of reservoir intervals into separate production objects based on their properties, and well pattern densification have proven effective under conditions of high geological heterogeneity. However, these methods lack universality and are insufficiently effective in formations where extensive waterflooding has already occurred. This fact explains the stabilization of oil recovery factors in modern waterflooding projects at the level of 0.3–0.5 of the geological reserves [15]. In this context, a key condition for further improving oil recovery is the restriction of water movement through high-permeability channels with established streamlines within the existing field development system.

Selective permeability reduction (SPR) methods, also known as water shutoff treatments (WST), are gaining increasing popularity today. According to their target technological effect, SPR methods can be classified into flow-diverting technologies, injectivity profile modification, and water inflow restriction. The essence of SPR technologies lies in selectively increasing the filtration resistance in water-swept, high-permeability zones of the reservoir, which leads to the redistribution of filtration flows within the formation. A water-blocking barrier is created in the water-flooded interval, which prevents further water movement and redirects it into oil-saturated, low-permeability layers. As a result, the sweep efficiency and the oil recovery factor (ORF) are increased.

Experience with the implementation of SPR technologies shows that their primary targets are predominantly terrigenous multilayer reservoirs characterized by a high degree of geological and physical heterogeneity, where permeability can vary by several orders of magnitude both vertically and laterally [68]. In particular, several sources describe the application of SPR methods in the northern Gulf of Mexico [69], in sandy reservoirs of Indonesia [70], in China's Daqing Oilfield, and at the Samotlor, Romashkinskoye, Vankor, and Novo-Elkhovskoye fields in Russia.

Flow-diverting technologies (FDT) have gained widespread application both in Russia and abroad. According to recent reviews [71], more than 100 different FDTs are currently known. In Russia, over the period from 2010 to 2020, more than 30,000 operations were carried out, resulting in a cumulative increase in oil production of 53 million tons. A confirmed example of the successful implementation of a flow-diverting technology using polymer systems is provided by the pilot field operations conducted at the Romashkinskoye field (reservoir D_0 of the Kynov horizon). In 2015, approximately 700 m^3 of a polymer–microgel composition (polymer + crosslinker + surfactant) was injected into an injection well. Following the treatment, a decrease in the water cut of producing wells from approximately 80% to 66% and an increase in oil production from about 500 to 700 tons per month were recorded. The achieved effect is attributed to selective permeability reduction. In 2020 alone, approximately 9000 SPR operations were performed, with additional oil production ranging from 0.3 to 1.6 thousand tons per well. The total contribution of FDTs to additional oil production across the country in 2020 is estimated at around 9.5 million tons, or approximately 8% of the total incremental production attributable to enhanced oil recovery methods. These figures emphasize the high effectiveness of selective permeability reduction technologies in addressing the growing problem of watercut and in localizing residual oil reserves at mature Russian oil fields [72].

To ensure an objective comparison of the effectiveness of various gel formulations used in chemical flooding technologies, real field application cases are summarized in the table below. Table 1, compiled by the authors, presents the key parameters: reservoir type and residual oil characteristics, the achieved effect—including production increase and water cut reduction—as well as the limitations under which the formulations can be applied, such as temperature, salt tolerance, rock sensitivity, and other conditions.

The technological approaches of flow-diverting technologies are classified into several main categories based on the mechanism by which the blocking barrier is formed. These include polymer gel-forming and viscoelastic systems, precipitate-forming compositions, surfactant-based foam systems, and curable (setting) systems. In real field applications, it is common to combine several mechanisms within a single treatment. For example, there are complex formulations that combine gel-forming and dispersed-phase components, which are injected sequentially to affect both the near-wellbore zone and deep reservoir intervals.

Table 1. Comparative characteristics of gel systems applied for residual oil recovery. [compiled by the authors].

Gel Type	Reservoir Type	Residual Oil	Effect	Limitations	Country of Application
Cr(III)–carboxylate/acrylamide gel	Sandstone, 2 D	RO zones with 100% water cut	Oil production increased from 9.2 to 20 m^3/day; water cut decreased to 60–70%	Applicable up to ~80–90 °C; sensitive to hard water (Ca^{2+}/Mg^{2+}); possible rock adsorption; requires gelation control using retarders (malonate, glycolates). Gelation accelerates at higher temperatures. Suitable for low- to medium-temperature reservoirs.	China
Al(III)–crosslinked gel	Carbonate, moderate T	RO in water-bearing intervals	Moderate water cut reduction; production rate increase (limited data)	Thermal stability up to 90 °C; sensitive to salt ions; difficult to control gelation time without retarders; crosslinking may occur too rapidly. Limited use under high-temperature conditions. Less toxic than Cr-based gels.	Iran
Organically crosslinked gel	Carbonate, 2–3 mD	RO in fractured zones	Water cut decreased by 42%; gas rate increased by 7.7 times.	Stable at T > 120 °C due to covalent bonds; thermally resistant up to 150 °C; effective in fractured and tight reservoirs. Requires shut-in period for gelation; inhibitors needed at high T. Chemically complex and costly.	China
Biopolymer gel	Sandstone, heterogeneous	RO in swept zones	Oil increase ≈ 3 t/day; water cut ↓	Thermal stability up to ~90 °C; insensitive to Ca^{2+}/Mg^{2+}; biodegradable (biocides required); high viscosity; stable under pH and salinity variations. Applicable in fresh and saline waters; environmentally friendly and low-corrosive.	Russia

Table 1. *Cont.*

Gel Type	Reservoir Type	Residual Oil	Effect	Limitations	Country of Application
Monomer-based gel	Various, high permeability	RO in deep zones	Historical success; increased oil recovery	Highly sensitive to temperature and oxygen; difficult to control free-radical polymerization. Acrylamide is toxic and requires degassing. Thermal stability up to 93 °C with composition modification. Largely replaced by pre-crosslinked gels.	USA
Synthetic polymer gel (PPG)	Sandstone/carbonate, >0.5 D	RO between interbeds	Oil rate increase up to 10 t/day; water cut reduction by 5–10%	Operable up to 140 °C; requires sufficient pore or fracture width (>0.1 mm); unsuitable for fine-pore rocks. Applicable in highly heterogeneous formations. Provides long-term effect; available in various particle sizes.	USA, China
Inorganic (silicate) gel	Carbonate, fractures >1 D	RO in matrix zones	Significant water reduction; oil production increase	Temperature range 20–120 °C; high salinity resistance; inexpensive and readily available; non-selective—may block oil-saturated zones. Brittle; possible breakthrough under pressure variation. Requires precise placement.	Russia

4.1. Gel-Forming Systems

Gels are gel-forming substances produced by the coagulation of a colloidal solution into a semi-solid phase. In modern petroleum engineering literature, the term "gel" encompasses elastic and semi-solid materials formed through the chemical crosslinking of water-soluble polymers in an aqueous solution.

An ideal gel injection technology for improving sweep efficiency should be applicable to both production and injection wells, for the purposes of displacement enhancement, inflow restriction, and flow control treatments. The technology must include a simple and effective gelation chemical system that can be used across a wide range of reservoir conditions, such as temperature, porosity, permeability, pressure, and others. The system should be stable over extended periods, provide sufficient gel strength across a broad range of conditions, and enable controlled and predictable thickening behavior. The gel should be formulated using relatively inexpensive and readily available chemical reagents, based on low-concentration compositions.

Field-applied gel systems for improving sweep efficiency exist in various forms and encompass a wide range of chemical compositions [73]. Below is a classification of gel systems most frequently referenced in contemporary literature.

Chromium(III) carboxylate/acrylamide-polymer gels are aqueous solutions of acrylamide-based polymer gels in which the chemical crosslinking agent is a carboxylate complex of Cr(III) [74]. These gels possess exceptionally stable chemical structures and are highly inert to surrounding environmental conditions. They are applicable across a wide pH range, making them suitable for use in both acidic and alkaline environments when properly formulated. Chromium triacetate, $Cr(CH_3COO)_3$, is often preferred as the crosslinking agent. To retard the gelation reaction, either low-molecular-weight hydrolyzed polyacrylamide is used, or strong carboxylate ligands are added, such as lactate ester or malonate ester [61].

Cr(IV) in redox reactions with acrylamide polymers represents one of the earliest technologies for creating blocking systems. Today, this technology is no longer used due to the high toxicity and carcinogenicity of Cr(IV), as well as the complexity of the crosslinking chemical reaction [75].

Aluminum(III)-crosslinked gels. There are numerous technologies that incorporate aluminum as a crosslinking agent; however, the most widespread involves the use of aluminum in combination with citric acid ester. According to several studies, gels based on high-molecular-weight hydrolyzed polyacrylamide encounter difficulties—and in some cases become completely immobile—in low-permeability reservoirs, although they exhibit good mobility in fractured formations. It is also noted that crosslinking with aluminum is a slow reaction, typically occurring over the course of several hours, and the viscosity of such systems does not exceed the viscosity of water-based polymer solutions without a crosslinking agent [76].

Polyacrylamide gels crosslinked with aluminum and acetic acid esters have recently gained popularity. These are referred to as colloidal dispersion gels and are primarily used in porous reservoirs. These systems are of particular interest, as crosslinking occurs via aluminum acetate, which provides thermal and chemical stability of the gel across a pH range of 3.5 to 8.5, as well as high stability in salinity levels up to 50 g/L [77].

An interesting direction is the use of aluminum alkoxides for modifying ethylene-vinyl acetate (EVA) copolymers, where transesterification reactions lead to the formation of three-dimensional crosslinked structures [78]. As a result of the interaction between the hydroxyl groups of the aluminum compounds and the acetate fragments of the polymer, a significant increase in viscosity is observed, along with the formation of a spatially crosslinked gel structure [79].

Gels crosslinked with other multivalent metal ions, such as Cu^{2+}, Zr^{4+}, Ni^{2+}, and Co^{2+}, are also being investigated. For example, the use of Zr^{4+} in gel systems significantly enhances their thermal stability and effectiveness when injected into high-temperature reservoirs. Modern modifications of such gels utilizing biocompatible components demonstrate stable behavior at temperatures up to 130 °C and effectively reduce permeability in high-permeability reservoir intervals, providing a reliable flow-diverting effect [80].

As part of the search for environmentally safe alternatives to toxic Cr^{3+} crosslinkers, gels based on divalent metals, particularly Co^{2+}, Ni^{2+}, and Cu^{2+}, are actively being studied. Research has shown that cobalt acetate forms strong and stable gels suitable for use at temperatures up to 150 °C and under high salinity conditions of formation water. Gel systems based on Ni^{2+} and Cu^{2+} have also demonstrated satisfactory stability and operational feasibility, effectively reducing water saturation and redistributing flow in heterogeneous reservoirs [81].

Gels crosslinked with organic crosslinking agents. This group of gels was originally developed in pursuit of strong and stable polymer gels that would eliminate the need for metallic crosslinkers. Most of the organic-crosslinked gels developed to date are based on the chemical properties of phenol and formaldehyde or their derivatives. However, phenol and formaldehyde are known to be highly toxic. In this regard, one of the modern solutions is the use of the amino acid L-lysine as a crosslinking agent for polyacrylamide. Gels developed on this basis demonstrate stable structure formation at temperatures up to 130 °C, high resistance to formation water salinity, and the ability to effectively reduce permeability in water-bearing zones [80].

Another actively researched direction involves organic gel systems based on hydroquinone and hexamethylenetetramine. These formulations allow for precise control of gelation time, which is critically important for the controlled placement of the gel within the pore space of the reservoir. Gels formed using these components exhibit high resistance to salinity and elevated temperatures, and are also effective in limiting water inflow in fractured formations.

Biopolymer gels are gels based on the crosslinking of biopolymers with either organic or inorganic crosslinking agents. One of the most established technologies in this category involves xanthan-based formulations crosslinked with inorganic Cr(III). In recent years, active attention has been given to the development of gels based on natural polysaccharides such as guar gum, scleroglucan, diutan-gamma, pectins, and gum arabic, owing to their biocompatibility, resistance to high salinity, and thermal stability. These polysaccharides act as the gel base, while crosslinking is achieved using agents such as Cr^{3+}, Zr^{3+}, and organoborate compounds, which effectively crosslink the hydroxyl groups of polysaccharides under alkaline pH conditions [82,83].

Guar gum is one of the most commonly used biopolymer bases for gels, especially under hydraulic fracturing conditions, due to its ability to form stable gel structures at low concentrations in the presence of organoborate crosslinkers. Scleroglucan and diutan-gamma are used under conditions of high temperature and high salinity, including in systems crosslinked with Cr^{3+} and Zr^{4+} ions [84].

Pectins, due to their carboxyl group content, interact with divalent cations (e.g., Ca^{2+}) to form ionic gels suitable for moderate temperature conditions [85].

Gum arabic, although it has limited gelation capability on its own, can enhance the solubility and thermal stability of other polysaccharides in mixed systems. It exhibits resistance to saline environments and maintains viscosity in the presence of Na^{+} and Ca^{2+} ions [86].

A promising direction is also the use of cellulose-based biopolymers, such as carboxymethyl cellulose (CMC) and hydroxyethyl cellulose, which have high water solubility and the ability to form gel-like structures. According to [87], such biopolymers are effectively used in environmentally friendly gel-forming systems aimed at enhancing oil recovery and controlling water inflow. Furthermore, experimental studies have shown that gelation of CMC may involve coordination crosslinking with multivalent cations, and the viscoelastic properties of such gels depend on both concentration and environmental conditions [88,89].

Another rapidly developing direction is the use of lignin as a natural biopolymer characterized by high chemical stability and environmental compatibility. Due to its polyfunctional structure, lignin can participate in the formation of gel-like networks, either as a base material or as a functional additive. Experimental studies have shown that lignin exhibits a high degree of adsorption on rock surfaces even under high salinity conditions, making it an effective flow-diverting material [90]. Moreover, combinations of lignin with inorganic components such as aluminum soaps allow the creation of durable gel-like materials with pronounced sorption and structural–mechanical properties, suitable for water-bearing zone isolation and enhanced oil recovery [91]. Additionally, hybrid gels based on lignin-containing nanocellulose have been developed, demonstrating thermal stability up to 150 °C and resistance to salinity levels up to 200,000 mg/L [92].

Monomeric gels are gels formed through the localized polymerization of organic monomers, with or without the inclusion of crosslinking monomers. The earliest developments in this area were based on the in situ polymerization of acrylamide monomer. However, due to its high toxicity, this method is now rarely used. Preference is currently given to the localized polymerization of less toxic monomers, such as acrylates. A key advantage of this type of gel is the low viscosity of the gel-forming solution, which facilitates injection. The main disadvantage is its high sensitivity to reservoir properties, which can limit the applicability of the method.

A promising direction is the use of multifunctional monomeric systems containing N,N-dimethylacrylamide, AMPSNa, and N-vinylpyrrolidone, which are polymerized in situ to enhance the thermochemical stability of gels under reservoir conditions [93]. In addition, technologies for controlled delayed polymerization are actively being developed. These are aimed at achieving deep flow diversion and uniform distribution of the gel front within the reservoir [94].

Gels based on synthetic organic polymers are gel-forming systems in which the primary component is a pre-synthesized polymer, such as polyacrylamide (PAM) and its copolymers. Unlike monomeric gels, which are formed in situ, these systems are injected into the reservoir in ready-to-use form, often in combination with crosslinking agents to stabilize the gel structure. A typical example is the PAM/PEI gel, which has demonstrated high efficiency at temperatures up to 120 °C, increasing the oil recovery factor by 24% in laboratory conditions [95]. To improve placement selectivity and blocking stability, such gels can be modified with inorganic nanoparticles, such as Fe_3O_4, without altering the organic nature of the base polymer. In particular, one study developed a magnetically responsive PAM gel modified with Fe_3O_4, capable of directed placement in porous media and thermal stability under reservoir conditions [96]. Synthetic organic polymers also serve as the basis for preformed particle gels (PPGs) in the micron size range [97]. The advantages of synthetic organic polymers include high salinity resistance, adaptability to various reservoir conditions, and a wide range of crosslinker options. However, limitations include increased viscosity prior to gelation and the potentially reduced depth of penetration compared to monomeric systems.

Inorganic gels. Throughout the study of inorganic gels, a wide variety of formulations have been examined and applied. These systems typically utilize gels based on silicates or aluminum ions, as well as gels formed from iron and magnesium hydroxides. Silicate gel is formed when an aqueous solution with a relatively high pH, containing a sufficient amount of monomeric orthosilicic acid ester or oligomers of orthosilicic acid, undergoes pH reduction or is exposed to hardness cations. Silicate gels may form under the influence of either an external or internal catalyst. An internal catalyst is typically an acid-generating compound that spontaneously lowers the pH after injection into the reservoir. Silicate gel formation via external catalysis occurs when the monomeric orthosilicic acid or oligomer solution comes into contact with formation water containing hardness cations, such as Ca^{2+}.

Such systems demonstrate high thermal stability and resistance to saline environments, as confirmed by numerous studies. In particular, interactions with Ca^{2+} and Mg^{2+} ions lead to the formation of strong CSH (calcium silicate hydrate) and MSH (magnesium silicate hydrate) gels, which maintain stability at temperatures up to 140 °C and provide an increase in oil recovery factor of up to 23.87% [98]. A similar approach was implemented in an earlier study, where a silicate gel was used for flow diversion and demonstrated high efficiency in selective water shutoff in permeable zones, while maintaining stability under reservoir temperature and salinity conditions [99]. More recent developments have shown that the use of inorganic reagents can achieve high residual resistance factors (RRF > 180) and confirm the promising potential of such systems for injection profile control, without the need for complex organic components [100]. Despite their advantages, the main limitations of inorganic gels include limited controllability of gelation time and possible instability under varying reservoir pH and water composition.

Subsequently, Table 2 presents a summary of laboratory studies and pilot field trials of various gel-forming compositions.

Table 2. Summary of studies on various gel-forming formulations. [compiled by the authors].

Gel Type	Composition	Country	Reservoir Type	Temperature, °C	Pressure, MPa	Permeability, mD	Water Salinity, g/L	Application Effect	Source
Cr(III)-carboxylate/PAM gel	Partially hydrolyzed polyacrylamide (PHPA) + Cr(III) acetate	USA Field (multiple applications)	Fractured formation	up to 124 (H_2S-resistant)	Not available	Not available	High (salt-tolerant)	Widely implemented: ~700 treatments, reduced water cut, increased oil recovery	[101, 102]
Al(III)-crosslinked gel	PAM + Al(III)-citrate complex (molar ratio ~2:1)	Not specified Laboratory tests	Porous (low-permeability)	~60–90 (moderate)	Not available	Not available	Depends on salinity (citrate selection at higher mineralization)	Effectively gels and blocks water at moderate temps	[103]
Organically crosslinked gel	PHPA (or AM/AMPS copolymer) + organic crosslinker (e.g., PEI)	Saudi Arabia (lab) Laboratory (core flooding)	Porous (sandstone)	up to 130 (stable ≥ 8 weeks)	3	Not available	Stable in salts/acids (no ionic precipitates)	At 90 °C and ΔP = 6.89 MPa, complete water shutoff for 3 weeks	[104]
Biopolymer gel	Polysaccharide (xanthan) + Cr(III) crosslinker + nano-silica	Iran (lab) Laboratory (bottle and core tests	Porous (sandstone)	60 (tested); stable at 80 (with 170 g/L salinity)	Not available	42.6	6.8 (simulated formation water); high salt tolerance	~72% water cut reduction after gel injection; high water selectivity	[102, 105]
Monomer-based gel	In situ free-radical polymerization of acrylamide/acrylate monomers (initiator + crosslinker)	China	Matrix (perforated intervals)	up to ~93 (controlled gelation); at 149 (stable for a year)	Not available	Not available	Insensitive (inorganic monomers)	Forms rigid gel-plug	[106]

Table 2. *Cont.*

Gel Type	Composition	Country	Reservoir Type	Temperature, °C	Pressure, MPa	Permeability, mD	Water Salinity, g/L	Application Effect	Source
Synthetic polymer gel	Pre-crosslinked gel (PPG) based on PHPA, particles 1.5–5 mm	China (Daqing, field) Field (pilot injections)	Porous (heterogeneous sandstone)	~37	~10 (est.)	up to ~1200 mD	~6.8 (low salinity)	In pilot: oil recovery ↑ by ~8% (after polymer flooding), water cut ↓	[107, 108]
Inorganic gel	Water glass (Na_2SiO_3) + Cr(III) salt (e.g., alum) + polyol (glycerin)	Russia (lab) Laboratory (fracture simulation)	Fractured/fractured-porous	≥100 (thermally stable)	Not specified (thermo-barostable)	Not available	Insensitive to salinity (no organics)	In core: RRF_water ≈ 180, RRF_oil ≈ 1.2—high water selectivity (>150)	[109]

4.2. Other Methods for Selective Permeability Reduction

Other technologies exist that are based on different mechanisms of barrier formation. For example, precipitate-forming systems are also effective tools under field conditions. These systems consist of chemical reagent compositions that rely on the precipitation of amphoteric or crystalline solids. An example of such a technology includes sodium carbonate and calcium chloride solutions. Other known approaches involve sulfate–sodium mixtures with calcium chloride, or sodium silicate injected together with inorganic acids. Injection is typically performed by displacing chemical slugs into the reservoir. Precipitation-based technologies were widely used in the past, and field experience has shown them to be both effective and economically accessible. The technology was actively applied by Surgutneftegaz, with over 250 wells treated between 2012 and 2014. However, field results indicated that the method was effective mainly during the second and third stages of field development, when reserve depletion was still relatively low. This limited applicability is likely due to the insufficient mobility of the precipitate barriers within the reservoir, as well as significant sensitivity to formation water salinity and pH conditions [103].

Curable systems, most commonly represented by resins, are also used in selective permeability reduction. The application of resins is currently proposed by a limited number of modern companies, yet several types are distinguished: epoxy resins, phenolic resins, furan resins, and crosslinked styrene–butadiene block copolymers [110,111]. The advantages of this technology include high mechanical strength, adhesion, thermal stability, and chemical resistance of the resins. Among the disadvantages, a key limitation is the short treatment radius from the wellbore due to the high viscosity of resin solutions. Additionally, the systems are chemically complex, which requires especially careful injection modeling. Resins are also expensive per unit volume, which often leads to the use of resin-transported dispersed phases to reduce consumption [112]. Field applications include: Furan resin treatments in production wells at the Kern River and San Ardo fields in California; Phenolic resin treatments at the Midway-Sunset field; Epoxy resin treatments at the Green Canyon field in the Gulf of Mexico.

Surfactant–foam systems currently have three main applications in oil production:

1. Mobility control agent during steam injection into the reservoir;
2. Mobility control agent during CO_2 injection;
3. Gas-blocking agent around production wells.

Despite their potential, foam systems have not seen widespread use, primarily because the injection process is technically complex and labor-intensive. Gel-based systems are generally preferred due to their greater operational simplicity and reliability.

Based on the information presented above regarding selective permeability reduction (SPR) compositions, as well as practical experience described in recent publications, a large number of modifications of various systems have been developed—many of which have now evolved into distinct classes of formulations. The main modifications that enhance the performance of flow-diverting technologies include:

1. Thermal stabilizers—chemical additives that prevent the degradation of chemical systems under high-temperature conditions;
2. Catalysts—substances that accelerate chemical reactions within the reservoir, for example, external and internal catalysts used in inorganic gel systems;
3. Stabilizers—compounds that ensure the integrity and functional performance of gel systems over time by providing resistance to salinity, pH fluctuations, shear stress, and filtration effects;
4. Fillers—solid particles introduced into the dispersed phase to increase mechanical strength and plugging capability. They enhance the mechanical properties of the

system and may provide additional filtration resistance. For instance, the combined addition of wood flour and clay powder significantly increases resistance to flow in water-swept zones: wood flour swells on contact with water, while clay flocculates with the polymer, forming a strong structural framework. Another example is the use of rubber crumb, where shredded recycled rubber is mixed with sodium silicate or bitumen; the crumb swells, blocking fractures and flow channels.

Thus, new subclasses of flow-diverting compositions are being developed, including thermotropic systems, polymer–dispersed systems, fiber–dispersed systems, gel–precipitate-forming systems, and others.

As previously noted, one of the criteria for an ideal flow-diverting system is the use of the most effective and relatively low-cost components. Under current conditions, a particularly relevant area of focus is the identification and implementation of industrial waste products as reagents for FDT applications. Such materials may include sludges from oil refineries, byproducts of metallurgical processes, and components from the pulp and paper or agricultural industries. Data from several projects involving the application of recycled materials have been summarized in Table 3.

Table 3. Case studies on the use of recycled products for profile control and water shutoff (PCWS). [compiled by the authors].

Waste Type	Composition	Country	Project Status	Reservoir Type	Reservoir Temperature, °C	Result	Source
Oil sludge (bottom sediment from oil storage tanks)	~40% oil, ~7% water, ~52% solids (silt, sand, etc.)	China	Field (injection into 15 injection wells)	Porous (sandstone, multilayered)	Not specified (effective at high temps)	Positive: after treatment, injection pressure increased, injection profile redistributed (absorption share in high-perm layer decreased from 71.3% to 44.5%; increased in low-perm layer from 9.7% to 45.5%). Additional oil recovery: ~12,186 tons from 15 wells; oil recovery factor increased by ~4.1%.	[113, 114]
Wood flour (finely ground sawdust; pulp and paper industry waste)	Cellulose and hemicellulose fibers, high lignin content (typical lignocellulosic material); possible additives (clay, reagents)	Russia	Field (pilot field trials)	Not specified (used in heterogeneous reservoir)	Not specified	Positive: field trials of wood flour-based system showed improved oil recovery and isolation of formation water influx	[115]
Cellulose flour (from agricultural waste, e.g., straw)	~40–58% microcellulose (from annual plants), ~40–60% fine hydromuscovite (clay filler), 1.5–5% thermally and salt-resistant polymer	Russia	Laboratory (patented formulation, tested on samples)	Not specified	Not specified (formulation stable at high temperatures)	Positive: the composition increases oil production and reduces water cut by selectively blocking water-bearing layers	[116]
Metallurgical sludge + ash and slag waste (blast furnace slag and thermal power plant ash in plugging formulation)	Mixture of powdered solid waste: granulated blast furnace slag, fly ash, clay component, perlite (total 50–85 wt.%); binder—phenol-formaldehyde resin (~10–45%); hardener—hexamethylenetetramine + benzoic acid (0.5–5%)	China	Laboratory (development and testing, patented)	Not specified	Not specified	Positive: composition is easy to prepare, reliably solidifies in formation, forms a durable long-term blocking barrier; low cost	[117]

Table 3. Cont.

Waste Type	Composition	Country	Project Status	Reservoir Type	Reservoir Temperature, °C	Result	Source
Blast furnace slag (metallurgical slag containing aluminum)	CaO (~39–41%), SiO_2 (~36%), Al_2O_3 (~10–13%)	China	Field	Fractured	High (≈80–150 °C)	Formation of a strong sealing structure; effective blocking of water-bearing fractures/channels (100% success in trials)	[118]
Oil sludge (from separators at TPPs, heavy oil)	~60–70% water, 10–15% oil, 15–30% solids (sand/clay; d ≈ 10–50 μm)	China	Laboratory + field (150 wells)	Porous	Not specified	After injection and solidification, sludge cemented large pores, redirecting flow to fine-pore zones; cumulative effect: +10,756 tons of additional oil recovered	[119]
Ground fly ash (coal)	Mainly SiO_2 (~92%) and Al_2O_3 (mullite ~8%)	China	Laboratory (patented)	Porous	Not specified	Mechanical "bridging" of pore channels; strong adhesion in pores, high thickening strength, low cost, long-term effect	[120]

The efficiency of gel-forming and flow-diverting compositions based on industrial waste is determined by their granulometric and chemical composition. Modified fly ash and blast furnace slag represent inorganic matrices with particle sizes of 5–50 μm and a predominance of SiO_2, Al_2O_3, and CaO. Activation with alkaline agents (NaOH 0.3–2%) and the use of borax or $CaCl_2$ as curing regulators enhance polymer adsorption and promote the formation of a dense silicate–polymer structure with a compressive strength of 0.8–1.5 MPa and thermal stability up to 300 °C [118].

Systems based on oily sludge contain both mineral and organic phases (10–25 wt.%), which determine the gelation kinetics. Finely dispersed particles (<20 μm) increase gel strength, while a moderate organic content imparts elasticity and hydrophobicity. When the organic phase is excessive, crosslinking slows down, reducing long-term stability. Optimization of the phase ratio allows regulation of the gelation time within 3–300 h, achieving sealing efficiency above 85% and thermal stability up to 350 °C [119].

Cellulose-based compositions [116] include microfibrous fillers and hydromicas, which provide reinforcement and salt tolerance at mineralization levels up to 200 g/L and temperatures of 120–150 °C. Natural biodegradable materials also exhibit selective water shutoff properties with minimal environmental impact, offering a sustainable alternative to synthetic systems.

Based on the analysis of available sources, several examples of industrial waste utilization as components for selective permeability reduction in reservoirs have been identified. Most case studies demonstrate a positive effect from implementation, and the diversity of waste products offers a wide range of chemical compositions, allowing their use in various modifications of base FDT systems. Bauxite residues deserve special mention, as they contain significant amounts of metal cation impurities, including Fe_2O_3, Al_2O_3, SiO_2, TiO_2, and Na_2O. In addition to their high content of reactive components, this byproduct is also highly alkaline due to the specifics of its processing. Other types of waste also contain multivalent metals, which, in turn, determine the direction of the authors' future research.

5. Results and Discussion

The current pace of field development inevitably leads to the formation of residual oil reserves. After the completion of primary recovery, up to 62.5% of oil reserves may remain unproduced in the reservoir. As noted by several authors, the low oil recovery factor observed at many Russian fields is a result of insufficient application of enhanced oil recovery methods, the implementation of which requires significant capital investment, as well as the lack of technologies capable of targeting the underlying causes of incomplete reservoir depletion [121].

Residual reserves in terrigenous reservoirs are formed at both the microscale, which accounts for approximately 30% of residual oil, and the macroscale, which accounts for about 70%. In the first case, as noted in fundamental sources, the primary cause of such accumulation is reservoir rock wettability, with oil existing in the form of films and dispersed droplets. In the second case, oil is present as accumulations—so-called pockets and lenses—formed due to reservoir heterogeneity, the viscoplastic properties of oil, and technological factors associated with the imperfections of the development system. There is no doubt that targeting oil reserves formed at the macroscale is a more universal and effective approach compared to attempting to recover capillary- and film-trapped oil. The predominance of macroscale residual reserves, as well as the availability of effective methods to act upon them, determines the authors' research focus on improving oil field development efficiency by involving previously undrained zones containing RRR.

At present, there are numerous field-proven methods for monitoring oil field development, many of which have long become routine practice for operating companies. Based on an analysis of modern advancements in this area, the authors have classified these monitoring methods into two groups: instrument-based methods and indirect (analytical) methods. The fundamental difference lies in how information is obtained regarding the current state of the development object. The instrumental methods group includes Vertical Seismic Profiling, well logging, well testing, and geochemical investigations. Each of these classes is actively evolving and, under current conditions, constitutes an essential part of field monitoring. Indirect or analytical methods refer to approaches for localizing and evaluating residual recoverable reserves. These include retrospective analysis, Voronoi diagram analysis, and the use of neural network algorithms, which—given the ongoing digital transformation of the industry—represent a promising direction. Among instrumental methods, the authors emphasize geochemical investigations, particularly the optical method for well analysis in the UV–visible–NIR spectrum. The optical properties of oil are extremely sensitive to changes in chemical composition during production, which, when properly implemented, can be effectively used as a tool for identifying and localizing zones containing RRR. Going forward, the authors plan to implement neural network approaches in combination with optical oil analysis to localize RRR zones—an approach that has the potential to become a highly effective technological solution.

The formation of dominant water-swept zones within the reservoir is an inevitable consequence of current field development systems, due to a number of factors. Although technological methods based on adjusting well operating parameters are effective tools for enhancing oil recovery, their application does not sufficiently enable the involvement of previously undrained reservoir zones. In this regard, given the substantial volumes of residual recoverable reserves at Russian oil fields, physicochemical enhanced oil recovery methods based on selective permeability reduction have proven highly effective. Injectivity profile control aimed at improving sweep efficiency, isolation of water-swept intervals, and redistribution of flow paths in the reservoir via the creation of remote blocking barriers are all recognized as efficient tools widely applied at oil fields around the world. Chemical systems for such applications can be classified by the physicochemical mechanism used to create the blocking barrier, namely into gel-forming systems, precipitate-forming systems, curable (setting) systems, and surfactant–foam systems.

Through source analysis, the authors have identified gel-forming systems as the most extensive, rapidly developing, and highly tunable group among selective permeability reduction technologies. Among gel systems, the most frequently applied are those based on Cr(III) and Al(III) crosslinked gels, biopolymer gels, inorganic gels based on sodium silicate, and gels based on synthetic organic polymers. Various modifications of these gels are also known, including the use of thermal stabilizers, catalysts, and the addition of dispersed fillers into the gel framework. Field applications have demonstrated the high efficiency of these compositions both in Russia and internationally. Under current technological and economic conditions, systems incorporating industrial waste products are gaining popularity. Numerous industrial sectors generate chemical components that can either serve as crosslinkers, form the base of a gel, or act as modifiers. Examples include byproducts from the pulp and paper, agricultural, metallurgical, and oil refining industries. Due to the presence of multivalent metals and natural polymers in production sludges, the authors propose this area as a promising direction for future research.

6. Conclusions

The present study outlines the principles of residual oil reserve formation at both the micro- and macroscale during oil field development, along with the methods for evaluation and localization of residual recoverable reserves and physicochemical enhanced oil recovery methods, specifically selective permeability reduction techniques, which represent one of the most promising directions in this field. Based on the results of this study, the following conclusions can be drawn:

1. At large oil fields in Russia, up to 62.5% of oil reserves remain unrecovered in the reservoir, of which 30% correspond to oil retained at the microscale and 70% to oil accumulated at the macroscale.
2. The formation of macroscale residual oil is caused by various types of reservoir heterogeneity, differences in oil and water properties, and imperfections in the pressure maintenance system.
3. A wide range of monitoring methods is used in the field to control reservoir development. These include both instrumental and indirect (analytical) methods. The key distinction between them lies in the manner of acquiring information about the reservoir.
4. A promising direction is the integration of optical well investigation methods in the UV–visible–NIR spectrum with neural network algorithms operating in situ. The combined implementation of these methods will improve mapping by enabling the localization of residual oil that differs significantly in its optical properties. However, the question of the feasibility and necessity of large-scale pilot projects for integrating neural network algorithms in field operations remains relevant.
5. Chemical methods for residual oil recovery based on selective permeability reduction have proven effective. These include gel-forming, precipitate-forming, curable, and surfactant–foam systems for creating blocking barriers. Gel systems, in particular, offer the highest degree of tunability.
6. A promising direction in the field of flow-diverting technologies is the use of waste products from various industries, including pulp and paper, agricultural, metallurgical, and petroleum refining sectors. These recycled materials may contain useful components such as crosslinking agents, gel-forming substances, and thermal stabilizers.

Author Contributions: I.R.: Conceptualization, Methodology, Investigation, Writing—Original Draft, Visualization, Project administration; M.R.: Validation, Writing—Review and Editing, Supervision; E.S.: Formal Analysis, Resources, Data Curation, Writing—Original Draft. All authors have read and agreed to the published version of the manuscript.

Funding: This research received no external funding.

Data Availability Statement: Not applicable.

Conflicts of Interest: The authors declare no conflict of interests.

References

1. Ministry of Natural Resources and Environment of the Russian Federation, Federal Agency for Subsoil Use. *State Report on the Condition and Use of Mineral Resources of the Russian Federation in 2022*; Ministry of Natural Resources and Environment of the Russian Federation: Moscow, Russia, 2023; 626p. Available online: http://rosnedra.gov.ru/ (accessed on 1 June 2025). (In Russian)
2. PJSC "LUKOIL". *Key Trends in the Global Liquid Hydrocarbon Market Development up to 2035*; PJSC "LUKOIL": Moscow, Russia, 2019; 110p, Available online: https://lukoil.ru/Business/Futuremarkettrends (accessed on 1 June 2025). (In Russian)
3. Litvinenko, V.S.; Petrov, E.I.; Vasilevskaya, D.V.; Yakovenko, A.V.; Naumov, I.A.; Ratnikov, M.A. Assessment of the role of the state in the management of mineral resources. *J. Min. Inst.* **2023**, *259*, 95–111. [CrossRef]
4. FGBU "VSEGEI". *Report on the State and Prospects for the Use of the Mineral Resource Base of the Republic of Tatarstan as of 15.03.2021*; Federal State Budgetary Institution "VSEGEI": Saint Petersburg, Russia, 2021. Available online: https://rosnedra.gov.ru/data/Fast/Files/202104/0d7902ff118308f4bc272cdb94c47d02.pdf (accessed on 1 June 2025). (In Russian)
5. Dvoynikov, M.V.; Minaev, Y.D. Mathematical Model of Non-Pressurized Flow for Calculating Killing of Gas Wells with Abnormally Low Reservoir Pressures. *Int. J. Eng.* **2025**, *38*, 1677–1684. [CrossRef]

6. Melnikov, П.Н.; Varlamov, A.I.; Fortunatova, N.; Poroskun, В.И.; Soloviev, A.V.; Skvortsov, M.B.; Kravchenko, M.N.; Kanev, A.S.; Sotnikova, A.G. Results of Quantitative Estimation of Unconventional Oil Resources of the Russian Federation. *Geol. Geophys.* **2024**, *65*, 8–25. [CrossRef]
7. Wang, L.; Zhu, L.; Cao, Z.; Liu, J.; Xue, Y.; Wang, P.; Cao, X.; Liu, Y. Thermo-Mechanical Degradation and Fracture Evolution in Low-Permeability Coal Subjected to Cyclic Heating–Cryogenic Cooling. *Phys. Fluids* **2025**, *37*, 086617. [CrossRef]
8. Wang, L.; Zhu, L.; Xue, Y.; Cao, X.; Liu, G. Multiphysics Modeling of Thermal–Fluid–Solid Interactions in Coalbed Methane Reservoirs: Simulations and Optimization Strategies. *Phys. Fluids* **2025**, *37*, 076649. [CrossRef]
9. Shishlyannikov, D.; Zvonarev, I.; Rybin, A.; Zverev, V.; Ivanchenko, A. Assessment of Changes in the Abrasiveness of Solid Particles in Hydraulic Mixtures Pumped with ESPs. *Appl. Sci.* **2023**, *13*, 1885. [CrossRef]
10. Burkhanov, R.N.; Lutfullin, A.A.; Raupov, I.R.; Maksyutin, A.V.; Valiullin, I.V.; Farrakhov, I.M.; Shvydenko, M.V. Localization and Involvement in Development of Residual Recoverable Reserves of a Multilayer Oil Field. *J. Min. Inst.* **2024**, *268*, 599–612, EDN DKXZSP. Available online: https://pmi.spmi.ru/pmi/article/view/15996 (accessed on 1 June 2025).
11. Burkhanov, R.N.; Lutfullin, A.A.; Ibragimov, I.I.; Maksyutin, A.V. Estimation and Localization of Residual Recoverable Oil Reserves by the Complex of Filtration, Optical and Field Research. In Proceedings of the Society of Petroleum Engineers—SPE Russian Petroleum Technology Conference 2020, RPTC 2020, Virtual, 26–29 October 2020. [CrossRef]
12. Pshenin, V.; Sleptsov, A.; Dukhnevich, L. Prospects of Improving the Vibroacoustic Method for Locating Buried Non-Metallic Pipelines. *Eng* **2025**, *6*, 121. [CrossRef]
13. Surguchev, M.L. *Methods of Monitoring and Regulating the Oil Field Development Process*; Nedra: Moscow, Russia, 1968; Available online: https://www.geokniga.org/books/33112 (accessed on 1 June 2025). (In Russian)
14. Zheltov, Y.V.; Surguchev, M.L.; Simkin, E.M. *Physicochemical Microprocesses in Oil and Gas Reservoirs*; Nedra: Moscow, Russia, 1984; Available online: https://www.geokniga.org/books/32008 (accessed on 1 June 2025). (In Russian)
15. Gazizov, A.A. *Enhanced Oil Recovery from Heterogeneous Reservoirs at the Late Stage of Development*; Nedra-Businesscenter LLC.: Moscow, Russia, 2002; ISBN 5-8365-0119-X. Available online: https://www.geokniga.org/books/2656 (accessed on 1 June 2025). (In Russian)
16. Gazizov, A.S.; Gazizov, A.A. *Improving the Efficiency of Oil Field Development Based on Water Movement Control in Reservoirs*; Nedra-Businesscenter LLC.: Moscow, Russia, 1999; ISBN 5-8365-0022-3. Available online: https://www.geokniga.org/bookfiles/geokniga-povyshenieeffektivnostirazrabotki.pdf (accessed on 1 June 2025). (In Russian)
17. Devlikamov, V.V.; Kabirov, M.M.; Khabibullin, Z.A. *Anomalous Oils*; Nedra: Moscow, Russia, 1975; Available online: https://www.geokniga.org/books/16600 (accessed on 1 June 2025). (In Russian)
18. Craig, F.F. *The Reservoir Engineering Aspects of Waterflooding*; SPE: Richardson, TX, USA, 1980; ISBN 0-89520-202-6. Available online: https://www.google.nl/books/edition/The_Reservoir_Engineering_Aspects_of_Wat/aThQAQAAIAAJ?hl=ru&gbpv=0&bsq=TheReservoirEngineeringAspectsofWaterfloodingcraigF.F (accessed on 1 June 2025).
19. Willhite, G.P. *Waterflooding: Textbook 3 (Spe Textbook Series)*; Society of Petroleum Engineers: Houston, TX, USA, 2011; ISBN 978-1-55563-005-8. Available online: https://www.google.ru/books/edition/Waterflooding/AVpQAQAAIAAJ?hl=ru&gbpv=0&bsq=Waterflooding:%20Textbook%203%20(Spe%20Textbook%20Series)%20-%20Softcover (accessed on 1 June 2025).
20. Rose, S.C.; Buckwalter, J.F.; Woodhall, R.J. *The Design Engineering Aspects of Waterflooding*; Society of Petroleum Engineers: Houston, TX, USA, 1989; ISBN 978-1-55563-016-4. Available online: https://www.google.nl/books/edition/The_Design_Engineering_Aspects_of_Waterf/KzlQAQAAIAAJ?hl=ru&gbpv=0&bsq=designengineeringaspectsofwaterflooding (accessed on 1 June 2025).
21. Lake, W.L. *Enhanced Oil Recovery*; Prentice Hall: Saddle River, NJ, USA, 2010; Available online: https://books.google.ru/books/about/Enhanced_Oil_Recovery.html?id=oERUDQEACAAJ&redir_esc=y (accessed on 1 June 2025).
22. Putilov, I.S.; Chizhov, D.B.; Kochergin, E.A. Appearance of Capillary End Effects in Filtration Studies. *Perm J. Pet. Min. Eng.* **2021**, *21*, 23–27. [CrossRef]
23. Morenov, V.A.; Shchigolev, K. Assessment of the Possibility of Carbon Dioxide Utilization for Enhanced Oil Recovery in Oil Field Conditions. *Neft. Khozyaystvo Oil Ind.* **2025**. [CrossRef]
24. Gabrielyants, G.A. *Geology of Oil and Gas Fields*, 2nd ed.; Nedra: Moscow, Russia, 1979; Available online: https://www.geokniga.org/books/16574 (accessed on 1 June 2025). (In Russian)
25. Ghanbarian, B.; Liang, F.; Liu, H. Unconventional Reservoirs. In *Physics of Fluid Flow and Transport in Unconventional Reservoir Rocks*; John Wiley & Sons: Hoboken, NJ, USA, 2023; pp. 1–14. [CrossRef]
26. Wu, J.; Ansari, U. From CO_2 Sequestration to Hydrogen Storage: Further Utilization of Depleted Gas Reservoirs. *Reserv. Sci.* **2025**, *1*, 19–35. [CrossRef]
27. Wang, F.; Kobina, F. The Influence of Geological Factors and Transmission Fluids on the Exploitation of Reservoir Geothermal Resources: Factor Discussion and Mechanism Analysis. *Reserv. Sci.* **2025**, *1*, 3–18. [CrossRef]

28. Wang, Z.; Liu, K.; Zhang, C.; Yan, H.; Yu, J.; Yu, B.; Liu, J.; Jiang, T.; Dan, W.; Hu, C. Integral Effects of Porosity, Permeability, and Wettability on Oil–Water Displacement in Low-Permeability Sandstone Reservoirs—Insights from X-Ray CT-Monitored Core Flooding Experiments. *Processes* **2023**, *11*, 2786. [CrossRef]
29. *GB/T 29172-2012*; Practices for Core Analysis. China Standards Press: Beijing, China, 2012.
30. *ASTM D4520-18*; Standard Practice for Determining Water Injectivity Through the Use of On-Site Floods. ASTM International: West Conshohocken, PA, USA, 2018.
31. *GOST 26450.1-85*; Rocks. Method for Determination of Open Porosity Coefficient by Fluid Saturation. Publishing House of Standards: Moscow, USSR, 1985.
32. *GOST 26450.2-85*; Rocks. Method for Determination of Absolute Gas Permeability Coefficient by Steady and Unsteady Filtration. Publishing House of Standards: Moscow, USSR, 1985.
33. Fan, H.; Liu, X.; Li, G.; Li, X.; Radwan, A.E.; Yin, S. Study on Characteristics, Efficiency, and Variations of Water Flooding in Different Stages for Low Permeability Oil Sandstone. *Energy Sci. Eng.* **2024**, *12*, 5245–5265. [CrossRef]
34. Zhou, Y.; Yang, W.; Yin, D. Experimental Investigation on Reservoir Damage Caused by Clay Minerals after Water Injection in Low Permeability Sandstone Reservoirs. *J. Pet. Explor. Prod. Technol.* **2022**, *12*, 915–924. [CrossRef]
35. He, J.; Liu, X.; Zhu, X.; Jiang, T.; He, H.; Zhou, L.; Liu, Q.; Zhu, Y.; Liu, L. Water-Flooding Characteristics of Lithologic Reservoir in Ordos Basin. *Sci. Rep.* **2021**, *11*, 2503. [CrossRef]
36. Sun, X.; Suo, L.; Huang, Y.; Wang, H.; Yu, H.; Xu, C.; Xu, J.; Qin, X.; Sun, W.; Cao, Y.; et al. Study on the Occurrence Characteristics of the Remaining Oil in Sandstone Reservoirs with Different Permeability after Polymer Flooding. *Polymers* **2024**, *16*, 1902. [CrossRef]
37. Liu, Y.; Ge, L.; Ma, K.; Chen, X.; Zhu, Z.; Hou, J. Study on Surfactant–Polymer Flooding after Polymer Flooding in High-Permeability Heterogeneous Offshore Oilfields: A Case Study of Bohai S Oilfield. *Polymers* **2024**, *16*, 2004. [CrossRef]
38. Chen, X.; Zhang, Z.; Liu, Q.; Sun, L.; Xiao, H.; Gao, J.; Kang, H. Physical Modeling of Development Adjustment Mechanism for Heterogeneous Thick Oil Reservoir. *Front. Energy Res.* **2023**, *10*, 1094697. [CrossRef]
39. Wang, H.; Torabi, F.; Zeng, F.; Xiao, H. A Novel Visualization Approach for Foamy Oil Non-Equilibrium Phase Behavior Study of Solvent/Live Heavy Oil Systems. *Fuel* **2020**, *272*, 117648. [CrossRef]
40. Lee, J.; Babadagli, T. Comprehensive Review on Heavy-Oil Emulsions: Colloid Science and Practical Applications. *Chem. Eng. Sci.* **2020**, *228*, 115962. [CrossRef]
41. Zheng, Z. Hydraulic Fractures of the Cusp Shape in a Hele-Shaw Cell. *J. Fluid Mech.* **2024**, *995*, A17. [CrossRef]
42. Howison, S. *Practical Applied Mathematics Modelling, Analysis, Approximation*; Cambridge University Press: Cambridge, UK, 2005; ISBN 978-0-521-84274-7. Available online: https://www.google.ru/books/edition/Practical_Applied_Mathematics/A2Hy_54Y1MsC?hl=ru&gbpv=0 (accessed on 1 June 2025).
43. Aime, S.; Sabato, M.; Xiao, L.; Weitz, D.A. Dynamic Speckle Holography. *Phys. Rev. Lett.* **2021**, *127*, 088003. [CrossRef]
44. Yadali Jamaloei, B. Effect of Wettability on Immiscible Viscous Fingering: Part I. Mechanisms. *Fuel* **2021**, *304*, 120726. [CrossRef]
45. Singh, A.; Pandey, K.M.; Singh, Y. Numerical Investigation of Immiscible Liquid-Liquid Displacement in Hele-Shaw Cell. *Mater. Today Proc.* **2020**, *45*, 7151–7155. [CrossRef]
46. Guerrero, F.; Bryan, J.; Kantzas, A. Visualization of Chemical Heavy Oil Eor Displacement Mechanisms in a 2d System. *Energies* **2021**, *14*, 950. [CrossRef]
47. Raza, S.; Gates, I.D. Effect of Cellulose Nanocrystal Nanofluid on Displacement of Oil in a Hele-Shaw Cell. *J. Pet. Sci. Eng.* **2021**, *196*, 108068. [CrossRef]
48. An, B.; Solorzano, D.; Yuan, Q. Viscous Fingering Dynamics and Flow Regimes of Miscible Displacements in a Sealed Hele-Shaw Cell. *Energies* **2022**, *15*, 5798. [CrossRef]
49. Ratanpara, A.; Kim, M. Wettability Alteration Mechanisms in Enhanced Oil Recovery with Surfactants and Nanofluids: A Review with Microfluidic Applications. *Energies* **2023**, *16*, 8003. [CrossRef]
50. Dvoynikov, M.V.; Nikitin, V.I.; Kopteva, A.I. Analysis of Methodology for Selecting Rheological Model of Cement Slurry for Determining Technological Parameters of Well Casing. *Int. J. Eng.* **2024**, *37*, 2042–2050. [CrossRef]
51. Savenok, O.V.; Chuikova, E.P.; Agaguena, A. Justification of Waterless Hydraulic Fracturing Methods. *Int. J. Eng.* **2025**, *38*, 1462–1475. [CrossRef]
52. Kang, H.; Li, G.; Gao, J. Study on Permeability Characteristics of Tight Oil Reservoir through High-Pressure Mercury Injection. *Front. Energy Res.* **2023**, *10*, 1023099. [CrossRef]
53. Jia, D.; Zhang, J.; Li, Y.; Wu, L.; Qiao, M. Recent Development of Smart Field Deployment for Mature Waterflood Reservoirs. *Sustainability* **2023**, *15*, 784. [CrossRef]
54. Jia, D.; Zhang, J.; Sun, Y.; Wang, S.; Gao, S.; Qiao, M.; Li, Y.; Qu, R. Collaboration between Oil Development and Water/Power Consumption in High-Water-Cut Oilfields. *Sustainability* **2023**, *15*, 11405. [CrossRef]
55. Smirnova, E.; Saychenko, L. Hydrodynamic Modeling and Evaluation of Partial Substitution of Cushion Gas During Creation of Temporary Underground Gas Storage in an Aquifer. *Int. J. Eng.* **2024**, *37*, 1221–1230. [CrossRef]

56. Zheng, Y.; Wang, Y.; Chang, X. VSP Imaging Using Free-Surface Multiples with Wavefield Decomposition: Synthetic and Field Data Examples. *Front. Earth Sci.* **2021**, *9*, 730184. [CrossRef]
57. Isaenkov, R.; Tertyshnikov, K.; Yurikov, A.; Shashkin, P.; Pevzner, R. Effect of Source Mispositioning on the Repeatability of 4D Vertical Seismic Profiling Acquired with Distributed Acoustic Sensors. *Sensors* **2022**, *22*, 9742. [CrossRef]
58. Bayamirova, R.; Togasheva, A.; Saduakasov, D.; Zholbasarova, A.; Tabylganov, M.; Gusmanova, A.; Sarbopeeva, M.; Nauyryzova, B.; Nugumarov, S. Reservoir Development and Well Operation Control Methods: Practical Application. *Processes* **2025**, *13*, 1541. [CrossRef]
59. Zhou, W.; Lin, Y.; Gao, G.; Wang, P. Enhanced Carbon/Oxygen Ratio Logging Interpretation Methods and Applications in Offshore Oilfields. *Processes* **2024**, *12*, 2301. [CrossRef]
60. Luo, R.; Liu, J.; Yang, D.; Deng, Q. Application and Analysis of Array Production Logging Technology for Multiphase Flow in Horizontal Wells. *Processes* **2023**, *11*, 3421. [CrossRef]
61. Gao, J.; Wang, X.; Zhang, D.; Wang, J. The Formation Mechanism of Residual Oil and Methods of Enhanced Oil Recovery in a Fractured Low-Permeability Metamorphic Rock Reservoir in Bohai Bay. *Processes* **2024**, *12*, 2281. [CrossRef]
62. Patidar, A.K.; Joshi, D.; Dristant, U.; Choudhury, T. A Review of Tracer Testing Techniques in Porous Media Specially Attributed to the Oil and Gas Industry. *J. Pet. Explor. Prod. Technol.* **2022**, *12*, 3339–3356. [CrossRef]
63. Yang, H.; Guo, K.; Lin, L.; Zhang, S.; Wang, Y. Application of Micro-Substance Tracer Test in Fractured Horizontal Wells. *J. Pet. Explor. Prod. Technol.* **2024**, *14*, 1235–1246. [CrossRef]
64. Bouguer, P. *Essay on Optics on the Gradation of Light*; Claude Jombert: Paris, France, 1729; Available online: https://archive.org/details/UFIE003101_TO0324_PNI-2703_000000/page/74/mode/2up (accessed on 1 June 2025).
65. Gomaa, S.; Soliman, A.A.; Mansour, M.; El Salamony, F.A.; Salem, K.G. Machine Learning Models for Estimating the Overall Oil Recovery of Waterflooding Operations in Heterogenous Reservoirs. *Sci. Rep.* **2025**, *15*, 14619. [CrossRef]
66. Maghsoudian, A.; Izadpanahi, A.; Bahmani, Z.; Avvali, A.H.; Esfandiarian, A. Utilizing Deterministic Smart Tools to Predict Recovery Factor Performance of Smart Water Injection in Carbonate Reservoirs. *Sci. Rep.* **2025**, *15*, 537. [CrossRef]
67. Sudakov, V.A.; Safuanov, R.I.; Kozlov, A.N.; Porivaev, T.M.; Zaikin, A.A.; Zinykov, R.A.; Lutfullin, A.A.; Farhutdinov, I.Z.; Tylyakov, I.Z. Localization and Development of Residual Oil Reserves Using Geochemical Studies Based on Neural Network Algorithms. *Georesursy* **2022**, *24*, 50–64. [CrossRef]
68. Galkin, S.; Krivoshchekov, S.; Kozyrev, N.; Kochnev, A.; Mengaliev, A. Accounting of Geomechanical Layer Properties in Multi-Layer Oil Field Development. *J. Min. Inst.* **2020**, *244*, 408–417. [CrossRef]
69. Gruenenfelde, M.A.; Zaitoun, A.; Kohler, N.; Ali, S.A.; Linser, T.M. *Implementing New Permeability Selective Water Shutoff Polymer Technology in Offshore, Gravel-Packed Wells*; Society of Petroleum Engineers: Tulsa, OK, USA, 1994.
70. Stanley, F.O.; Tanggu, P.S.; Marnoch, E. Amphoteric Polymer Improves Hydrocarbon/Water Ratios in Producing Wells—An Indonesian Case Study. *SPE Prod. Facil.* **1997**, *12*, 181–186. [CrossRef]
71. Malozyomov, B.V.; Martyushev, N.V.; Kukartsev, V.V.; Tynchenko, V.S.; Bukhtoyarov, V.V.; Wu, X.; Tyncheko, Y.A.; Kukartsev, V.A. Overview of Methods for Enhanced Oil Recovery from Conventional and Unconventional Reservoirs. *Energies* **2023**, *16*, 4907. [CrossRef]
72. Kashirina, K.O.; Elov, I.N. Review of Domestic and Foreign Applications of Conformance Control Technologies. *Nat. Sci.* **2016**, *85*, 10. Available online: https://www.elibrary.ru/item.asp?id=26361702 (accessed on 1 June 2025). (In Russian).
73. Ketova, Y.; Bai, B.; Khizhnyak, G.; Gladkikh, Y.; Galkin, S. Testing of Preformed Particles Polymer Gel Technology on Core Filtration Models to Limit Water Inflows. *J. Min. Inst.* **2020**, *241*, 91. [CrossRef]
74. Pu, J.; Bai, B.; Schuman, T. Systematic Evaluation of a Novel Self-Healing Poly(Acrylamide-Co-Vinyl Acetate)/Alginate Polymer Gel for Fluid Flow Control in High Temperature and High Salinity Reservoirs. *Polymers* **2021**, *13*, 3616. [CrossRef]
75. Kawai, K.; Zhao, W.; Nishimura, S.; Yamada, A. High-Voltage Cr 4+/Cr 3+ Redox Couple in Polyanion Compounds. *ACS Appl. Energy Mater.* **2018**, *1*, 928–931. [CrossRef]
76. Neto, V.J.S.; Dores, F.G.L.; Oliveira, P.F.; Mansur, C.R.E. Evaluation of the Influence of Adding Clay in Polymeric Hydrogels Based on Partially Hydrolyzed Polyacrylamide and Aluminum Citrate. *J. Appl. Polym. Sci.* **2023**, *141*, e55019. [CrossRef]
77. Shamlooh, M.; Hussein, I.; Nasser, M.; Magzoub, M.; Salehi, S. Development of pH-Controlled Aluminum-Based Polymeric Gel for Conformance Control in Sour Gas Reservoirs. *ACS Omega* **2020**, *5*, 24504–24512. [CrossRef] [PubMed]
78. Galkin, S.V.; Rozhkova, Y.A. Analysis of Experience in the Use of Preformed Particle Polymer Gels in the Development of High-Water-Cut Production Facilities in Low-Temperature Oil Reservoirs. *J. Min. Inst.* **2024**, *265*, 55–64, EDN 20CNCFIW. Available online: https://pmi.spmi.ru/pmi/article/view/14686 (accessed on 1 June 2025). (In Russian).
79. Rátzsch, M.; Schade, H.; Pabst, P. The Conversion of Ethylene–Vinyl Acetate Copolymers with Aluminum Alcoholates. *J. Appl. Polym. Sci.* **1971**, *15*, 589–595. [CrossRef]
80. Song, T.; Bai, B.; Eriyagama, Y.; Schuman, T. Lysine Crosslinked Polyacrylamide—A Novel Green Polymer Gel for Preferential Flow Control. *ACS Appl. Mater. Interfaces* **2023**, *15*, 4419–4429. [CrossRef]

81. Nimir, H.; Hamza, A.; Hussein, I. Development of Greener D-Metal Inorganic Crosslinkers for Polymeric Gels Used in Water Control in Oil and Gas Applications. *Energies* **2020**, *13*, 4262. [CrossRef]
82. Mao, J.; Mao, J.; Liu, B.; Xiao, Y.; Yang, X.; Lin, C.; Zhang, Y.; Wang, Q.; Zhang, Q. Study of Crosslinker Size on the Rheological Properties of Borate Crosslinked Guar Gum. *Int. J. Biol. Macromol.* **2023**, *231*, 123284. [CrossRef]
83. Zhang, L.; Chen, F.; Wang, H.; He, J.; Luo, M.; Chen, H.; Zhong, C. Starch Microsphere Silicon-Boron Crosslinker for Low Concentration Hydroxypropyl Guar Gum Based Fracturing Fluid. *Int. J. Biol. Macromol.* **2024**, *264*, 130575. [CrossRef]
84. Gao, X.; Huang, L.; Xiu, J.; Yi, L.; Zhao, Y. Evaluation of Viscosity Changes and Rheological Properties of Diutan Gum, Xanthan Gum, and Scleroglucan in Extreme Reservoirs. *Polymers* **2023**, *15*, 4338. [CrossRef] [PubMed]
85. Afdhol, M.K.; Hidayat, F.; Erfando, T.; Purnama, D.P. Pectin Extraction of Jackfruit Peel as a Biopolymer Potential with Microwave Assisted Extraction Method. *Sci. Contrib. Oil Gas* **2024**, *47*, 125–133. [CrossRef]
86. Ezeh, O.S.; Ikiensikimama, S.S.; Akaranta, O. Comparative Analysis of the Effects of Monovalent and Divalent Ions on Imported Biopolymer-Xanthan Gum and Locally Formulated Biopolymers-Gum Arabic and Terminalia Mantaly. *J. Energy Res. Rev.* **2021**, *9*, 33–46. [CrossRef]
87. Abou-alfitooh, S.A.M.; El-hoshoudy, A. Eco-Friendly Modified Biopolymers for Enhancing Oil Production: A Review. *J. Polym. Environ.* **2023**, *32*, 2457–2483. [CrossRef]
88. Medvedeva, V.V.; Myasnikova, L.; Semchikov, Y.; Rogovina, L. Kinetics of Coordination Networks in the Sodium Carboxymethyl Cellulose-Cr^{3+} Salt System. *Polym. Sci. Ser. B* **1998**, *40*, 80–84. Available online: https://www.researchgate.net/publication/286654559_Kinetics_of_coordination_networks_in_the_sodium_carboxymethyl_cellulose-Cr3_salt_system?_tp=eyJjb250ZXh0Ijp7ImZpcnN0UGFnZSI6InB1YmxpY2F0aW9uIiwicGFnZSI6InNjaWVudGlmaWNDb250cmlidXRpb25zIiwicHJldmlvdXNQYWdlIjoicHVibGljYXRpb24ifX0 (accessed on 1 June 2025).
89. Medvedeva, V.V.; Semchikov, Y. Effect of pH on Gelation in the Sodium Carboxymethyl Cellulose-Cr^{3+} Salt System. *Polym. Sci. Ser. B* **2001**, *43*, 162–165. Available online: https://www.researchgate.net/publication/288978541_Effect_of_pH_on_Gelation_in_the_Sodium_Carboxymethyl_Cellulose-Cr3_Salt_System (accessed on 1 June 2025).
90. Abbas, A.H.; Moslemizadeh, A.; Sulaiman, W.; Jaafar, M.; Pourafshary, P. Lignin Adsorption in Sandpacks during Horizontal Flow Using UV–Vis Spectrometry as a Quantification Tool. *Arab. J. Geosci.* **2022**, *15*, 501. [CrossRef]
91. Cui, S.; Qi, B.; Liu, H.; Sun, X.; He, R.; Lian, J.; Li, Y.; Lu, J.; Bao, M. Aluminum Soap Nanoparticles-Lignin Powder Form Phase-Selective Gelator as an Efficient Sorbent for Oils/Water Separation. *Chemosphere* **2023**, *340*, 139803. [CrossRef]
92. Yang, Z.; Gao, F.; Guo, J.; Gao, Q.-Y.; Xu, Y.-Y.; Yang, H.; Xu, Z.-X.; Leng, L. Improving Temperature and Salinity Tolerance of Polymer Gels by Modifying with Lignin-Containing Nanocellulose. *Fuel* **2025**, *384*, 134019. [CrossRef]
93. Saghafi, H.; Naderifar, A.; Gerami, S.; Emadi, M. Improvement in Thermo-chemical Stability of Nanocomposite Preformed Particle Gels for Conformance Control in Harsh Oil Reservoir Conditions. *Can. J. Chem. Eng.* **2016**, *94*, 1880–1890. [CrossRef]
94. AlSofi, A.; Dokhon, W. Toward Deep Diversion for Waterflooding and EOR: From Representative Delayed Gelation to Practical Field-Trial Design. In Proceedings of the SPE Improved Oil Recovery Conference, Virtual, 25 April 2022. [CrossRef]
95. Amir, Z.; Saaid, I.; Jan, B.M.; Patah, M.F.A.; Khalil, M.; Bakar, W.Z.W. PAM/PEI Polymer Gel for Water Control in High-Temperature and High-Pressure Conditions: Core Flooding with Crossflow Effect. *Korean J. Chem. Eng.* **2022**, *39*, 605–615. [CrossRef]
96. Wang, K.; Guo, Y.; Wen, J.; Yang, H.; Zhang, H. Magnetic Smart Polymer Gel with Directional Plugging for Conformance Control in Oil Reservoirs. *J. Mol. Liq.* **2024**, *405*, 125046. [CrossRef]
97. Zhao, Y.; Leng, J.; Lin, B.; Wei, M.; Bai, B. Experimental Study of Microgel Conformance-Control Treatment for a Polymer-Flooding Reservoir Containing Superpermeable Channels. *SPE J.* **2021**, *26*, 2305–2317. [CrossRef]
98. Cao, W.; Xie, K.; Cao, B.; Lu, X.; Tian, Z. Inorganic Gel Enhanced Oil Recovery in High Temperature Reservoir. *J. Pet. Sci. Eng.* **2021**, *196*, 107691. [CrossRef]
99. Tang, X.; Yang, L.; Liu, Y.; Li, Z.; Chang, Z.; Li, Y.; Wei, F. A New In-Depth Fluid Diverting Agent of Inorganic Gel Coating. *Pet. Explor. Dev.* **2012**, *39*, 82–87. [CrossRef]
100. Zulkifli, N.; Hendraningrat, L.; Razali, N.; Bakri, C.N.; Zainal, S.; Kechut, N.I. Evaluation of Flow Diverter Chemical to Improve Waterflood Performance as Conformance Control for Heterogeneous Reservoir and High-Temperature Field Application: An Innovative Experimental Design. In Proceedings of the International Petroleum Technology Conference, Bangkok, Thailand, 1 November 2022. [CrossRef]
101. Portwood, J.T.; Romero, J.L. Waterflood Conformance Improvement—Practical Considerations & Lessons Learned. *Rev. Fuentes El Reventón Energético* **2018**, *16*, 7–21. Available online: https://revistas.uis.edu.co/index.php/revistafuentes/article/view/9156/8998 (accessed on 1 June 2025).
102. Al Brahim, A.; Bai, B.; Schuman, T. Comprehensive Review of Polymer and Polymer Gel Treatments for Natural Gas-Related Conformance Control. *Gels* **2022**, *8*, 353. [CrossRef]

103. Zemtsov, Y.V.; Mazaev, V.V. *Current State of Physicochemical Methods for Enhanced Oil Recovery: A Literature and Patent Review*; Publishing Solutions LLC.: Yekaterinburg, Russia, 2021; Available online: http://eues.ru/sites/default/files/2022-02/%D0%9C%D0%BE%D0%BD%D0%BE%D0%B3%D1%80%D0%B0%D1%84%D0%B8%D1%8F%20%D0%A4%D0%A5%20%D0%9C%D0%A3%D0%9D_2021_%D0%97%D0%B5%D0%BC%D1%86%D0%BE%D0%B2-%D0%9C%D0%B0%D0%B7%D0%B0%D0%B5%D0%B2.pdf (accessed on 1 June 2025). (In Russian)
104. Al-Muntasheri, G.A.; Nasr-El-Din, H.A.; Zitha, P.L.J. Gelation Kinetics and Performance Evaluation of an Organically Crosslinked Gel at High Temperature and Pressure. *SPE J.* **2008**, *13*, 337–345. [CrossRef]
105. Dastjani-Farahani, R.; Asadzadeh, N.; Jafari, A.; Vafaie-Sefti, M.; Baghban-Salehi, M. Enhancing Xanthan Hydrogels for Water Shutoff Treatment in Sandstone Oil Reservoirs through Core Flooding and CCD Approach. *Sci. Rep.* **2025**, *15*, 14016. [CrossRef]
106. Han, J.; Sun, J.; Lv, K.; Yang, J.; Li, Y. Polymer Gels Used in Oil–Gas Drilling and Production Engineering. *Gels* **2022**, *8*, 637. [CrossRef]
107. Pi, Y.; Liu, J.; Cao, R.; Liu, L.; Ma, Y.; Gu, X.; Li, X.; Fan, X.; Zhao, M. Visualized Study on a New Preformed Particle Gels (PPG) + Polymer System to Enhance Oil Recovery by Oil Saturation Monitoring Online Flooding Experiment. *Gels* **2023**, *9*, 81. [CrossRef]
108. Liu, H.; Han, H.; Li, Z.; Wang, B. Granular-Polymer-Gel Treatment Successful in the Daqing Oil Field. *SPE Prod. Oper.* **2006**, *21*, 142–145. [CrossRef]
109. Duryagin, V.; Nguyen Van, T.; Onegov, N.; Shamsutdinova, G. Investigation of the Selectivity of the Water Shutoff Technology. *Energies* **2023**, *16*, 366. [CrossRef]
110. Telin, A.; Vezhnin, S.; Safarov, F.E.; Lenchenkova, L.; Yakubov, R.N.; Safuanova, R.M. Application of Hydrogels for Conformance Control and Flow Diversion. *Pet. Eng.* **2023**, *21*, 58–77. [CrossRef]
111. Meng, X.; Zhang, G.; Wu, J.; Zhao, X.; Wang, L.; Zhang, F. Experimental Study on Phenol-Formaldehyde Resin Aggregates as In-Depth Conformance Control Agents Stabilized by Polymer. *Polymers* **2022**, *14*, 3159. [CrossRef] [PubMed]
112. Wu, Q.; Ge, J.; Ding, L.; Wei, K.; Deng, X.; Liu, Y. Identification and Characterization of a Proper Gel System for Large-Volume Conformance Control Treatments in a Fractured Tight Reservoir: From Lab to Field. *J. Pet. Sci. Eng.* **2020**, *198*, 108199. [CrossRef]
113. Liu, Y.; Bai, B.; Wang, Y. Applied Technologies and Prospects of Conformance Control Treatments in China. *Oil Gas Sci. Technol.—Rev. IFP Energ. Nouv.* **2010**, *65*, 859–878. [CrossRef]
114. Xu, C.; Bi, W.; Li, X.; Yang, H.; Xue, F.; Yi, P. *Application Prospects for Harmless Treatments of Oily Sludge in Changqing Oilfield*; National Engineering Laboratory for Exploration and Development of Low Permeability Oil and Gas Field, Oil & Gas Technology Research Institute of Changqing Oilfield Company: Xi'an, China, 2015; Available online: https://www.atlantis-press.com/article/19576.pdf (accessed on 1 June 2025).
115. Meshalkin, V.; Akhmetov, A.; Lenchenkova, L.; Nzioka, A.; Politov, A.; Strizhnev, V.; Telin, A.; Fakhreeva, A. Application of Renewable Natural Materials for Gas and Water Shutoff Processes in Oil Wells. *Energies* **2022**, *15*, 9216. [CrossRef]
116. Farid Alfredovich, G.; Kashayev, R.A. Cellulose Flour for Isolation of Water-Bearing or Watered Formations for Enhanced Oil Recovery and Method of Its Production. RU2575488C2, 20 February 2016. Available online: https://patents.google.com/patent/RU2575488C2/ru (accessed on 1 June 2025).
117. Liu, M.; Wu, B.; Du, Y.; Leng, Q.; Gong, J. Oil and Water Well Plugging Agent and Its Producing Process. CN1335348A, 15 November 2006. Available online: https://patents.google.com/patent/CN1335348A/en (accessed on 1 June 2025).
118. Pan, J.; Deng, W.; Li, Y. Slag Plugging Agent for Heavy Oil Thermal Recovery Steam Injection Well. CN101353570A, 28 January 2009. Available online: https://eureka.patsnap.com/patent-CN101353570A (accessed on 1 June 2025).
119. Lv, B.; Sun, P.; Wu, Y.; Yang, Z.; Liu, P.; Wang, C.; Liu, Q. Study and Application of Oily Sludge Profile Control Technology in Heavy Oil Reservoir. *Energies* **2023**, *16*, 5064. [CrossRef]
120. Lu, X. Use of Modified Flyash as Profile Control Agent. CN1317481C, 23 May 2007. Available online: https://patents.google.com/patent/CN1317481C/en?oq=CN1317481C (accessed on 1 June 2025).
121. Prischepa, O.M.; Nefedov, Y.V.; Kochneva, O.E. Raw Material Base of Hard-to-Extract Oil Reserves of Russia. *Periódico Tchê Química* **2020**, *17*, 915–924. [CrossRef]

energies

MDPI

Article

Stimulation Effect Evaluation of Boundary Sealing and Reservoir Fracturing on Offshore Challenging Gas Hydrates

Shuaishuai Nie [1], Ke Liu [1,*] and Xiuping Zhong [2]

[1] School of Rail Transit, Chengdu Vocational & Technical College of Industry, Chengdu 610213, China; niess20@mails.jlu.edu.cn

[2] College of Construction Engineering, Jilin University, Changchun 130026, China; zhongxp_jlu@163.com

* Correspondence: lkcdpc@hotmail.com

Abstract

Depressurization combined with thermal stimulation based on injection-production well patterns is considered promising for gas hydrate development. Nevertheless, its direct application to Shenhu challenging hydrates may be problematic due to the presence of low reservoir permeability and permeable boundaries. The present study proposes to improve the development potential of Shenhu hydrate by reservoir reconstruction, including boundary sealing and reservoir fracturing, and numerically investigates the production performance. The results showed that water intrusion, hot loss, and gas leakage can be effectively addressed by boundary sealing. Nevertheless, it cannot enhance productivity as thermal decomposition gas accumulated around the injection well. Conversely, reservoir fracturing can significantly improve extraction efficiency as substantial amounts of hydrates dissociate along the fractures, and the gas can be well recovered through the fractures. However, reservoir fracturing was not conducive to water control and energy utilization as it induced more severe water flooding and gas leakage. Under the synergistic effect of the two, there was no methane leakage, and the gas production rate increased with increasing fracture conductivity, while the gas-to-water ratio and energy ratio presented the opposite trend. To obtain a favorable production performance, a fracture with a conductivity of 1–10 D·cm was recommended. Therefore, the combination of boundary sealing and reservoir fracturing makes it feasible for safe and efficient extraction of offshore challenging hydrate under the injection-production mode.

Keywords: gas hydrate; boundary sealing; reservoir fracturing; productivity enhancement; water control

1. Introduction

Natural gas hydrate (NGH) is a crystalline compound in which guest gas molecules (mainly methane) are encapsulated in a cage of water molecules [1]. Due to the wide distribution and huge reserves, with a carbon content of about 1/3 of the global total [2,3], NGH is widely recognized as a promising new clean energy resource that will play a significant role in future energy supply [4].

At present, Japan, China, the European Union, as well as other countries and regions, have been vigorously advancing the commercial development and utilization of hydrates [5]. However, more than 97% of NGH is hosted in deep-sea shallow sediments, facing enormous challenges for development due to the complex marine environment, unique occurrence conditions, and a range of potential geo-environmental issues [6]. The Shenhu

Area of the South China Sea is currently one of the most active NGH trial production sites in the world, where NGH is endowed in low-permeability clayey silty sediments without impermeable boundaries [7–9]. According to the classification of hydrates by Moridis et al. (2011), this falls under the category of challenging hydrates due to the great extraction difficulty [10]. Furthermore, the two rounds of trial production of Shenhu hydrates have proven the feasibility of recovering gas from this deposit [11,12]. Especially in the second trial production, horizontal wells were used for productivity enhancement, with 86.14×10^4 m^3 of gas produced within 30 days [12]. Nevertheless, it is still far below the accepted commercial productivity of 5–50 m^3/d [13,14]. Consequently, enhancing the productivity of challenging hydrates has become a hot topic.

The feasible NGH extraction methods include depressurization, thermal stimulation, gas displacement (mainly CO_2), and inhibitor injection [15]. Among them, the combination of depressurization and thermal stimulation is generally considered highly promising as it can provide sufficient heat for hydrate decomposition [16–19]. Particularly, the injection-production mode based on a multi-well shows tantalizing potential because of the efficient mass and heat transfer and has been proposed for the commercial development of NGH by Japan [20]. However, this development mode may be problematic if applied directly to challenging hydrates. The most pronounced distinction is that the permeability of the clayey silty deposits in the Shenhu Area (mostly less than 10 mD) is much lower than the sandy deposits in the Nankai Trough (hundreds to thousands of mD), which is unfavorable for temperature–pressure transfer and fluid flow. Li et al. (2010) and Su et al. (2012) found that thermal stimulation was confined to a limited area [21,22]. Feng et al. (2015) and Jin et al. (2016) revealed that the low permeability cuts off the inter-well hydraulic connection, resulting in the thermal decomposition gas being hard to recover [23,24]. Additionally, the absence of an impermeable boundary results in significant heat fluid losses, which are highly detrimental to energy utilization [25]. Most importantly, thermal decomposition gas tends to escape to the cover layers driven by the high injection pressure, significantly increasing the geo-environmental risk [26]. These issues need to be well addressed in challenging hydrate development under the injection-production mode.

Reservoir fracturing (RF) is an important strategy for unconventional oil and gas development [27–29]. Because of this, RF has been proposed to improve the injection-production performance of challenging hydrates. Furthermore, the fracability of hydrate deposits and the stimulation potential have been initially evaluated [30–36]. Ju et al. (2020) found that RF improved the hot water wave area, thereby facilitating hydrate decomposition [34]. Zhong et al. (2011, 2022) discovered that RF greatly increased the mass and heat transfer between injection and production wells, and was expected to reach commercial productivity [13,37]. However, the hot water loss and gas leakage induced by the open boundary were not considered. According to our previous assessment, uncontrolled hydrate decomposition caused by RF can significantly increase the risk of gas leakage [38]. Against this background, the current study attempts to address the negative effects induced by RF and permeable boundaries.

The permeable cover layer of the NGH deposits is similar to the boundary conditions of gas reservoirs containing edge or bottom water. The primary solution technique is boundary sealing (BS), that is, the use of sealing and plugging materials (e.g., cement) to create impermeable artificial barriers that prevent water intrusion. Yue et al. (2012) developed a flow model for horizontal wells in bottom-water-driven reservoirs with impermeable barriers, demonstrating that the barrier can enhance the critical rate and delay water breakthrough [39]. Liu et al. (2023) constructed an improved artificial barrier by injecting oil-soluble resin as the selective water shut-off agent during the crack generation phase [40]. Bai et al. (2024) reviewed the macroscopic effects and microscopic mechanisms of gel

https://doi.org/10.3390/en19010120

agents during the "injection, migration, plugging, and stabilization" process [41]. Similarly, this technology can be integrated into hydrate development. Our previous research has demonstrated that the hydraulic connection between the reservoir and caprock can be severed by the BS, thereby addressing issues including injected hot loss, boundary water intrusion, and methane leakage [42]. However, the low reservoir permeability hindered the migration of thermal decomposition gas to production wells, resulting in no contribution to productivity during the first three years of hot water injection. Given all the considerations above, the combination of BS and RF may be a novel and promising development mode for offshore challenging hydrates.

This study intends to improve the development potential of challenging hydrates under the injection-production mode through reservoir reconstruction, including BS and RF. The production performance, including NGH decomposition, gas–water production, energy utilization, as well as hot loss and gas leakage, was numerically analyzed, with Shenhu challenging hydrates as the geological background. Furthermore, the synergistic effects between BS and RF were thoroughly evaluated. The novel findings of this study could provide valuable insights for challenging hydrate extraction.

2. Modeling

2.1. Development Plan

The development plan is designed based on the geological characteristics of hydrates at the Shenhu SH2 site, as shown in Figure 1. Geological surveys indicate that the hydrate deposits at the Shenhu SH2 site belong to Class 2 [43,44]. Furthermore, both the overlayer (OL) and underlayer (UL) are permeable with the same permeability as the hydrate-bearing layer (HBL) of 10 mD. Consequently, this represents a classic challenging hydrate characterized by the absence of impermeable boundaries and low reservoir permeability [10]. In the development plan, a five-spot horizontal well pattern with a rectangular distribution was adopted, including one hot water injection well located in the middle of the reservoir and four production wells distributed at the top and bottom of the HBL. Previous studies have demonstrated that this type of well deployment is efficient for gas recovery [45,46]. One feasible reservoir reconstruction scheme is given as follows: firstly, a high-conductivity artificial fracture is created in the middle of the HBL by RF through the injection well; then, impermeable artificial barriers are formed at the top and bottom of the HBL by injecting sealing materials (e.g., cement) through the production wells, thereby sealing open boundaries.

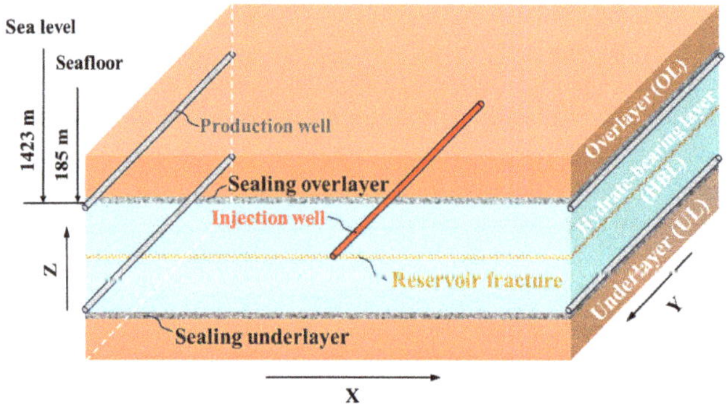

Figure 1. Proposed development plan for the NGH reservoir at the Shenhu SH2 site, wherein the injection and production horizontal wells were arranged along the X-axis; all horizontal sections were drilled along the Y-axis; and the Z-axis represents the vertical direction.

https://doi.org/10.3390/en19010120

2.2. Geometric Modeling

The HBL thickness at Shenhu SH2 is reported to be 0–44 m [43], and 40 m is taken here. Several researchers have demonstrated that a 30 m thick cover layer is sufficient for the simulation of reservoir temperature–pressure evolution [47–49]. While considering the permeable characteristics in this case, the thickness of the OL and UL was set to be 80 m. The horizontal well spacing between the injection and production wells was 100 m, assuming that the reservoir was homogeneous along the Y-axis, and thus only one unit length of the horizontal well section was modelled. As a result, the model size is 200 m (X) × 1 m (Y) × 200 m (Z).

The fracture width and half-length were set as 0.01 and 1 m, with a conductivity of 1–50 D·cm. The 1 m-thick layer at the top and bottom of the HBL was modeled as the barrier, and the permeability was set as 0 to simulate the sealing effect. In the X direction, the grids were refined from 1 m to 0.2 m toward the wellbore, resulting in 203 subdivided meshes. In the Y direction, only one grid was subdivided. In the Z-direction, the grids were refined from the model boundary toward the fracture to 0.01 m, resulting in 106 subdivided meshes. As a result, the total number of meshes is 21,518.

2.3. Simulation Code

Tough + Hydrate (T + H, v1.0), developed by Lawrence Berkeley, was used here because of its reliability in describing phase transitions, heat exchange, and fluid flow in complex media. Furthermore, its accuracy has been fully tested on both experimental and mine scales [50–52], which ensures the accuracy of the simulation results.

2.4. Initial and Boundary Conditions

Reservoir pressure and temperature were initialized using hydrostatic pressure and geothermal gradients due to the permeable cover layers. Dirichlet boundary conditions were imposed at the top and bottom model boundaries, as sufficient cover layers have been considered. Model parameters are presented in Table 1. Notably, to ensure the reliability of simulation results, the selected reservoir parameters in the model, including porosity, permeability, and hydrate saturation, as well as the properties of the OL and UL, were consistent with the real geological conditions of Shenhu SH2 Site.

Table 1. Model parameters.

Reservoir Parameters	**Value and Unit**
Sediment and seawater densities	2650 kg/m^3, 1024 kg/m^3
Sediment porosity	0.38
Sediment permeability	10 mD
Hydrate saturation	0.55
Sediment specific heat	1000 J/(kg·K)
Dry and wet thermal conductivity of sediments	1 W/(m·K), 3.1 W/(m·K)
Initial average reservoir pressure	14.70 MPa
Seafloor temperature	3.60 °C
Thermal gradient	50 °C/km
Injection and production parameters	Value & Unit
Production pressure	4.50 MPa
Injection pressures	20 MPa
Hot water temperature	60 °C

https://doi.org/10.3390/en19010120

Table 1. *Cont.*

Constitutive parameters	Value & Unit
Capillary pressure P_{cap}	$P_{cap} = -P_{thr}\left[\left(\frac{S_A - S_{irA}}{S_{mxA} - S_{irA}}\right)^{-1/\lambda} - 1\right]^{1-\lambda}$
Gas phase relative permeability k_{rG}	$k_{rG} = \left(\frac{S_G - S_{irG}}{1 - S_{irA}}\right)^{n_G}$
Aqueous phase relative permeability k_{rA}	$k_{rA} = \left(\frac{S_A - S_{irA}}{1 - S_{irA}}\right)^{n}$
P_{thr}	1×10^5 Pa
λ	0.45
S_{mxA}	1.00
n	3.50
n_G	3.50
S_{irA}	0.30
S_{irG}	0.03

2.5. Simulation Scheme

In our simulations, the production behaviors under various BS cases including sealing OL (BS_O), UL (BS_U), or both (BS_{OU}) were investigated; subsequently, various RF cases were analyzed, including C_f of 1 (RF_1), 5 (RF_5), 10 (RF_{10}), and 50 (RF_{50}) D·cm, respectively; and finally, the synergistic effect between RF and BS was evaluated. Notably, the unreconstructed reservoir serves as a base case for comparative analyses. Consequently, the final simulation scheme in Table 2 was developed.

Table 2. Simulation scheme.

Cases	Sealed Layer	C_f, D·cm	Cases	Sealed Layer	C_f, D·cm
Base case	None	None	RF_{10}	None	10
BS_{OU}	OL, UL	None	RF_{50}	None	50
BS_O	OL	None	BS_{OU} + RF_1	OL, UL	1
BS_U	UL	None	BS_{OU} + RF_5	OL, UL	5
RF_1	None	1	BS_{OU} + RF_{10}	OL, UL	10
RF_5	None	5	BS_{OU} + RF_{50}	OL, UL	50

3. Results and Discussion

3.1. Production Performance Under BS

3.1.1. Reservoir Physical Field Distribution

The reservoir physical field distribution at 1800 days in the base, BS_{OU}, BS_O, and BS_U cases is shown in Figure 2.

The high injection pressure extended to the cover layer in the base case (Figure 2a), and considerable injected hot water flowed into the OL and UL (Figure 2b). Furthermore, the propagation of low production pressures was suppressed, similar to the boundary water intrusion phenomenon in the development of conventional gas reservoirs containing edge or bottom water. This interlayer permeability difference is detrimental to hydrate dissociation and gas recovery. It can be observed that the thermal decomposition gas escaped into cover layers (Figure 2c) and a significant amount of NGH was reserved (Figure 2d). In the BS_{OU} case, the high injection pressure can be successfully maintained in the HBL, and the low production pressure was also efficiently propagated, creating a satisfactory pressure gradient between the injection and production wells (Figure 2e). In addition, there was no hot water loss (Figure 2f) and gas escape (Figure 2g). Regrettably, gas was hard to transport to the production wells due to the low reservoir permeability. This causes a large amount of gas that is trapped in the vicinity of the injection wells

https://doi.org/10.3390/en19010120

(Figure 2g), thus inhibiting the injection and diffusion of hot water (Figure 2f). As a result, the NGH decomposition efficiency (Figure 2h) was lower relative to the base case. In the BS_O case, high injection pressure maintenance failed (Figure 2i). The mechanism is that the injected hot water penetrated through the bottom of the HBL and migrated downward into the permeable UL, causing partial thermal decomposition gas to escape into the UL (Figure 2j–l). In the BS_U case, high injection pressure primarily propagated towards the permeable OL (Figure 2m). This results in hot water breaking through the top of the HBL and thermal decomposition gas escaping into the permeable OL (Figure 2n–p). Thus, partial sealing does not fully prevent hot water loss and gas escape.

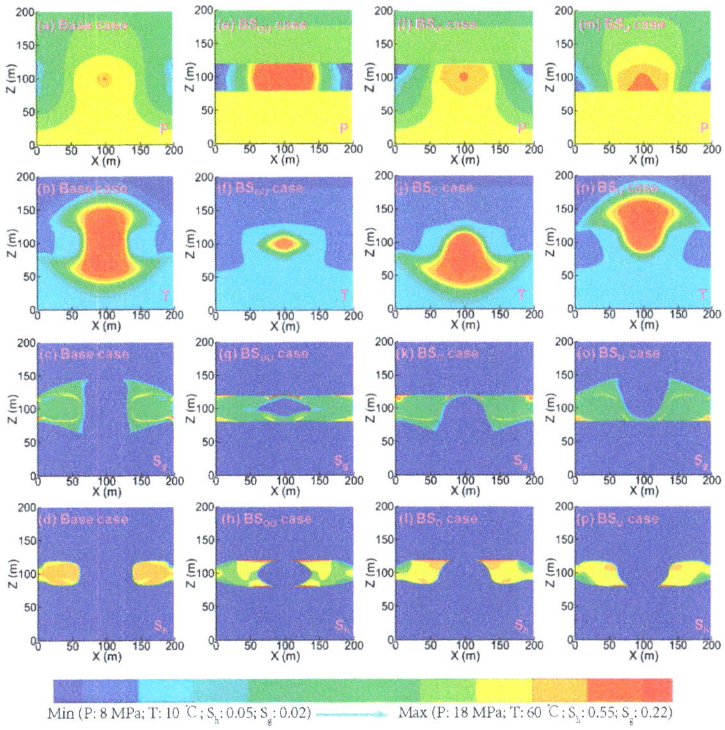

Figure 2. Reservoir pressure (P), temperature (T), hydrate saturation (S_h), and gas saturation (S_g) in the base, BS_{OU}, BS_O, and BS_U cases.

3.1.2. Hydrate Dissociation Behavior

The variation of NGH dissociation rate (Q_h, gas released volume per day) and NGH decomposition ratio (R_{hd}, defined by the ratio of decomposed hydrate mass to the in situ hydrate mass) in the base, BS_{OU}, BS_O, and BS_U cases is shown in Figure 3. The variation of Q_h in the base case presents four different phases. In the first phase (~330 days), Q_h increased greatly with injected hot water spreading and low production pressure propagation and peaked at 113 m^3/d. Subsequently (~1000 days), Q_h decreased sharply, with a minimum value of 50 m^3/d. Mechanisms contributing to this phenomenon include a significant amount of hot water leaked into UL and OL, the slower and slower forward advance of the thermal decomposition leading edge, as well as the boundary water intrusion that inhibits the propagation of low production pressures. In the third phase (~2340 days), inter-well interaction gradually intensified with NGH decomposition, and therefore Q_h

https://doi.org/10.3390/en19010120

increased slowly and peaked at 71 m^3/d. After that, Q_h dropped sharply as there was less and less hydrate remaining in the reservoir.

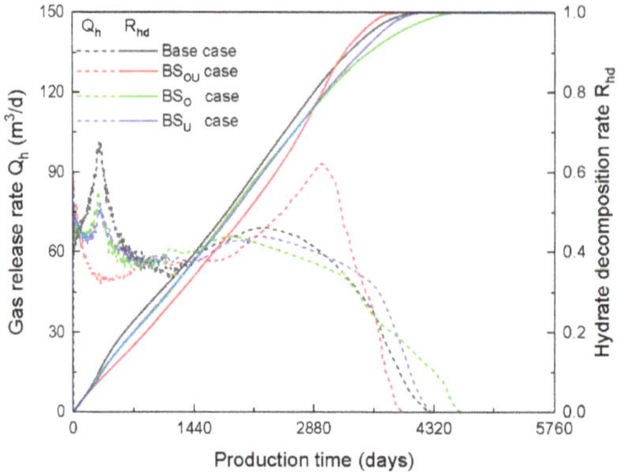

Figure 3. Variation of Q_h and V_{hd} in the base, BS_{OU}, BS_O, and BS_U cases.

Unlike the base case, the initial Q_h in the BS_{OU} case decreased rapidly with a minimum value of 45 m^3/d at 327 days. This is because the gas accumulated around the injection well suppressed the injection and diffusion of hot water. Then, a piston-like displacement gradually formed under a stable inter-well pressure gradient, and thus Q_h stabilized at 55 m^3/d. Afterward, Q_h began to rise at about 1800 days and peaked at 98 m^3/d at 3090 days. This is because the gas is recovered by the production wells, which promotes the injection and diffusion of hot water, thus increasing the hydrate decomposition efficiency. After that, Q_h decayed dramatically as the inter-well NGH was nearly completely decomposed. The variation of Q_h in BS_O and BS_U cases is similar to the base case. In addition, the NGH decomposition cycle T_h for the base, BS_U, and BS_O cases is 4130–4530 days, suggesting that sealing the UL or OL alone has a limited effect on hydrate decomposition. Nevertheless, simultaneously sealing both of them decreased the initial Q_h, but the mid- and post-Q_h were more substantial, resulting in the shortest T_h of 3800 days.

3.1.3. Gas Production Behavior

The variation of gas production rate Q_g and cumulative gas production V_g in the base, BS_{OU}, BS_O, and BS_U cases is presented in Figure 4.

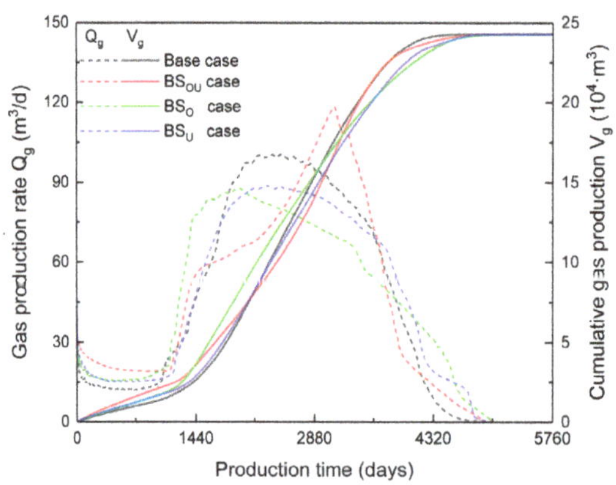

Figure 4. Variation of Q_g and V_g in the base, BS_{OU}, BS_O, and BS_U cases.

https://doi.org/10.3390/en19010120

As shown in Figure 4, the variation of Q_g presented three phases in all cases. Firstly, there was a low gas production phase, where Q_g stabilized after a slight decrease, and the produced gas was contributed by depressurized decomposed gas. Furthermore, Q_g in the BS_{OU} case is the maximum, about 20 m^3/d, followed by those in BS_O and BS_U cases, about 15 m^3/d, and the minimum is 12 m^3/d in the base case. The second phase was entered at approximately 1000 days, where Q_g increased significantly as more and more thermal decomposition gas was recovered by production wells. Affected by the evolutions of Q_h, the peak values of Q_g in the BS_{OU} case are the maximum (118 m^3/d), followed by 100 m^3/d in the base case, 88 m^3/d in the BS_O case, and 86 m^3/d in the BS_U case. Ultimately, there is a gas production depletion phase due to the insufficient gas supply. Additionally, the final V_g within 5760 days in all cases was consistent, and boundary conditions barely affected the gas recovery cycles T_g, as they were about 4700–4900 days.

3.1.4. Water Production Behavior

Figure 5 shows the variation of water production rate Q_w and cumulative water production V_w with gas recovery ratio R_{gr} (defined by the ratio of V_g to total gas available in the reservoir) in the base, BS_{OU}, BS_O, and BS_U cases. It can be seen that Q_w exhibits three phases in all cases, including an initial slight rise, followed by a brief decline, and another rapid rise. The mechanism is that more and more injected water was recovered by the production wells as inter-well interactions increased. This is similar to the water flooding phenomenon in water-driven reservoir development, since open boundaries could induce severe water intrusion. The Q_w in the base case is the highest, followed by the BS_O, BS_U, and BS_{OU} cases, respectively. Specifically, the V_w at R_{gr} of 0.7 (acceptable for water-driven development of China's offshore low-permeability gas reservoirs) in the base case is 3.13×10^4 m^3, which is 5.89, 1.38, and 1.61 times that of BS_O, BS_U, and BS_{OU} cases, respectively. Consequently, boundary intrusion can be well addressed by simultaneously sealing the upper and lower boundaries, which is an effective means of water control.

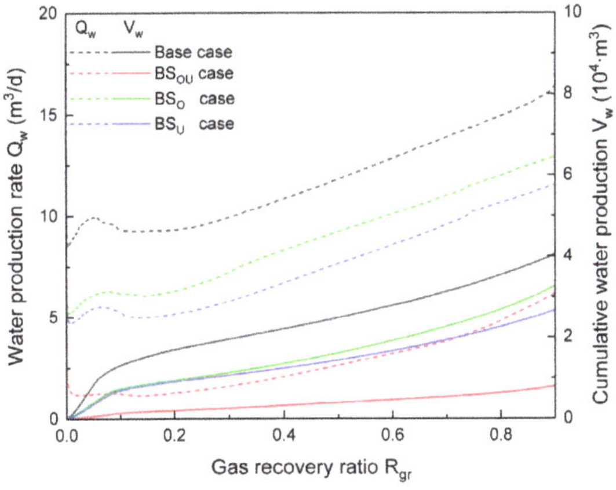

Figure 5. Variation of Q_w and V_w in the base, BS_{OU}, BS_O, and BS_U cases.

Figure 6 shows the variation of gas-to-water ratio R_{gw} (defined by the ratio of V_g to V_w) in the base, BS_{OU}, BS_O, and BS_U cases. It can be seen that R_{gw} in the base case is the lowest (<5), and its change presents three stages of fall-rise-fall, which correspond to the initial stage of low gas and low water production, the middle stage of increasing water and gas production, and the later stage of low gas and high water production, respectively. Compared to the base, the evolution of R_{gw} in BS_O and BS_U cases is similar, and its enhancement is not significant. Encouragingly, R_{gw} in the BS_{OU} case is satisfactory,

https://doi.org/10.3390/en19010120

reaching 32 at an R_{gr} of 0.7, 5.87, 4.25, and 3.63 times that of the base, BS_O, and BS_U cases, respectively. Therefore, gas–water production behavior can be improved by sealing OL and UL.

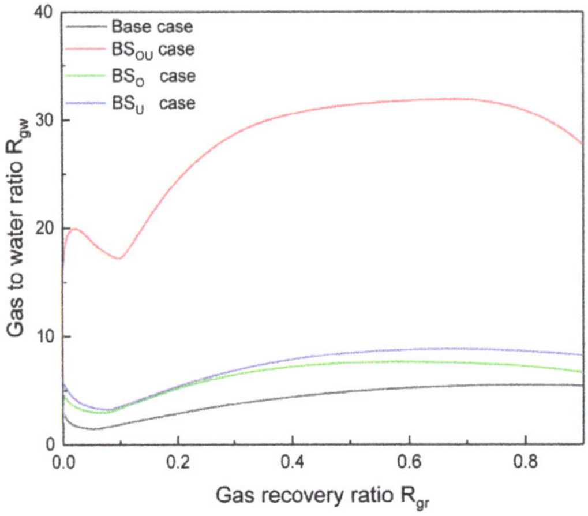

Figure 6. Variation of R_{gw} in the base, BS_{OU}, BS_O, and BS_U cases.

3.2. Injection-Production Performance Under Reservoir Fracturing

3.2.1. Reservoir Physical Field Distribution

Figure 7 shows the reservoir physical field distribution at 720 days in the RF_1, RF_5, RF_{10}, and RF_{50} cases. It can be seen that as C_f increases from 1 to 50 D·cm, the high injection pressure propagates faster (Figure 7a–d) and a wider wave of hot water in the horizontal direction is observed (Figure 7e–h). This improves the reservoir heating efficiency, thus promoting hydrate decomposition (Figure 7i–l). Furthermore, more hydrate dissociates along the fracture surface as C_f increases, which leads to a shift in hydrate decomposition mode from piston-type to volumetric-type. However, increasing C_f could not avoid gas leakage (Figure 7m–p), which is consistent with previous studies. It should be specifically pointed out that the current model assumes a constant C_f value throughout the production process. In reality, given the weakly cemented nature of NGH deposits, fractures may buckle, become plugged with sand proppant, or interact with natural heterogeneities. This may impact long-term forecasts. Consequently, the variation pattern of C_f as well as its influence needs to be further investigated.

3.2.2. Hydrate Dissociation Behavior

Figure 8 shows the variation of Q_h and R_{hd} in the base, RF_1, RF_5, RF_{10}, and RF_{50} cases. As can be seen, the variation of Q_h in the RF cases is similar to that in the base case. The difference is that the initial Q_h increases with the increase in C_f. Specifically, the peak value increases from 130 to 317 m^3/d as C_f increases from 1 to 50 D·cm, which is 1.16–2.83 times higher than that in the base case. The mechanism is that a greater C_f is more favorable for hot water migration, allowing more hydrate to dissociate along the fracture. Additionally, at greater C_f, the leading edge of hydrate decomposition moves away from the fracture faster, which leads to a higher post-peak Q_h decay rate. The T_h reduces from 3700 to 2170 days as C_f increases from 1 to 50 D·cm, a reduction of 430–1960 days compared to the base case. As a result, increasing C_f is helpful for hydrate decomposition.

https://doi.org/10.3390/en19010120

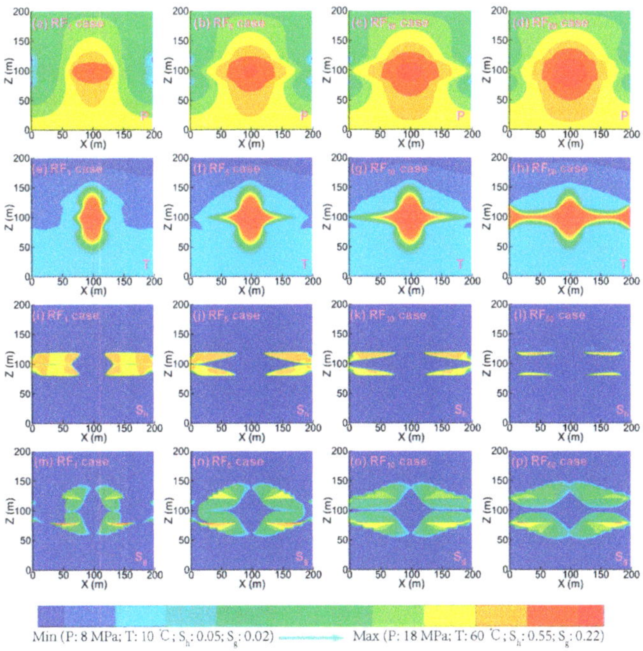

Figure 7. Reservoir pressure (P), temperature (T), hydrate saturation (S_h), and gas saturation (S_g) in the RF_1, RF_5, RF_{10}, and RF_{50} cases.

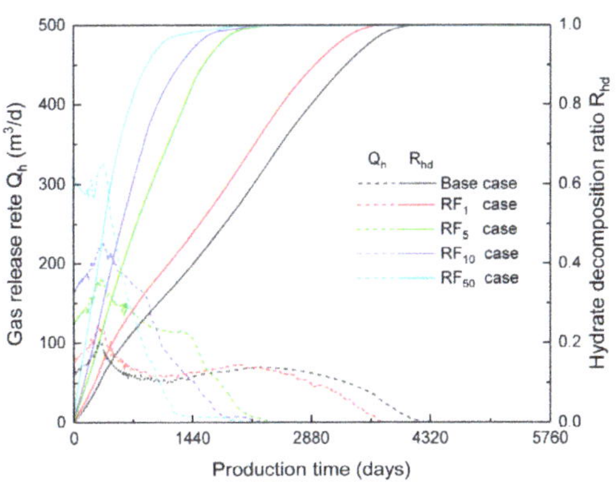

Figure 8. Variation of Q_h and V_{hd} in the base, RF_1, RF_5, RF_{10}, and RF_{50} cases.

3.2.3. Gas Production Behavior

Figure 9 shows the variation of Q_g and V_g in the base, RF_1, RF_5, RF_{10}, and RF_{50} cases. With C_f increasing, the initial low productivity phase is greatly shortened, and the corresponding time for Q_h to peak is earlier. Specifically, this time is shortened from 2160 days to 400 days as C_f increases from 1 to 50 D·cm, a reduction of 260–1790 days relative to the base case. The mechanism is that the gas flow rate inside the fracture is higher with a greater C_f, and therefore, thermal decomposition gas can be extracted more rapidly from production wells. However, the peak value of Q_g does not always increase as C_f increases, with a maximum value of 161 m^3/d occurring at C_f of 10 D·cm. This is attributed to the fact that increasing C_F simultaneously increases the hot water injection rate, thereby reducing the gas-phase relative permeability in the fracture. Additionally,

https://doi.org/10.3390/en19010120

it is observed that the V_g within 5760 days decreases from 24.13 to 22.12 $\times$ 10^4 m^3 as C_F increases from 1 to 50 D·cm, a reduction of 0.18–2.18 $\times$ 10^4 m^3 compared to the base case. This suggests that gas leakage occurs in the RF case and that the amount of leaked gas increases with increasing C_f. Thus, increasing C_f can substantially increase the gas recovery efficiency, but also induce more serious gas leakage.

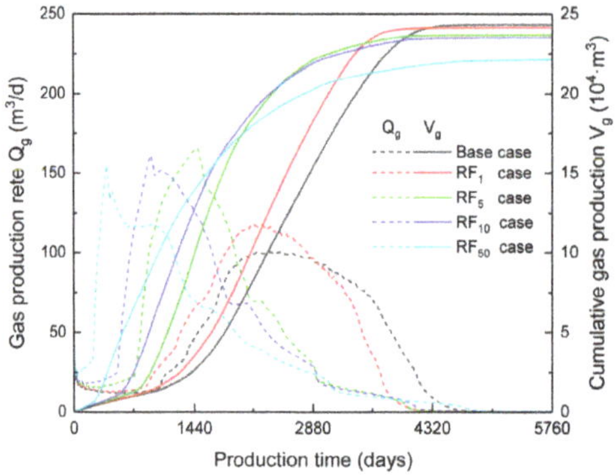

Figure 9. Variation of Q_g and V_g in the base, RF_1, RF_5, RF_{10}, and RF_{50} cases.

3.2.4. Water Production Behavior

Figure 10 shows the variation of Q_w and V_w in the base, RF_1, RF_5, RF_{10}, and RF_{50} cases. At the early production stage (R_{gr} < 0.1), the influence of C_f on Q_w is not significant. Interestingly, Q_w and V_w do not always increase with increasing C_f. Specifically, when R_{gr} is 0.7, V_w decreases from 2.84 to 1.94 $\times$ 10^4 m^3 as C_f increases from 1 to 5 D·cm, a reduction of 0.29–1.19 $\times$ 10^4 m^3 compared to the base case. In contrast, V_w increases from 1.94 to 3.60 as C_f increases from 5 to 50 D·cm, indicating that increasing C_f in this interval is more favorable for water production compared to gas production. Consequently, properly controlling the C_f value of 1–10 D·cm is conducive to water control.

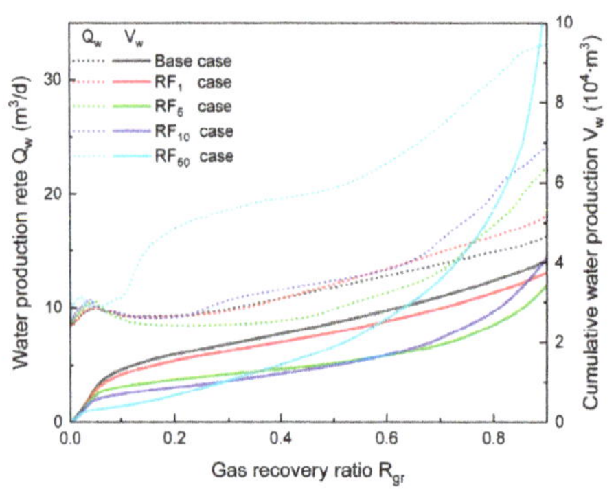

Figure 10. Variation of Q_w and V_w in the base, RF_1, RF_5, RF_{10}, and RF_{50} cases.

Figure 11 shows the variation of R_{gw} in the base, RF_1, RF_5, RF_{10}, and RF_{50} cases. The R_{gr} corresponding to the peak value of R_{gw} reduced from 0.72 to 0.20 as C_f increases from 1 to 50 D·cm, a reduction of 0.08–0.60 relative to the base case. The reason is that increasing C_f can significantly improve Q_g, but also induce premature water flooding. The R_{gw} at R_{gr} of 0.7 increases first and then decreases with increasing C_f, and the maximum value of 8.82 appears at C_F of 5 D·cm, 1.61 times that in the base case. As a result, RF can improve the gas and water production behavior, but needs to control C_f reasonably.

https://doi.org/10.3390/en19010120

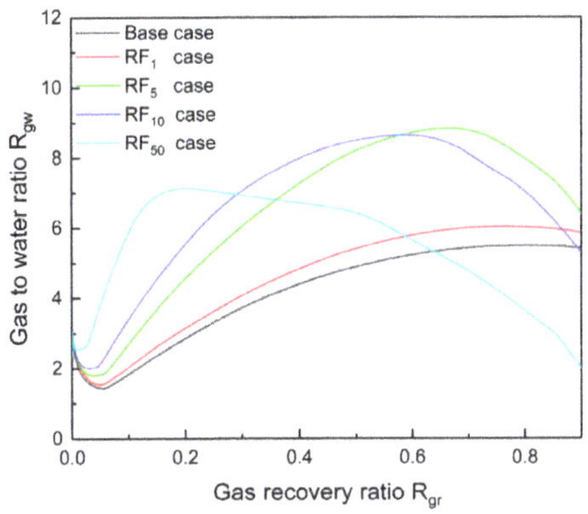

Figure 11. Variation of R_{gw} in the base, RF_1, RF_5, RF_{10}, and RF_{50} cases.

3.3. Production Performance Under BS and RF

3.3.1. Reservoir Physical Field Distribution

Figure 12 shows the reservoir physical field distribution at 1800 days in the BS_{OU} + RF_1, BS_{OU} + RF_5, BS_{OU} + RF_{10}, and BS_{OU} + RF_{50} cases. The high injection pressure only propagated in the HBL, and ideal pressure gradients between the injection and production wells were generated (Figure 12a–d). Similar to the RF cases in Figure 6, the hot water spread area increases as C_F increases (Figure 12e–h), and therefore, more hydrates are decomposed along the fracture (Figure 12i–l). The difference is that there is no hot water loss (Figure 12e–h) and gas leakage (Figure 12i–l). As a result, BS can improve the utilization efficiency of hot water while effectively addressing gas leakage induced by RF.

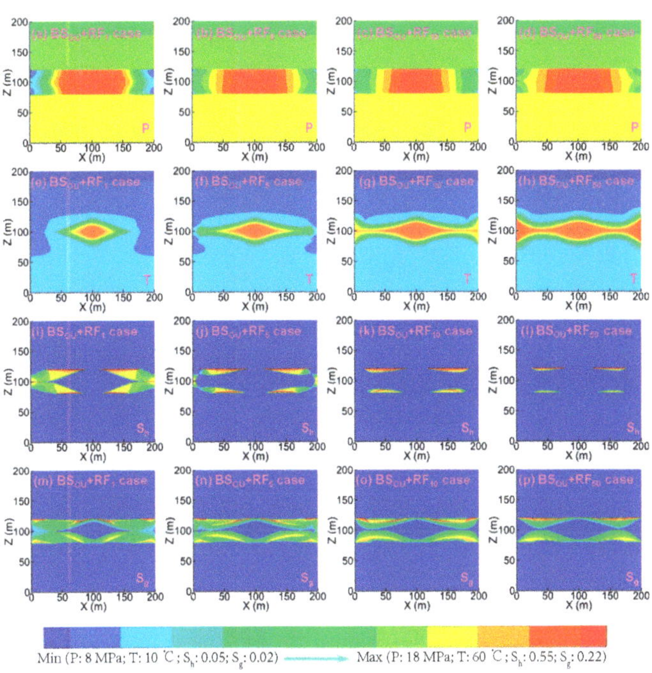

Figure 12. Reservoir pressure (P), temperature (T), hydrate saturation (S_h), and gas distribution (S_g) in the BS_{OU} + RF_1, BS_{OU} + RF_5, BS_{OU} + RF_{10}, and BS_{OU} + RF_{50} cases.

https://doi.org/10.3390/en19010120

3.3.2. Hydrate Dissociation Behavior

The variation of Q_h and V_{hd} in the base, BS_{OU} + RF_1, BS_{OU} + RF_5, BS_{OU} + RF_{10}, and BS_{OU} + RF_{50} cases is shown in Figure 13. As can be seen, the peak value of Q_h increases from 122 to 204 m^3/d as C_f increases from 1 to 50 D·cm, which is 1.72–2.87 times that of the base case. Additionally, the T_h reduced from 3370 to 2380 days as C_f increased from 1 to 50 D·cm, a reduction of 760–1750 days relative to the base case. As a result, increasing C_f in BS + RF cases is still an effective method for promoting hydrate decomposition.

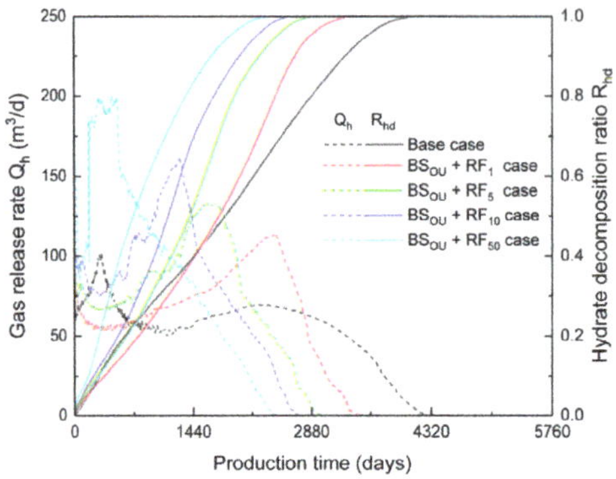

Figure 13. Variation of Q_h and R_{hd} in the base, BS_{OU} + RF_1, BS_{OU} + RF_5, BS_{OU} + RF_{10}, and BS_{OU} + RF_{50} cases.

3.3.3. Gas Production Behavior

Figure 14 shows the variation of Q_g and V_g in the base, BS_{OU} + RF_1, BS_{OU} + RF_5, BS_{OU} + RF_{10}, and BS_{OU} + RF_{50} cases. The Q_g in the BS + RF cases is much higher than that in the base case. Furthermore, as C_F increases from 1 to 50 D·cm, the peak value of Q_g increases from 142 to 186 m^3/d, and the T_{gr} is shortened from 4210 to 3720 days, indicating that the extraction efficiency is greatly improved by increasing C_f. The final V_g within 5760 days in all cases is consistent, indicating there is no gas leakage. As a result, BS can greatly improve gas recovery efficiency while addressing gas leakage induced by RF.

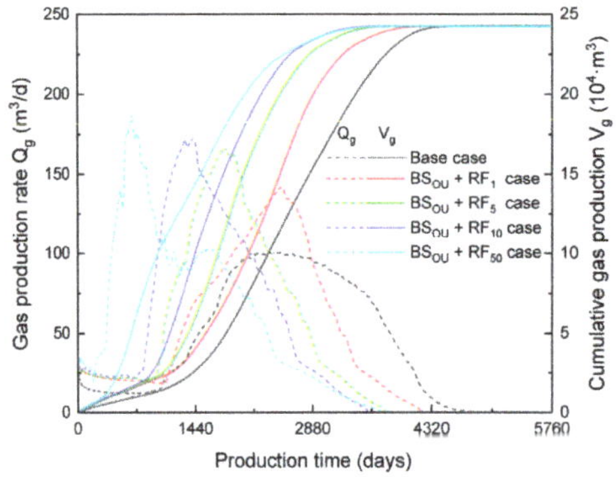

Figure 14. Variation of Q_g and V_g in the base, BS_{OU} + RF_1, BS_{OU} + RF_5, BS_{OU} + RF_{10}, and BS_{OU} + RF_{50} cases.

3.3.4. Water Production Behavior

The variation of Q_w and V_w in the base, BS_{OU} + RF_1, BS_{OU} + RF_5, BS_{OU} + RF_{10}, and BS_{OU} + RF_{50} cases is shown in Figure 15. The Q_w in C_f of 1–10 D·cm cases is much lower than that in the base case. For C_f of 50 D·cm, Q_w is higher than that of the base case in the

https://doi.org/10.3390/en19010120

mid and late periods. In addition, Q_w and V_w increase with increasing C_f. At R_{gr} of 0.7, V_w increases from 0.55 to 1.61 × 10^4 m^3 as C_f increases from 1 to 50 D·cm, accounting for 17–51% of the base case. Thus, significant reductions in water production are achieved.

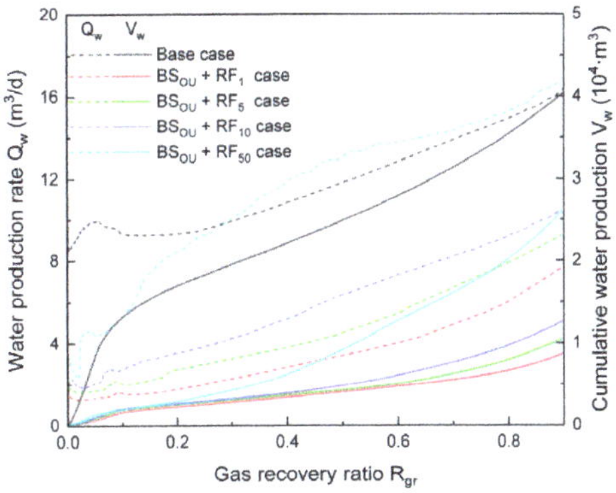

Figure 15. Variation of Q_w and V_w in the base, BS_{OU} + RF_1, BS_{OU} + RF_5, BS_{OU} + RF_{10}, and BS_{OU} + RF_{50} cases.

Figure 16 shows the variation of R_{gw} in the base, BS_{OU} + RF_1, BS_{OU} + RF_5, BS_{OU} + RF_{10}, and BS_{OU} + RF_{50} cases. The R_{gw} in BS + RF cases is much higher than in the base case, and it decreases with increasing C_f. Specifically, the R_{gw} at R_{gr} of 0.7 decreases from 30 to 10 as C_F increases from 1 to 50 D·cm, 1.84–5.53 times higher than the base case. Therefore, combining BS with RF greatly improves gas–water production performance.

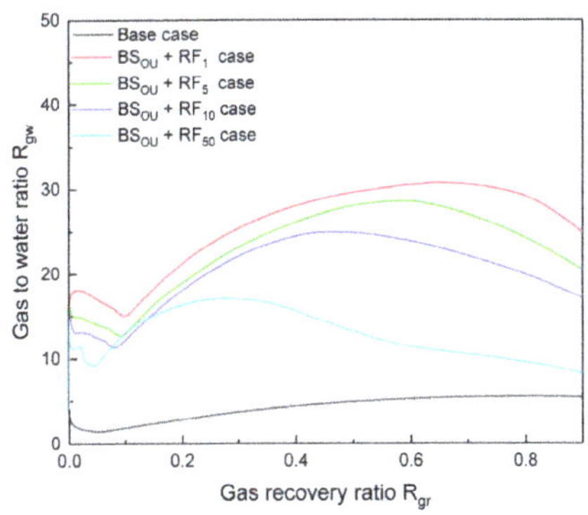

Figure 16. Variation of R_{gw} in the base, BS_{OU} + RF_1, BS_{OU} + RF_5, BS_{OU} + RF_{10}, and BS_{OU} + RF_{50} cases.

4. Synergistic Effects of BS and RF

From the analysis of the above results, it is found that the hot water loss and boundary water intrusion can be addressed by BS; RF can significantly improve extraction efficiency, but also exacerbate the gas leakage, whereas the combination of the two seems to result in a satisfactory production behavior. Here, the synergistic effects of BS and RF on gas and water production are analyzed with the average gas production rate Q_{ag} (Figure 17a) and R_{gw} (Figure 17b) at R_{gr} of 0.7. In addition, for the multi-well injection development mode, energy ratio ER, defined by $V_g \cdot C_g / [C_w \cdot M_i \cdot (T_i - T_0)]$, is a crucial parameter to measure the economy and feasibility of exploitation [53]. As a result, the ER at R_{gr} of 0.7 is employed as a key indicator for synergistic effect analysis (Figure 17c). Notably, the ER used here did not account for the energy required to create artificial barriers and perform reservoir fracturing.

Given the significant challenges in reconstructing hydrate deposits, more comprehensive economic feasibility assessments need to be conducted based on systems analysis methods before field application [54].

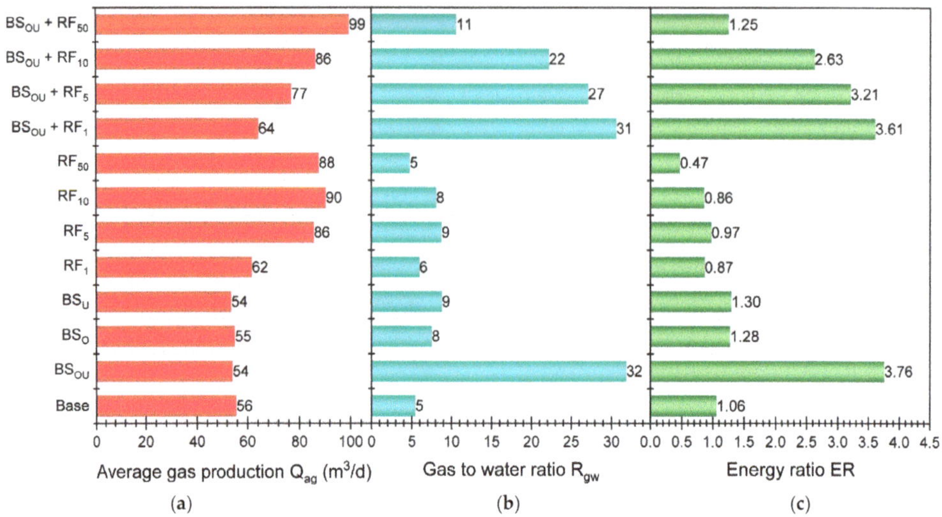

Figure 17. The Q_{ag} (**a**), R_{gw} (**b**), and ER (**c**) at R_{gr} of 0.7 in all simulation cases.

As can be observed from Figure 17a, the Q_{ag} in the BS_O, BS_U, and BS_{OU} cases is slightly lower than that in the base case. The mechanism involves the thermal decomposition gas accumulating around the injection well, which suppresses hot water injection and thereby reduces the thermal stimulation area. This indicates that boundary sealing alone does not enhance productivity. Surprisingly, Q_{ag} can be significantly increased by RF, especially at $C_f \geq 5$ D·cm. Furthermore, the synergistic effect of BS and RF shifts with changes in C_f. That is, they exhibit positive synergy at C_f of 1 and 50 D·cm, and negative synergy at C_f of 5 and 10 D·cm. However, this synergistic effect is not significant as the variation interval of Q_{ag} is only $[-9, 11]$. This indicates that in fractured reservoirs, the effect of BS on productivity is very limited. Furthermore, under the synergistic effect of BS and RF, Q_{ag} increases significantly with the increase in C_f, specifically from 64 to 99 m^3/d as C increases from 1 to 50 D·cm, an enhancement of 10–45 m^3/d relative to the BS_{OU} case. Given that BS can address RF-induced methane leakage and has a limited impact on Q_{ag}, it is suggested here to increase productivity by adjusting C_f.

As can be seen from Figure 17b, relative to the base case, there is very limited improvement in R_{gw} by RF alone or sealing one open boundary. However, simultaneously sealing the upper and lower boundaries obtains the highest R_{gw} of 32, 6.40 times that of the base case. At different C_f values, the R_{gw} after boundary sealing is all enhanced, which demonstrates the excellent water control ability of the BS. However, the increase in C_f weakens the water control ability of BS as a considerable amount of injected water induces severe water flooding. Nevertheless, the final water yield under the synergistic effect of BS and RF is still much lower than that of the base case.

As can be seen from Figure 17c, the ER of the base case is 1.06, indicating that it is not economically feasible to develop challenging hydrates directly using the multi-well injection-production mode. After the application of BS, the ER reaches 1.28, 1.30, and 3.76 for the BS_O, BS_U, and BS_{OU} cases, respectively, indicating that only the simultaneous sealing of the upper and lower boundaries could significantly improve ER because hot

https://doi.org/10.3390/en19010120

water loss is effectively addressed. The ER values in the RF case are all lower than 1.0, indicating that the application of RF alone is still not economically feasible. In addition, ER decreases significantly with increasing C_f under the synergistic effect of BS and RF. However, they are still greater than 1.0, 1.18–3.41 times that of the base case.

To summarize, BS can effectively address boundary water intrusion and hot water loss, thereby substantially increasing R_{gw} and ER, but barely affecting Q_{ag}, whereas RF can significantly increase Q_{ag}, but also induce severe gas leakage and decrease R_{gw} and ER. Nevertheless, with the synergistic effect of the two, safe extraction and the relatively ideal production behavior can be obtained by regulating the C_f.

5. Conclusions

The present research numerically investigated the injection-production performance under various BS and/or RF cases and thoroughly analyzed their synergistic stimulation mechanisms. The following conclusions can be drawn.

1. BS can effectively address boundary water intrusion, hot loss, and methane leakage, thereby reducing water production and enabling the full utilization of injection energy. However, it cannot speed up hydrate decomposition and gas production. Additionally, partial sealing does not fully prevent hot water loss and gas escape.
2. RF is very effective in facilitating hydrate decomposition and gas recovery, and the extraction efficiency improves greatly with increasing C_f. However, the final ER is lower than 1.0 due to the severe water flooding induced by the fracture, making it unfeasible for economic extraction. Additionally, RF exacerbates the risk of gas leakage, which poses a great challenge for the safe exploitation of marine hydrates.
3. There is no methane leakage under the synergistic effect of RF and BS. Furthermore, Q_h and Q_g improve significantly as C_f increases, whereas R_{gw} and ER show the opposite trend. Therefore, rational regulation of C_f value is essential to obtain satisfactory multi-well injection-production behavior, with 1–10 D·cm recommended.
4. BS is an indispensable technology for the economic and safe extraction of offshore challenging hydrates under the multi-well injection-production mode. Although RF is helpful for hydrate decomposition and gas production, its induced water flooding and methane leakage should not be neglected in future applications.

The current simulations indicate that the combination of BS and RS can significantly enhance the performance of the injection-production well pattern. Nevertheless, it should be noted that the fracture and barrier settings in our model are relatively ideal to maximize the desired effect, and the injection-production parameters employed also need further evaluation and optimization. Additionally, a homogeneous model without interlayers was employed because the hydrate distribution remains unclear. Consequently, follow-up research should fully account for these issues to advance and support engineering implementation.

Author Contributions: Conceptualization, S.N. and K.L.; data curation, K.L.; software, X.Z. and S.N.; writing—original draft, S.N. All authors have read and agreed to the published version of the manuscript.

Funding: This research was Funded by Science Research Project of Hebei Education Department, grant number QN2024286.

Data Availability Statement: The original contributions presented in this study are included in the article. Further inquiries can be directed to the corresponding author.

Conflicts of Interest: The authors declare no conflicts of interest. The funders had no role in the design of the study, in the collection, analyses, or interpretation of data, in the writing of the manuscript, or in the decision to publish the results.

https://doi.org/10.3390/en19010120

References

1. Soga, K.; Lee, S.L.; Ng, M.Y.A.; Klar, A. Characterisation and engineering properties of Methane hydrate soils. In Proceedings of the 2nd International Workshop on Characterisation and Engineering Properties of Natural Soils, Singapore, 29 November–1 December 2006.
2. Milkov, A.V. Global estimates of hydrate-bound gas in marine sediments: How much is really out there? *Earth-Sci. Rev.* **2004**, *66*, 183–197. [CrossRef]
3. Klauda, J.B.; Sandler, S.I. Global Distribution of Methane Hydrate in Ocean Sediment. *Energy Fuels* **2005**, *19*, 459–470. [CrossRef]
4. McGinnis, D.F.; Greinert, J.; Artemov, Y.; Beaubien, S.E.; Wüest, A. Fate of rising methane bubbles in stratified waters: How much methane reaches the atmosphere? *J. Geophys. Res.* **2006**, *111*, 09007. [CrossRef]
5. Chong, Z.R.; Yang, S.H.B.; Babu, P.; Linga, P.; Li, X. Review of natural gas hydrates as an energy resource: Prospects and challenges. *Appl. Energy* **2016**, *162*, 1633–1652. [CrossRef]
6. Xie, Y.; Li, R.; Wang, X.; Zheng, T.; Cui, J.; Yuan, Q.; Qin, H.; Sun, C.; Chen, G. Review on the accumulation behavior of natural gas hydrates in porous sediments. *J. Nat. Gas Sci. Eng.* **2020**, *83*, 103520. [CrossRef]
7. Su, M.; Yang, R.; Wang, H.; Sha, Z.; Liang, J.; Wu, N.; Qiao, S.; Cong, X.; Yang, R. Gas hydrates distribution in the Shenhu Area, northern South China Sea: Comparisons between the eight drilling sites with gas-hydrate petroleum system. *Geol. Acta* **2016**, *14*, 79–100.
8. Kang, D.; Lu, J.A.; Zhang, Z.; Liang, J.; Kuang, Z.; Lu, C.; Kou, B.; Lu, Q.; Wang, J. Fine-grained gas hydrate reservoir properties estimated from well logs and lab measurements at the Shenhu gas hydrate production test site, the northern slope of the South China sea. *Mar. Pet. Geol.* **2020**, *122*, 104676. [CrossRef]
9. Zhang, W.; Liang, J.; Wei, J.; Lu, J.A.; Su, P.; Lin, L.; Huang, W.; Guo, Y.; Deng, W.; Yang, X.; et al. Geological and geophysical features of and controls on occurrence and accumulation of gas hydrates in the first offshore gas-hydrate production test region in the Shenhu area, Northern South China Sea. *Mar. Pet. Geol.* **2020**, *114*, 104191. [CrossRef]
10. Moridis, G.J.; Reagan, M.T.; Boyle, K.L.; Zhang, K. Evaluation of the Gas Production Potential of Some Particularly Challenging Types of Oceanic Hydrate Deposits. *Transp. Porous Media* **2011**, *90*, 269–299. [CrossRef]
11. Li, J.; Ye, J.; Qin, X.; Qiu, H.; Wu, N.; Lu, H.; Xie, W.; Lu, J.; Peng, F.; Xu, Z.; et al. The first offshore natural gas hydrate production test in South China Sea. *China Geol.* **2018**, *1*, 5–16. [CrossRef]
12. Ye, J.; Qin, X.; Xie, W.; Lu, H.; Ma, B.; Qiu, H.; Liang, J.; Lu, J.; Kuang, Z.; Lu, C.; et al. The second natural gas hydrate production test in the South China Sea. *China Geol.* **2020**, *3*, 197–209. [CrossRef]
13. Zhong, X.; Pan, D.; Zhu, Y.; Wang, Y.; Tu, G.; Nie, S.; Ma, Y.; Liu, K.; Chen, C. Commercial production potential evaluation of injection-production mode for CH-Bk hydrate reservoir and investigation of its stimulated potential by fracture network. *Energy* **2022**, *239*, 122113. [CrossRef]
14. Konno, Y.; Fujii, T.; Sato, A.; Akamine, K.; Naiki, M.; Masuda, Y.; Yamamoto, K.; Nagao, J. Key Findings of the World's First Offshore Methane Hydrate Production Test off the Coast of Japan: Toward Future Commercial Production. *Energy Fuels* **2017**, *31*, 2607–2616. [CrossRef]
15. Li, X.; Xu, C.; Zhang, Y.; Ruan, X.; Li, G.; Wang, Y. Investigation into gas production from natural gas hydrate: A review. *Appl. Energy* **2016**, *172*, 286–322. [CrossRef]
16. Wan, Q.; Si, H.; Li, B.; Li, G. Heat transfer analysis of methane hydrate dissociation by depressurization and thermal stimulation. *Int. J. Heat Mass Transf.* **2018**, *127*, 206–217. [CrossRef]
17. Feng, J.; Wang, Y.; Li, X. Large Scale Experimental Evaluation to Methane Hydrate Dissociation Below Quadruple Point by Depressurization Assisted with Heat Stimulation. *Energy Procedia* **2017**, *142*, 4117–4123. [CrossRef]
18. Li, B.; Liu, S.; Liang, Y.; Liu, H. The use of electrical heating for the enhancement of gas recovery from methane hydrate in porous media. *Appl. Energy* **2018**, *227*, 694–702. [CrossRef]
19. Ma, X.; Sun, Y.; Liu, B.; Guo, W.; Jia, R.; Li, B.; Li, S. Numerical study of depressurization and hot water injection for gas hydrate production in China's first offshore test site. *J. Nat. Gas Sci. Eng.* **2020**, *83*, 103530. [CrossRef]
20. Consortium M.R. 2017. Available online: http://www.mh21japan.gr.jp/mh21wp/wp-content/uploads/mh21form2017_doc01.pdf (accessed on 1 December 2024).
21. Li, G.; Moridis, G.J.; Zhang, K.; Li, X. Evaluation of Gas Production Potential from Marine Gas Hydrate Deposits in Shenhu Area of South China Sea. *Energy Fuels* **2010**, *24*, 6018–6033. [CrossRef]
22. Su, Z.; Moridis, G.J.; Zhang, K.; Wu, N. A huff-and-puff production of gas hydrate deposits in Shenhu area of South China Sea through a vertical well. *J. Pet. Sci. Eng.* **2012**, *86–87*, 54–61. [CrossRef]
23. Feng, J.-C.; Wang, Y.; Li, X.-S.; Li, G.; Chen, Z.-Y. Production behaviors and heat transfer characteristics of methane hydrate dissociation by depressurization in conjunction with warm water stimulation with dual horizontal wells. *Energy* **2015**, *79*, 315–324. [CrossRef]

https://doi.org/10.3390/en19010120

24. Jin, G.; Xu, T.; Xin, X.; Wei, M.; Liu, C. Numerical evaluation of the methane production from unconfined gas hydrate-bearing sediment by thermal stimulation and depressurization in Shenhu area, South China Sea. *J. Nat. Gas Sci. Eng.* **2016**, *33*, 497–508. [CrossRef]
25. Nie, S.; Zhong, X.; Ma, Y.; Pan, D.; Liu, K.; Wang, Y.; Li, X.; Chen, C. Numerical simulation of a new methodology to exploit challenging marine hydrate reservoirs without impermeable boundaries. *J. Nat. Gas Sci. Eng.* **2021**, *96*, 104249. [CrossRef]
26. Ma, S.; Zheng, J.; Zhao, J.; Yang, M.; Song, Y. Experimental analysis on thermodynamic stability and methane leakage during solid fluidization process of methane hydrate. *Fuel* **2021**, *284*, 119020. [CrossRef]
27. Xie, L.; Min, K.-B.; Song, Y. Observations of hydraulic stimulations in seven enhanced geothermal system projects. *Renew. Energy* **2015**, *79*, 56–65. [CrossRef]
28. Nianyin, L.; Yu, J.; Daocheng, W.; Chao, W.; Jia, K.; Pingli, L.; Chengzhi, H.; Ying, X. Development status of crosslinking agent in high-temperature and pressure fracturing fluid: A review. *J. Nat. Gas Sci. Eng.* **2022**, *107*, 104369. [CrossRef]
29. Liu, W.; Zeng, Q.; Yao, J.; Liu, Z.; Li, T.; Yan, X. Numerical Study of Elasto-Plastic Hydraulic Fracture Propagation in Deep Reservoirs Using a Hybrid EDFM–XFEM Method. *Energies* **2021**, *14*, 2610. [CrossRef]
30. Nie, S.; Liu, K.; Zhong, X.; Wang, Y.; Yang, B.; Song, J. Research on Hydraulic Fracture Propagation Patterns in Multilayered Gas Hydrate Reservoirs Using a Three-Dimensional XFEM-Based Cohesive Zone Method. *Energy Fuels* **2024**, *38*, 5106–5123. [CrossRef]
31. Ito, T.; Igarashi, A.; Suzuki, K.; Nagakubo, S.; Matsuzawa, M.; Yamamoto, K. Laboratory Study of Hydraulic Fracturing Behavior in Unconsolidated Sands for Methane Hydrate Production. In Proceedings of the Offshore Technology Conference (OTC19324), Houston, TX, USA, 5–8 May 2008.
32. Chen, C.; Yang, L.; Jia, R.; Sun, Y.; Guo, W.; Chen, Y.; Li, X. Simulation Study on the Effect of Fracturing Technology on the Production Efficiency of Natural Gas Hydrate. *Energies* **2017**, *10*, 1241. [CrossRef]
33. Guo, W.; Zhang, P.; Wang, Y.; Jia, R.; Li, B. Evolution on the Gas Production from Low Permeability Gas Hydrate Reservoirs by Depressurization Combined with Reservoir Stimulation. *Energy Fuels* **2020**, *34*, 15819–15828. [CrossRef]
34. Ju, X.; Liu, F.; Fu, P.; White, M.D.; Settgast, R.R.; Morris, J.P. Gas Production from Hot Water Circulation through Hydraulic Fractures in Methane Hydrate-Bearing Sediments: THC-Coupled Simulation of Production Mechanisms. *Energy Fuels* **2020**, *34*, 4448–4465. [CrossRef]
35. Li, B.; Ma, X.; Zhang, G.; Guo, W.; Xu, T.; Yuan, Y.; Sun, Y. Enhancement of gas production from natural gas hydrate reservoir by reservoir stimulation with the stratification split grouting foam mortar method. *J. Nat. Gas Sci. Eng.* **2020**, *81*, 103473. [CrossRef]
36. Ma, X.; Sun, Y.; Guo, W.; Jia, R.; Li, B. Numerical simulation of horizontal well hydraulic fracturing technology for gas production from hydrate reservoir. *Appl. Ocean Res.* **2021**, *112*, 102674. [CrossRef]
37. Zhong, X.; Pan, D.; Zhu, Y.; Wang, Y.; Zhai, L.; Li, X.; Tu, G.; Chen, C. Fracture network stimulation effect on hydrate development by depressurization combined with thermal stimulation using injection-production well patterns. *Energy* **2021**, *228*, 120601. [CrossRef]
38. Nie, S.; Chen, C.; Chen, M.; Song, J.; Wang, Y.; Ma, Y. Numerical Evaluation of a Novel Development Mode for Challenging Oceanic Gas Hydrates Considering Methane Leakage. *Sustainability* **2022**, *14*, 14460. [CrossRef]
39. Yue, P.; Du, Z.; Chen, X.; Liang, B. The critical rate of horizontal wells in bottom-water reservoirs with an impermeable barrier. *Pet. Sci.* **2012**, *9*, 223–229. [CrossRef]
40. Liu, K.; Zhu, W.; Pan, B. Laboratory evaluation on oil-soluble resin as selective water shut-off agent in water control fracturing for low-permeability hydrocarbon reservoirs with bottom aquifer. *Geoenergy Sci. Eng.* **2023**, *225*, 211672. [CrossRef]
41. Bai, Y.; Pu, W.; Jin, X.; Shen, C.; Ren, H. Review of the micro and Macro mechanisms of gel-based plugging agents for enhancing oil recovery of unconventional water flooding oil reservoirs. *J. Mol. Liq.* **2024**, *399*, 124318. [CrossRef]
42. Nie, S.; Liu, K.; Xu, K.; Zhong, X.; Tang, S.; Song, J.; Zhang, H.; Li, J.; Wang, Y. Numerical study on the stimulation effect of boundary sealing and hot water injection in marine challenging gas hydrate extraction. *Sci. Rep.* **2024**, *14*, 15280. [CrossRef]
43. Wu, N.; Zhang, H.; Yang, S.; Zhang, G.; Liang, J.; Lu, J.A.; Su, X.; Schultheiss, P.; Holland, M.; Zhu, Y. Gas Hydrate System of Shenhu Area, Northern South China Sea: Geochemical Results. *J. Geol. Res.* **2011**, *2011*, 370298. [CrossRef]
44. Moridis, G.J.; Collett, T.S. *Strategies for Gas Production from Hydrate Accumulations Under Various Geological and Reservoir Conditions*; Report LBNL-52568; Lawrence Berkeley National Laboratory: Berkeley, CA, USA, 2002.
45. Liang, Y.; Li, X.; Li, B. Assessment of Gas Production Potential from Hydrate Reservoir in Qilian Mountain Permafrost Using Five-Spot Horizontal Well System. *Energies* **2015**, *8*, 10796–10817. [CrossRef]
46. Wang, Y.; Feng, J.; Li, X.; Zhang, Y.; Li, G. Analytic modeling and large-scale experimental study of mass and heat transfer during hydrate dissociation in sediment with different dissociation methods. *Energy* **2015**, *90*, 1931–1948. [CrossRef]
47. Moridis, G.J. Numerical Studies of Gas Production from Methane Hydrates. *SPE J.* **2003**, *8*, 359–370. [CrossRef]
48. Moridis, G.J. Numerical Studies of Gas Production from Class 2 and Class 3 Hydrate Accumulations at the Mallik Site, Mackenzie Delta, Canada. *SPE Reserv. Eval. Eng.* **2004**, *7*, 175–183. [CrossRef]

https://doi.org/10.3390/en19010120

49. Moridis, G.J.; Kowalsky, M.B.; Pruess, K. Depressurization-Induced Gas Production from Class-1 Hydrate Deposits. *SPE Reserv. Eval. Eng.* **2007**, *10*, 458–481. [CrossRef]
50. Yin, Z.; Moridis, G.; Chong, Z.R.; Tan, H.K.; Linga, P. Numerical analysis of experimental studies of methane hydrate dissociation induced by depressurization in a sandy porous medium. *Appl. Energy* **2018**, *230*, 444–459. [CrossRef]
51. Feng, Y.; Chen, L.; Suzuki, A.; Kogawa, T.; Okajima, J.; Komiya, A.; Maruyama, S. Numerical analysis of gas production from layered methane hydrate reservoirs by depressurization. *Energy* **2019**, *166*, 1106–1119. [CrossRef]
52. Chen, L.; Feng, Y.; Okajima, J.; Komiya, A.; Maruyama, S. Production behavior and numerical analysis for 2017 methane hydrate extraction test of Shenhu, South China Sea. *J. Nat. Gas Sci. Eng.* **2018**, *53*, 55–66. [CrossRef]
53. Feng, J.-C.; Wang, Y.; Li, X.-S.; Li, G.; Zhang, Y. Three dimensional experimental and numerical investigations into hydrate dissociation in sandy reservoir with dual horizontal wells. *Energy* **2015**, *90*, 836–845. [CrossRef]
54. Ilyushin, Y.; Nosova, V.; Krauze, A. Application of Systems Analysis Methods to Construct a Virtual Model of the Field. *Energies* **2025**, *18*, 1012. [CrossRef]

https://doi.org/10.3390/en19010120

energies

MDPI

Article

Thermal Displacement with CO_2 for E-CBM Recovery: Mechanisms and Efficacy of Temperature–Pressure Synergy in Permeability Enhancement

Xiaohu Xu [1,2,3], Tengze Ge [2,3], Ersi Gao [2,3], Shuguang Li [2,3], Kai Wei [4], Yulong Liu [4,*] and Ao Wang [4]

1 School of Chemical Engineering, University of Chinese Academy of Sciences, Beijing 100049, China
2 China United Coalbed Methane National Engineering Research Center Co., Ltd., Beijing 100095, China
3 PetroChina Coalbed Methane Co., Ltd., Beijing 100028, China
4 State Key Laboratory of Low Carbon Catalysis and Carbon Dioxide Utilization (Yangtze University), School of Petroleum Engineering, Yangtze University, Wuhan 430100, China
* Correspondence: yulongliu@yangtzeu.edu.cn

Abstract

The efficient development of coalbed methane (CBM) faces persistent challenges due to low recovery rates. While CO_2 thermal displacement offers a promising approach, the pore–fracture structure (PFC) evolution and gas displacement mechanisms under temperature–pressure coupling remain insufficiently clear. To address this knowledge gap, the in situ, dynamic quantification of pore–fracture evolution during CO_2 displacement was achieved by an integrated system with NMR and CT scanning, revealing the expansion, connection, and reconfiguration of coal PFC under temperature–pressure synergy and establishing the intrinsic relationship between supercritical CO_2 ($ScCO_2$)-induced permeability enhancement and methane displacement efficiency. Experimental results identify an observed transition in permeability near 80 °C under the tested conditions as a critical permeability transition point: below this value, permeability declines from 0.61 mD to 0.49 mD, reflecting pore structure adjustment; above it, permeability rises markedly to 1.18 mD, indicating a structural shift toward fracture-dominated flow. A "pressure-dominated, temperature-assisted" mechanism is elucidated, wherein pressure acts as the primary driver in creating macro-fractures and forming percolation pathways, while temperature—mainly via thermal stress—promotes micro-fracture development and assists gas desorption, offering only limited direct contribution to permeability. Although elevated injection pressure enhances permeability and establishes fracture networks, displacement efficiency eventually reaches a physical limit. To transcend this constraint, a synergistic production mechanism is proposed in which pressure builds flow channels while temperature activates microporous desorption. This study provides an integrated, in situ quantification of the pore–fraction evolution under high-temperature $ScCO_2$ conditions. The elucidated synergy between pressure and temperature offers insights and an experimental basis for the design of deep CBM recovery and CO_2 storage strategies.

Keywords: CO_2 thermal flooding; coalbed methane; pore–fracture structure; temperature–pressure synergy; permeability enhancement

1. Introduction

China has vast CBM resources, with approximately 36.8 trillion cubic meters within 2000 m depth, ranking third globally [1–3]. Current extraction mainly uses hydraulic fracturing and dewatering for pressure reduction, creating single seepage channels to boost gas production. However, the generally low permeability of coal reservoirs severely hinders gas desorption, diffusion, and flow, leading to low recovery rates [4–6]. Furthermore, about two-thirds of China's CBM resources lie deeper than 1000 m, where increased in situ stress and gas content further reduce permeability, raising extraction difficulties [7]. Therefore, effectively enhancing permeability in low-permeability coal seams is a critical challenge for China's CBM development [8].

Extensive research shows that $ScCO_2$ exhibits properties of both gas and liquid phases, characterized by low viscosity, high diffusivity, and excellent fluidity, enabling it to rapidly penetrate micro-fractures and pores in porous media and enhance permeability [9,10]. When injected into low-permeability coal seams, $ScCO_2$ not only displaces CH_4 like non-supercritical CO_2 but also significantly improves permeability by extracting organic matter, disrupting the molecular structure and ordered aromatic lamellae within coal, and inducing matrix swelling and deformation [11,12]. These processes alter the PFC evolution, mechanical strength, and gas adsorption capacity of coal, thereby increasing permeability and enhancing CBM recovery [13,14]. Additionally, elevated coal seam temperature activates CH_4 molecules, promoting desorption and migration while also improving permeability, thereby strengthening gas displacement efficiency and CO_2 sequestration [15–17]. Thus, heated $ScCO_2$ injection combines the dual advantages of heating and CO_2 injection, emerging as a promising technique for enhanced CBM recovery.

In recent years, extensive experimental research has been conducted to understand the evolution of seepage pathways and underlying mechanisms in coal reservoirs following CO_2 injection. Liu et al. [18] observed a continuous decline in coal permeability with increasing temperature during supercritical CO_2 ($ScCO_2$) stimulation at 35–65 °C and 7–10 MPa, with reported reductions ranging from approximately 20% to 60% depending on coal type and conditions, attributing this to pore compression caused by thermal expansion. In contrast, under non-supercritical conditions (0.21–1.8 MPa, 20–50 °C), Zhang et al. [19] reported a V-shaped permeability trend: decreasing by up to ~40% below 40 °C due to thermal expansion and increasing by ~15–30% above 40 °C as thermal stress-induced fractures developed. Li et al. [20] identified a similar but higher transition temperature near 55 °C under comparable conditions, with permeability variations in the order of 35–50%, reflecting variability across coal ranks. Further diverging from these trends, Liu et al. [21] monitored permeability evolution during pyrolysis up to 400 °C under unconfined conditions, revealing an N-shaped trend: a slight increase of ~5–10% up to 100 °C from water evaporation and gas desorption, a decline of ~20–25% at 100–200 °C due to asphaltene clogging, and a sharp rise exceeding 200% at 300–400 °C resulting from extensive fracture formation. While these studies highlight diverse thermal effects on permeability, variations in experimental conditions, coal types, and control methods have hindered a unified understanding of PFC evolution during heating [22–24]. In particular, the behavior of coal under $ScCO_2$ injection at temperatures above 70 °C remains unclear, warranting further systematic investigation to elucidate the governing mechanisms.

Furthermore, the dynamic evolution of coal's micro-fracture and pore structure remains unclear, with no consensus established [25,26]. For instance, Zhang et al. [27], using high-pressure gas injection coupled with micro-CT, found that under 7–10 MPa injection pressure, the proportion of micropores (<2 nm) initially increased and then decreased with injection time, while mesopore (2–50 nm) volume expanded significantly during early stages before stabilizing. They attributed this reorganization to $ScCO_2$'s extraction of organic matter. In contrast, Liu et al. [28] observed under near-critical conditions (31 °C, 7.2 MPa) that fracture aperture initially decreased due to matrix swelling, reducing permeability by approximately 20%, but after 6 h of continuous injection, average aperture

https://doi.org/10.3390/en19020496

increased by 15% from mineral dissolution and structural relaxation, revealing time-varying and nonlinear characteristics. Li et al. [29] further highlighted rank-dependent responses: medium-rank bituminous coal exhibited a 12.8% increase in total pore volume after $ScCO_2$ injection, whereas high-rank anthracite showed only a 3.5% increase, with newly formed fractures mainly isolated [30–32]. These studies collectively demonstrate limited systematic understanding of the evolution pathways, spatial reconfiguration, and subsequent impacts on seepage capacity of coal fractures and pores under $ScCO_2$ [33–35]. Particularly within the 70–150 °C range, the response under thermo–mechanical–chemical coupling remains unclear, and it is still debated whether the dominant factor in PFC evolution is temperature or pressure. Clarifying the dominant control mechanism and advancing in situ observation and multi-scale characterization are urgently needed to reveal the underlying mechanisms.

This study investigated the evolution of coal micro-fractures and pores under supercritical CO_2 at 35–150 °C through high-pressure thermal stimulation experiments. The fundamental novelty of this work lies in the in situ, same-sample, multi-modal characterization (NMR and CT) of pore–fracture evolution, enabling a quantitative comparison of temperature versus pressure effects within a controlled, high-temperature $ScCO_2$ experimental framework. By comparing NMR and CT scans of identical samples before and after treatment, this work tracks structural changes and clarifies the temperature–pressure coupling effects on pore–fracture evolution. The analysis clarifies temperature–pressure coupling effects on fracture propagation and pore reconfiguration, contributing to a clearer understanding of the relationship between $ScCO_2$-induced permeability enhancement and gas displacement efficiency. The findings address key mechanisms of $ScCO_2$–coal interaction at elevated temperatures and provide data and mechanistic support for enhancing deep CBM recovery and CO_2 geological storage.

2. Experiments Methods

2.1. Coal Sample Preparation

The coal samples used in this experiment were collected from the No. 5 coal seam in the Sanjiao Block, Ordos Basin, China. This medium-rank bituminous coal was selected for its representativeness of a major CBM-producing reservoir in China and its suitability for investigating temperature–pressure induced pore–fracture transformations under $ScCO_2$ conditions. The samples were precision wire-cut into standard cylindrical specimens (φ25 mm × 50 mm) with parallel end faces and no visible lateral fractures.

The key parameters presented in Table 1 provide the essential baseline for interpreting the experimental outcomes. The vitrinite reflectance ($R_{0,\max}$) confirms the medium rank of the bituminous coal samples, which is representative of significant deep CBM resources in China. The high vitrinite content is relevant to gas adsorption and mechanical behavior during stimulation. The low initial porosity and permeability are characteristic of the tight coal reservoirs targeted for enhancement. The samples were specifically selected with closely matched properties to minimize the influence of sample heterogeneity, allowing the effects of temperature–pressure coupling to be isolated and analyzed. These property ranges are consistent with reported data for similar rank coals in the Ordos Basin, ensuring the relevance of our findings to the region.

Prior to thermal stimulation experiments, the initial porosity and permeability of all samples were measured using a gas porosimeter–permeameter [36]. To control systematic errors and ensure experimental reliability, six samples with closely matched porosity and permeability values were selected from the total set as valid experimental subjects for subsequent supercritical CO_2 thermal stimulation tests. The reported initial porosity and permeability values represent the mean of triplicate measurements, with standard deviations of less than 5%. Additionally, the reported permeability values are Klinkenbergcorrected absolute permeabilities, derived from measurements at multiple mean pore pressures to account for gas slippage effects, which are significant in this low-permeability range. Table 1 summarizes the maceral analysis, proximate analysis, and initial porosity–permeability data of the four coal cores.

Table 1. Maceral analysis, proximate analysis, and initial porosity–permeability of the coal.

Samples	$R_{0,max}$/%	Maceral Content/%			Porosity/%	Permeability (mD)	Density (g/cm^3)
		Vitrinite	Inertinite	Minerals			
CL-1	1.222	75.6	21.8	2.6	6.1	0.023	1389.6
CL-2	1.235	74.3	22.4	3.3	5.8	0.031	1392.1
CH-1	1.288	78.3	20.3	1.4	5.5	0.024	1401.4
CH-2	1.246	73.8	23.5	2.7	5.7	0.025	1396.3

2.2. Experimental Procedure

2.2.1. CO_2 Thermal Displacement with Online NMR Monitoring

The experiment employed a custom-integrated, in-house designed variable-temperature high-pressure nuclear magnetic resonance (NMR) imaging analysis system, as illustrated in Figure 1. The setup primarily consists of a constant-pressure/constant-flow pump, a high-low temperature and pressure circulation system, a heating and temperature control unit, pressure transducers, and a Suzhou Niumag variable-temperature high-pressure NMR imaging analysis system.

Figure 1. Variable-temperature high-pressure nuclear magnetic resonance (NMR) analysis system: (**a**) NMR system; (**b**) NMR chamber.

The temperatures of 35, 80, 120, and 150 °C were selected to achieve the following objectives: 35 °C establishes a supercritical baseline near common reservoir conditions; 80 °C allows for investigation of potential transition behavior; 120 °C and 150 °C enable the study of pronounced thermo-mechanical effects and enhanced desorption kinetics, exploring the potential of thermal stimulation for recovery enhancement.

(1) CH_4 Saturation and Adsorption

The dried coal sample was placed in the NMR holder and saturated with CH_4 at set temperatures (35, 80, 120, and 150 °C). The injection pressure was gradually increased from atmospheric pressure to the reservoir pressure of 8 MPa at a rate of 0.1 MPa/min. While maintaining the target temperature, CH_4 was replenished slowly whenever the pressure dropped by 0.015 MPa until equilibrium was achieved (pressure fluctuation $\leq$ 0.015 MPa

https://doi.org/10.3390/en19020496

over 8 h), and the total injected CH_4 volume was recorded. Confining pressure was increased synchronously with injection pressure to prevent sample damage, reaching 8 MPa simultaneously.

(2) Online NMR CO_2 Displacement under High Temperature and Pressure

① CO_2 was heated to the target temperature (35, 80, 120, or 150 °C) and injected at a constant pressure of 8 MPa. Injection began at a low flow rate of 0.01 mL/min, which was gradually adjusted based on outlet flow and pressure to avoid rapid pressure changes and sample failure. Injected CO_2 volume was recorded in real time.

② A back-pressure regulator maintained a constant outlet pressure of 6 MPa, stabilizing the core pressure at 8 MPa. Throughout the experiment, gas flow and pressure at both the injection and outlet ends were continuously monitored. The flow rate was carefully controlled to ensure stable, laminar flow through the low-permeability core. Prior to each NMR measurement, the system was allowed to reach steady-state flow under these constant pressure conditions. This ensured a stable fluid pressure distribution within the sample and minimized transient compression/decompression effects on the acquired T_2 signal.

2.2.2. Pore–Fracture Characterization via NMR Before and After Displacement

Nuclear Magnetic Resonance (NMR) provides an effective approach for quantitatively characterizing the internal pore and fracture structure of coal. This study employs the transverse relaxation time (T_2) distribution to reveal pore structure characteristics of coal samples. The principle relies on the fact that the relaxation rate of hydrogen nuclei in the pore fluid is governed by surface interactions, where the relaxation time correlates directly with pore size, as expressed by

$$\frac{1}{T_2} = \rho_2 \frac{S}{V} \tag{1}$$

Here, T_2 (in ms) is the measured transverse relaxation time, ρ_2 (in μm/ms) is the surface relaxivity of the coal, and S and V (in cm^2 and cm^3) represent the surface area and volume of a single pore, respectively. For spherical pores, the S/V ratio simplifies to $3/r$, while for slit-shaped fractures, it approximates $2/r$, where r denotes the characteristic pore radius.

The experiments were conducted using a variable-temperature high-pressure NMR imaging analysis system, with a main magnetic field strength of 0.3 T and an RF frequency range of 1–42 MHz. Prior to testing, standard coal rock cores were subjected to pretreatment: they were placed in a vacuum pressurized saturation device and saturated with distilled water for over 8 h to ensure full pore filling. The saturated samples were then positioned at the magnet center and scanned using the CPMG sequence to acquire raw echo train data. The T_2 spectrum, representing pore size distribution, was obtained by applying an inversion algorithm to the data. After all NMR measurements, the samples were transferred to a vacuum drying oven and dried for 12 h for subsequent experiments.

It is explicitly acknowledged that the transverse relaxation time (T_2) distribution is a non-unique indicator of pore structure. The measured T_2 values are influenced by a combination of factors including pore size, surface relaxivity, fluid properties, and diffusion effects. Therefore, in this study, absolute T_2 values are not used as direct, quantitative proof of specific pore-to-fracture conversion. Instead, the relative changes observed in the T_2 spectra of the same sample before and after treatment, under rigorously controlled experimental conditions, are interpreted as qualitative trends consistent with alterations in the pore–fracture architecture. These interpretations are further validated and substantiated by direct, three-dimensional visualization from X-ray micro-CT scans (Figure 2).

https://doi.org/10.3390/en19020496

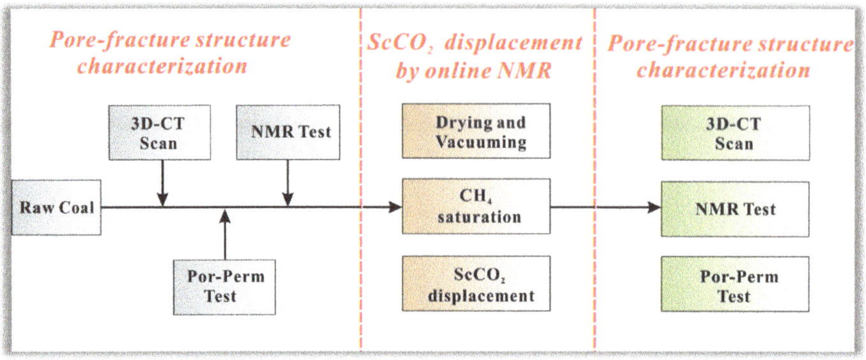

Figure 2. Schematic of PFC characterization pre- and post-CO_2 flooding.

2.2.3. Pre- and Post-Displacement 3D CT Scanning

The experiments were performed using a NanoVoxel-350E 3D computed tomography system (Tianjin, China). This instrument achieves a maximum spatial resolution of 0.5 μm, with a maximum scanning voltage of 190 kV and a maximum power of 25 kW. A stepwise scanning mode was employed (Figure 3).

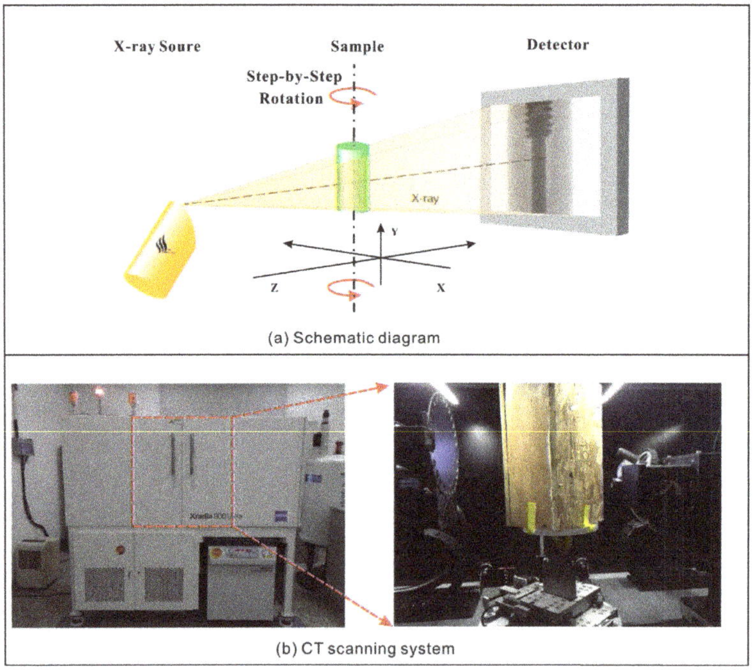

Figure 3. NanoVoxel-350E 3D computed tomography system.

To characterize the three-dimensional spatial distribution of mesopores and fractures within the sample, the two-dimensional slices underwent preprocessing and three-dimensional reconstruction. To enhance the quality of the original CT slices, a non-local means filtering method was applied for 3D denoising. This approach combines the advantages of Gaussian filtering in preserving image details and edges with the pronounced smoothing effect of mean filtering.

https://doi.org/10.3390/en19020496

After applying the median filter (Figure 4), the grayscale CT images with intensity values ranging from 0 to 255 were segmented using Equation (2):

$$v(i,i) = \begin{cases} 0 & l(i,j) \leqslant T_1 \\ 1 & T_1 < l(i,j) \leqslant T_2 \\ 2 & T_2 < l(i,j) \end{cases} \quad (2)$$

where $l\ (i, j)$ represents the grayscale value of the pixel, T_1 and T_2 are the segmentation thresholds. $v\ (i, j)$ denotes the three segmented primary components, with 0 representing voids, 1 representing matrix, and 2 representing minerals.

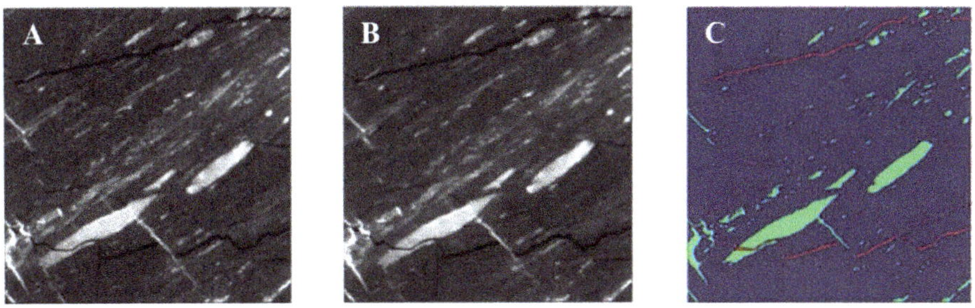

Figure 4. The image processing procedure: (**A**) original grayscale image; (**B**) after denoising; (**C**) after threshold segmentation.

During data analysis of the segmented images, the volume fraction of each component was calculated using Equation (3):

$$\varphi = N_c / N_t \times 100\% \quad (3)$$

where N_c is the number of voxels belonging to a specific component, and N_t is the total number of voxels.

3. Experimental Results and Analysis

3.1. Temperature-Dependent Permeability Enhancement During Thermal Flooding

3.1.1. PFC Evolution at Different Temperatures

To systematically investigate the impact of thermal CO_2 flooding on the pore structure of coal, this study conducted tests on coal samples before and after thermal CO_2 treatment using an NIUMAG™ (Shanghai, China) low-field nuclear magnetic resonance analysis system. Based on Hodot's pore classification scheme [37], the pore structures were categorized into four types: micropores (<10 nm), mesopores (10–100 nm), macropores (100–1000 nm), and fractures (>1000 nm).

Raw coal samples primarily displayed bimodal or trimodal T_2 distributions, with the main peak in the 0.1–1 ms range reflecting micropores, and a secondary peak in the 1–100 ms range indicating meso- and macropores. While minor variations existed among initial samples, all spectra consistently exhibited peaks within 10–100 ms, approximating a bimodal profile. These results indicate that despite subtle differences in peak features, the overall pore–fracture architecture remains broadly consistent across coal samples (Figure 5).

https://doi.org/10.3390/en19020496

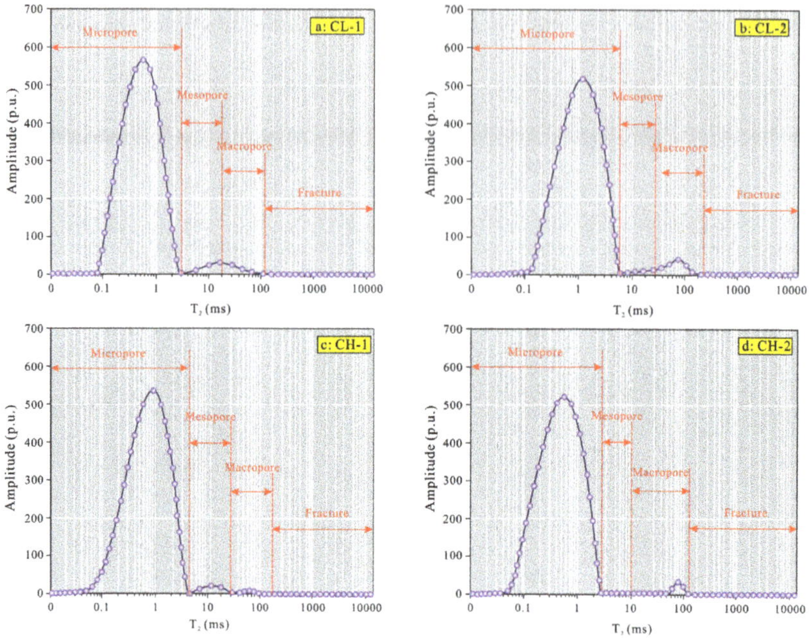

Figure 5. The T_2 spectra of different coal samples pre-CO_2 flooding: (**a**) CL-1 (35 °C injection), (**b**) CL-2 (80 °C injection), (**c**) CH-1 (120 °C injection), and (**d**) CH-2 (150 °C injection).

The T_2 spectrum of coal reveals distinct temperature-dependent evolution after CO_2–thermal treatment. At elevated temperatures (120 °C and 150 °C), the signal intensity in the 10–1000 ms range increases markedly (Figure 6a,c), demonstrating substantial growth in macropores and fractures. Concurrently, the peak in the 0.1–1 ms range weakens and shifts rightward, indicating micropore enlargement and a reduction in volume. Quantitative analysis of pore–fracture distribution further confirms a more pronounced increase in fractures proportion under higher temperatures. Specifically, samples CH-1 and CH-2 exhibited fractures proportion increases of 16.1% and 25.5%, respectively, accompanied by a decrease in micropores (32.0% and 41.8%) and an increase in macropores (10.7% and 8.9%). The observed conversion of micropores to macropores and the development of fractures at elevated temperatures are consistent with the effects of thermo-mechanical stress. We infer that non-uniform thermal expansion likely induces localized tensile stress, promoting micro-crack formation and propagation.

Compared to pre-treatment conditions, T_2 spectral changes in coal samples subjected to lower-temperature CO_2 thermal treatment were statistically insignificant (35 °C and 80 °C). The spectrum shifted slightly rightward, with reduced amplitude in the 0.1–1 ms range and lower peaks between 1 and 100 ms. This reflects general pore enlargement, a decrease in micropores, and a slight increase in meso- and macropores. Notably, no clear peak formed in the 100–1000 ms range—unlike under high temperatures—indicating no significant fractures development. These results confirm that weaker thermal stress at lower temperatures is inadequate to substantially modify the fractures system or induce extensive micro-fractures.

https://doi.org/10.3390/en19020496

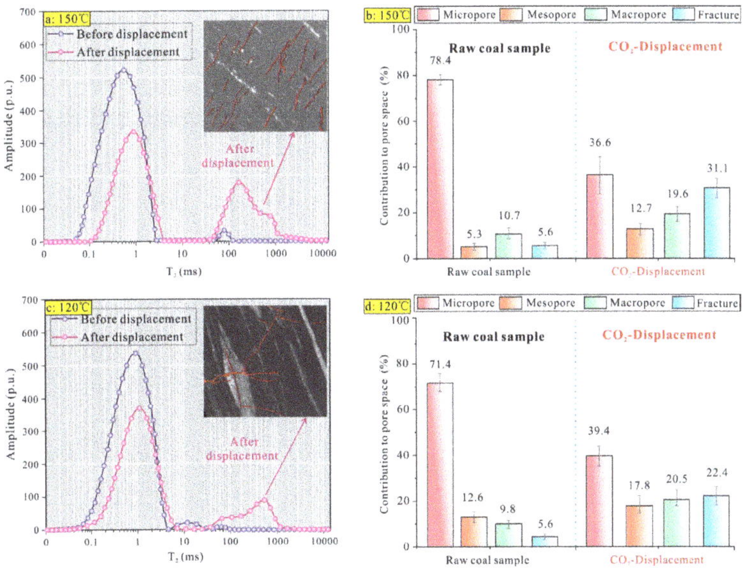

Figure 6. Evolution of T_2 spectra and PFC in response to CO_2 thermal flooding temperatures (120 °C and 150 °C).

Figure 7 illustrates the evolution of pore-fissure distribution following CO_2 thermal treatment. At lower injection temperatures (35 °C and 80 °C), the increase in fractures proportion remained limited, rising by only 10.3% and 8.6%, respectively. At lower treatment temperatures, the limited thermal stress applied to the coal is insufficient to induce significant tensile failure, resulting in a low degree of fracture development.

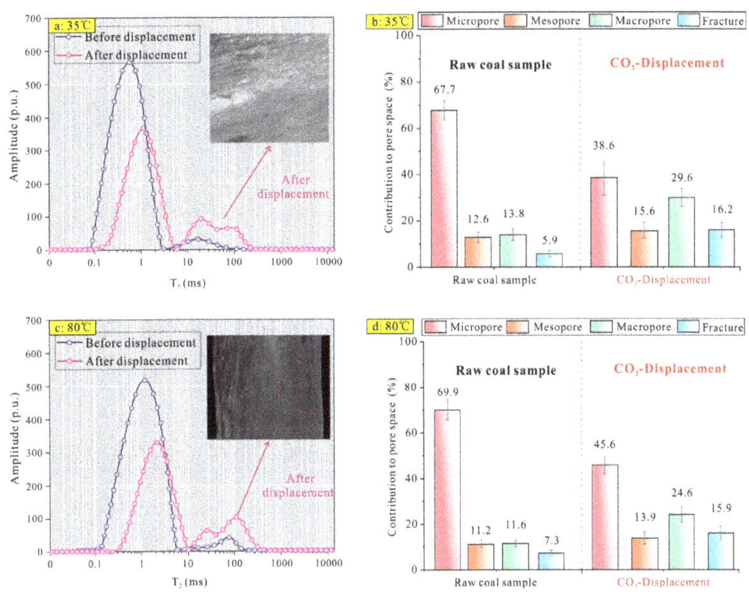

Figure 7. Evolution of T_2 spectra and PFC in response to CO_2 thermal flooding temperatures (35 °C and 80 °C).

https://doi.org/10.3390/en19020496

Compared to the fractures system, low-temperature CO_2 thermal treatment exerts a more consistent influence on the pore structure. After treatment, the proportion of micropores decreased by 29.1% and 24.3%, while that of macropores increased by 15.8% and 13.0%, respectively (Figure 8). These alterations in the pore system are primarily attributed to structural readjustments induced by thermal expansion and CO_2–coal matrix interactions. Under identical thermal treatment conditions, the expansive response of the coal matrix remains relatively uniform, leading to a consistent evolutionary trend in the pore system.

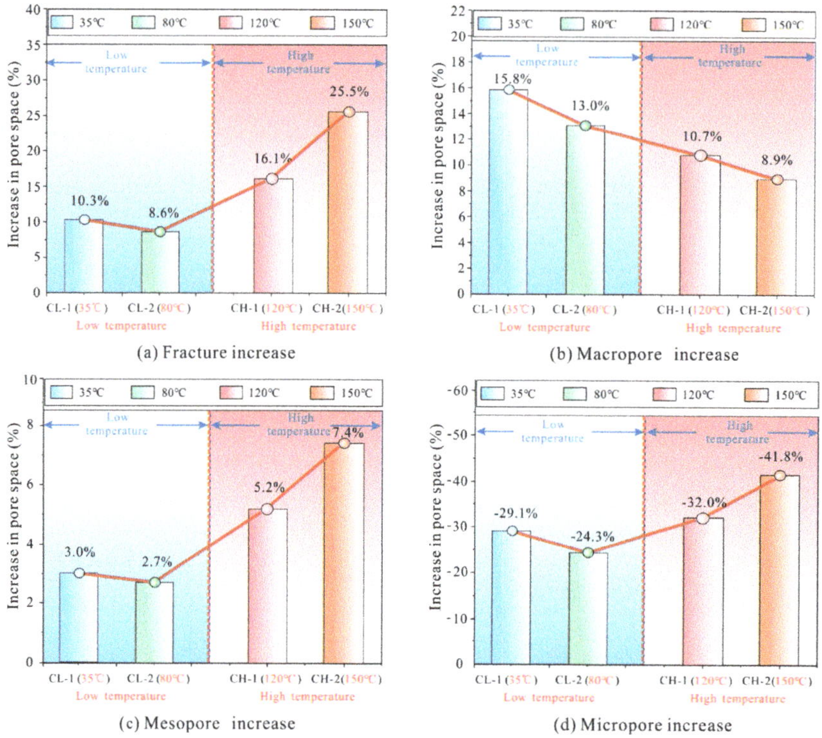

Figure 8. The evolution of pore-fissure distribution following CO_2 thermal treatment.

3.1.2. Evolution of Poro-Permeability Across Thermal Flooding Temperatures

Figure 9 illustrates the variations in NMR porosity of coal samples after CO_2 thermal treatment at different temperatures. Experimental results show a significant increase in total porosity after treatment, with the growth ranging from 94.8% to 238%. Under high-temperature conditions (120 °C and 150 °C), the coal matrix experienced substantial thermal stress, inducing numerous newly generated fractures and resulting in pronounced porosity increases of 193% and 238%, respectively. In contrast, under medium-low temperature treatments (35 °C and 80 °C), porosity also increased—by 120% and 94.8%, respectively—though the mechanism was dominated by macropore expansion, with limited fractures development induced by thermal stress.

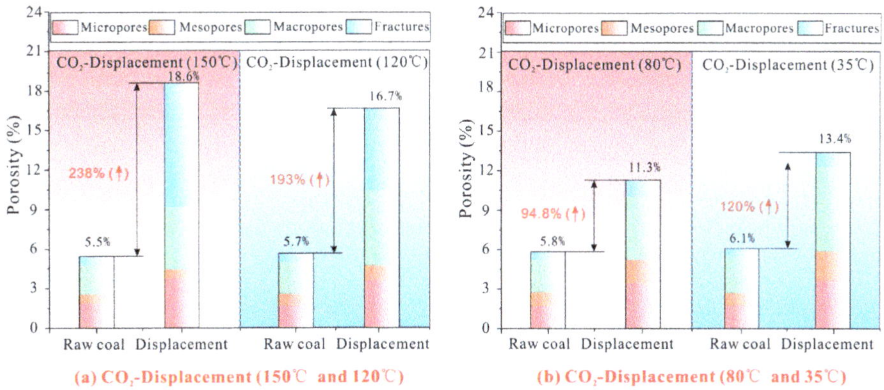

Figure 9. The variations in NMR porosity of coal samples after CO_2 thermal treatment.

Figure 10 further reveals that the porosity increase induced by high-temperature thermal treatment primarily originates from the development of fracture systems. Concurrently, as the thermal treatment temperature rises, the proportion of micropores in the coal decreases, reflecting the systematic impact of thermal transformation on the pore structure. In contrast, under medium-low temperature conditions, the enhancement in porosity is mainly attributed to the expansion of macropores, with fractures contributing minimally to the overall porosity increase (Figure 10).

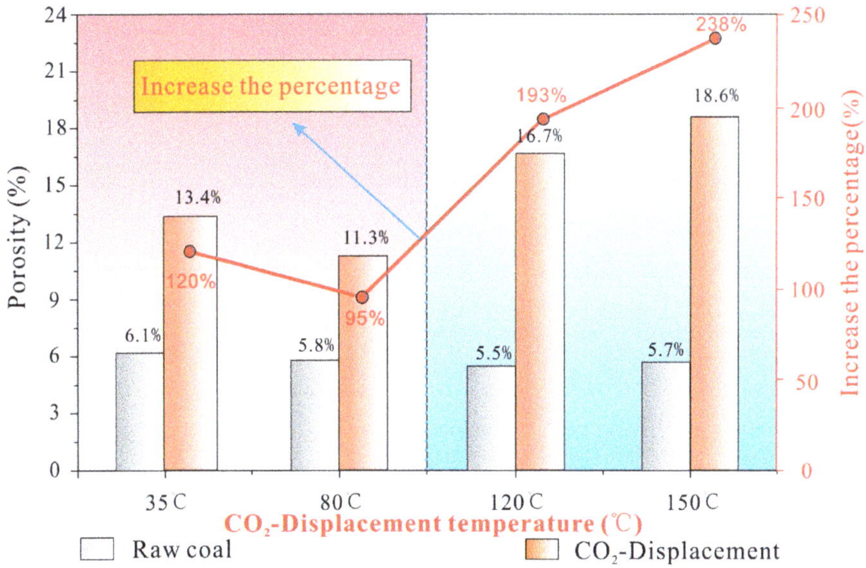

Figure 10. The porosity increases induced by different-temperatures thermal treatment.

These results demonstrate a distinct threshold effect of temperature on coal modification, with different dominant mechanisms across temperature ranges. Below 80 °C, CO_2 thermal treatment primarily induces pore structure rearrangement, characterized by a reduction in micropores and an increase in macropores. In contrast, at elevated temperatures (120 °C and 150 °C), intensified thermal stress leads to extensive fissure development, becoming the dominant factor enhancing permeability.

https://doi.org/10.3390/en19020496

This mechanistic transition corresponds to a nonlinear permeability evolution, with values of 0.61 mD (35 °C), 0.49 mD (80 °C), 0.97 mD (120 °C), and 1.18 mD (150 °C) displaying a V-shaped trend centered at 80 °C (Figure 11). These data reveal a clear non-monotonic, V-shaped trend in permeability as temperature rises, with a minimum value observed at 80 °C. Given the sample-specific nature of each temperature path, the observed V-shaped trend is presented as a clear and consistent observational trend across the systematically varied conditions. By extending the temperature range to 150 °C, this study confirms 80 °C as the critical transition temperature for pore-permeability evolution, resolving previous debates over temperature thresholds and providing key insights for optimizing thermal-enhanced coalbed methane recovery.

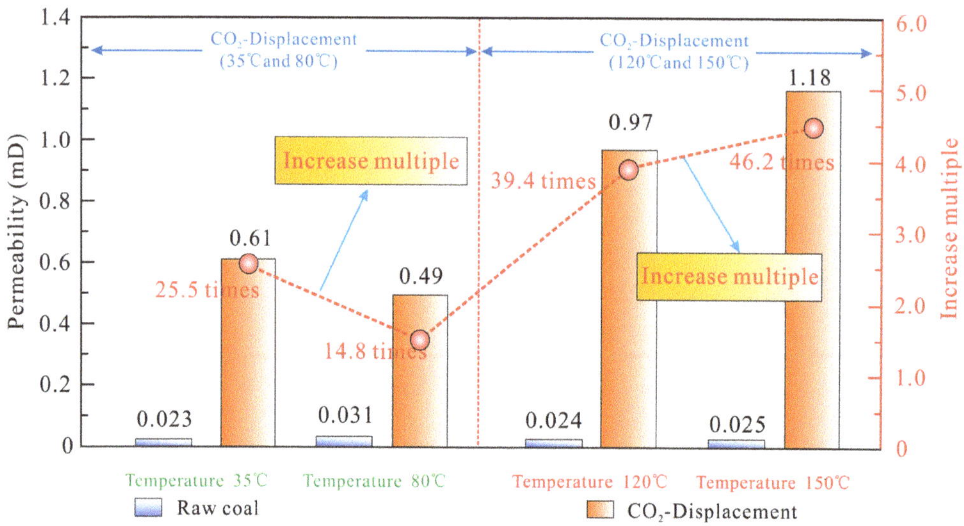

Figure 11. The permeability increases induced by different-temperatures thermal treatment.

3.2. *Pressure-Dependent Permeability Enhancement During Thermal Flooding*

3.2.1. Pore–Fracture Evolution Under Different Injection Pressures by NMR

The evolution of T_2 spectra and pore–fracture structure in coal under CO_2-thermal treatment was systematically compared at 150 °C with injection pressures increased from 8 MPa to 10 MPa. Enhanced injection pressure was observed to significantly strengthen the T_2 signal beyond 1000 ms, indicating the development of larger-scale macro-fractures. Correspondingly, the fracture proportion rose from 31.7% at 8 MPa to 47.4% under the elevated pressure condition (Figure 12).

This change is primarily attributed to the enhanced mechanical expansion effect under elevated injection pressure. As pressure increased, pre-existing pores and micro-fractures in the coal were progressively expanded and interconnected, gradually forming a macroscopic fracture system with certain aperture and extension. The development of such fractures not only directly increased the fracture proportion within the pore structure but also significantly improved the overall flow conductivity of the reservoir, serving as the key structural foundation for permeability enhancement.

https://doi.org/10.3390/en19020496

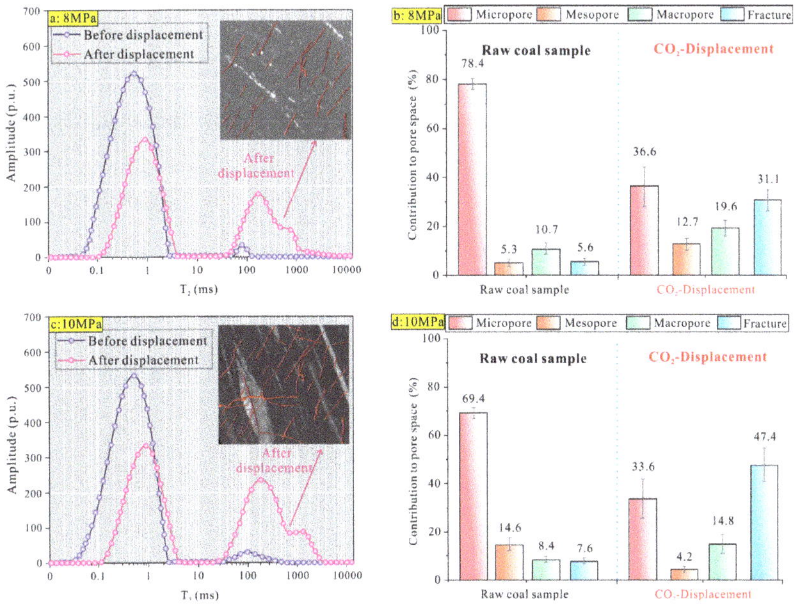

Figure 12. Evolution of T_2 spectra and PFC in response to CO_2 injection pressures.

3.2.2. Pore–Fracture- Evolution Under Different Injection Pressures by C

The development characteristics of coal fractures under CO_2–thermal treatment exhibit significant variations with increasing injection pressure. As shown in the 3D reconstruction results (Figure 13), at an injection pressure of 8 MPa, a dense network of short, morphologically complex fractures is predominantly formed. When the injection pressure rises to 10 MPa, the fracture system evolves further, developing more dominant macro-fracture channels on the basis of the existing short-fracture network (Figure 14).

The observed evolution is closely related to the action mechanisms of $ScCO_2$ in coal. At 8 MPa, CO_2 primarily induces microporous restructuring and short-fracture formation through physicochemical interactions with the coal matrix. When pressure increases to 10 MPa, the enhanced fluid potential and penetration pressure subject the coal to stronger tensile stress, which not only extends existing fractures but also promotes the generation of macro-fractures on the 10^4–10^5 nm scale. Although fracture directionality remains poor at 10 MPa, the significant increase in macro-fracture proportion substantially improves network connectivity, establishing multi-scale flow pathways from micro- to macroscopic levels.

A comparative analysis of pore–fracture structure evolution under 8 MPa and 10 MPa injection pressures reveals significant reorganization of the coal pore system as pressure increases. As pressure rises from 8 MPa to 10 MPa, the proportion of macro-fractures (10^4–10^5 nm) increases markedly from 4.7% to 24.5%, a rise of 19.8 percentage points (Figures 13 and 14). The extensive development of such fractures enhances and interconnects the existing fracture network, forming a more continuous and efficient flow pathway system, thereby significantly improving the overall permeability and fluid transport efficiency of the reservoir. The corresponding signal intensity in the NMR T_2 spectrum for the macro-fracture range also strengthens, further confirming these structural changes. This indicates that the macro-fracture system formed under high injection pressure is a key structural element controlling the enhancement of reservoir seepage capacity.

https://doi.org/10.3390/en19020496

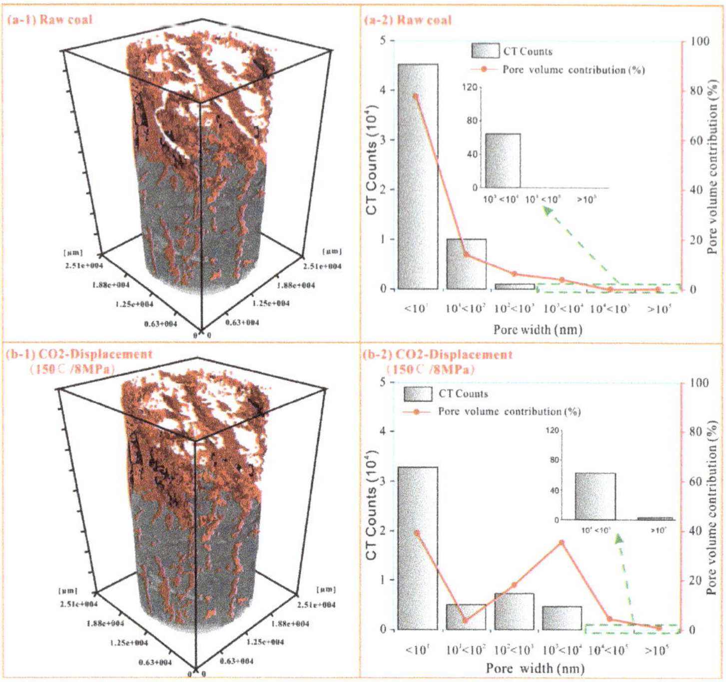

Figure 13. The 3D evolution of coal PFC from pre- to post-CO_2 thermal flooding (150 °C/8 MPa).

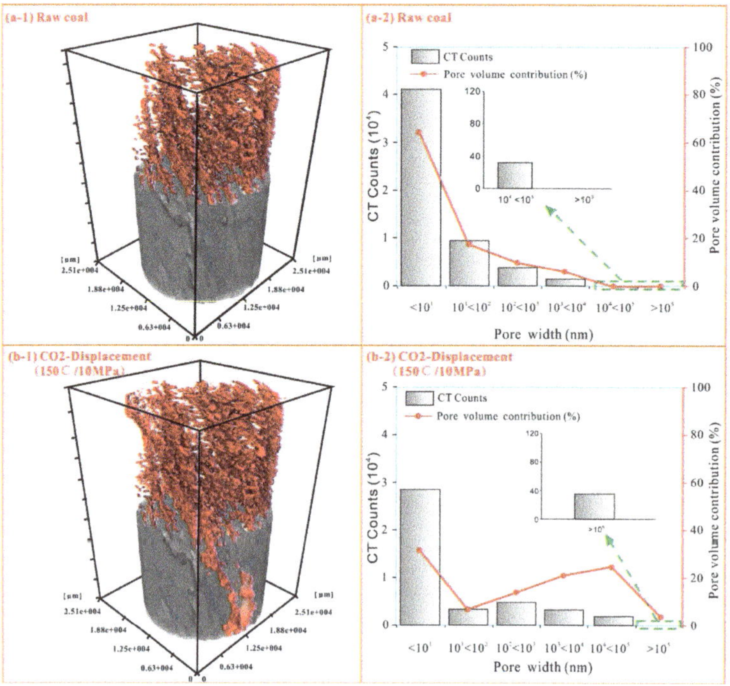

Figure 14. The 3D evolution of coal PFC from pre- to post-CO_2 thermal flooding (150 °C/10 MPa).

3.3. Permeability Enhancement Mechanism During $ScCO_2$ Thermal Flooding

Under a heating temperature of 150 °C, as the injection pressure increases gradually from 6 MPa to 8 MPa, and further to 10 MPa, the porosity and permeability of coal show a clear linear growth trend. Experimental data from pre- and post-flooding indicate that porosity increases linearly from 9.6% at 6 MPa to 16.7% at 8 MPa, reaching 34.8% at 10 MPa (Figure 15). Similarly, permeability exhibits a linear rise from 0.54 mD at 6 MPa to 1.18 mD at 8 MPa, and further to 2.37 mD at 10 MPa (Figure 16).

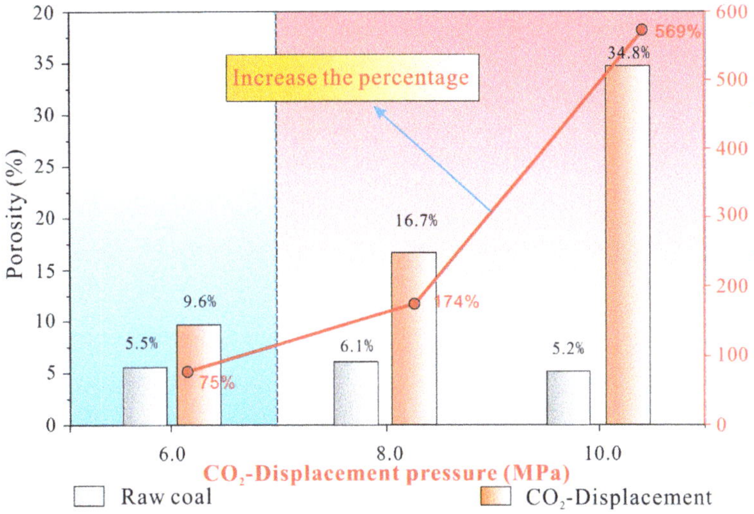

Figure 15. The porosity increases induced by thermal treatment at different pressures.

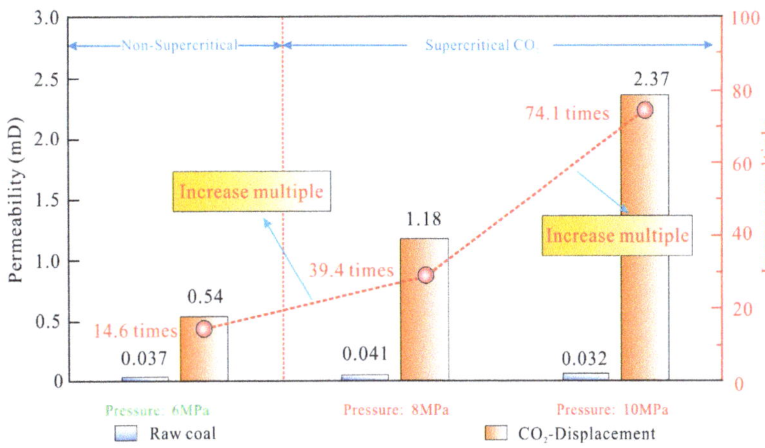

Figure 16. The permeability increases induced by thermal treatment at different pressures.

This linear growth pattern is primarily attributed to the direct correlation between injection pressure and macro-fracture development. As injection pressure increases, the permeation pressure and fluid wedging effect of CO_2 in coal are continuously enhanced, promoting the propagation and interconnection of micro-fractures to form a more developed macro-fracture network. These newly formed macro-fractures not only expand storage space but, more importantly, significantly improve pore connectivity, establishing efficient fluid transport pathways. The continuous expansion and enhanced connectivity

https://doi.org/10.3390/en19020496

of the macro-fracture system provide a reliable physical mechanism and structural basis for the linear improvement in coal porosity and permeability.

Based on systematic experimental data, the contribution rates of temperature and pressure to coal permeability were first quantified, revealing a "pressure-dominated, temperature-assisted" mechanism for CO_2-enhanced displacement. Between 80 and 150 °C, an 87.5% temperature increase raised permeability from 0.49 mD to 1.18 mD, corresponding to a 0.008 mD increase per 1% temperature rise (Figure 10). Over 6–10 MPa, a 66.7% pressure increase elevated permeability from 0.54 mD to 2.37 mD, equivalent to a 0.027 mD gain per 1% pressure increase (Figure 16).

The quantitative relationship confirms the dominant role of injection pressure in enhancing coal permeability. Increased pressure directly strengthens CO_2 permeation and fracturing effects, promoting macro-fracture propagation and interconnection. In contrast, temperature elevation mainly facilitates micro-fracture development through thermal stress, with limited contribution to permeability. The established "pressure-dominated, temperature-assisted" mechanism provides a theoretical basis for optimizing thermal-enhanced CBM extraction. Practical applications should prioritize injection pressure enhancement over temperature modulation to maximize energy efficiency and recovery performance.

4. Discussion

The identification of 80 °C as a critical transition point is supported by the confluence of permeability data and pore–fracture evolution. The V-shaped permeability trend (minimum at 80 °C) coincides with a shift in the dominant structural modification mechanism. At and below 80 °C, permeability reduction is primarily associated with pore restructuring (micropore decrease, macropore increase) without significant fracture development, likely due to thermal expansion effects. Above 80 °C, the sharp permeability increase correlates with the onset of thermal stress-induced fracturing, where the generation and connectivity of fractures become the principal factor enhancing flow capacity. We interpret that, under our experimental conditions, the thermo-mechanical stress surpasses a threshold near 80 °C, initiating the transition from a pore-adjusted to a fracture-dominated flow regime.

CO_2 displacement tests conducted at 8 MPa and various temperatures demonstrate a linear increase in methane desorption efficiency between 35 °C and 150 °C. The efficiency rose from 57% at 35 °C to 68% at 80 °C, and further reached 76% at 120 °C and 82% at 150 °C. This represents a 44% improvement relative to the baseline at 35 °C, establishing 150 °C as the optimal injection temperature within the 0–150 °C range (Figure 17).

Based on linear regression of experimental data, the temperature dependence of desorption efficiency was quantified, producing corresponding linear equations. These equations allow for accurate prediction of methane desorption efficiency and volume at any temperature between 0 and 150 °C, offering a practical model for optimizing field injection temperature and forecasting production.

Further analysis at 150 °C across varying injection pressures shows that raising pressure from 6 MPa to 8 MPa markedly increased methane desorption efficiency from 59% to 82%. However, further increasing pressure to 10 MPa only raised efficiency marginally to 84%, revealing a distinct efficiency plateau in the 8–10 MPa range. This result contradicts the conventional view that higher pressure consistently yields significant efficiency gains (Figure 18).

https://doi.org/10.3390/en19020496

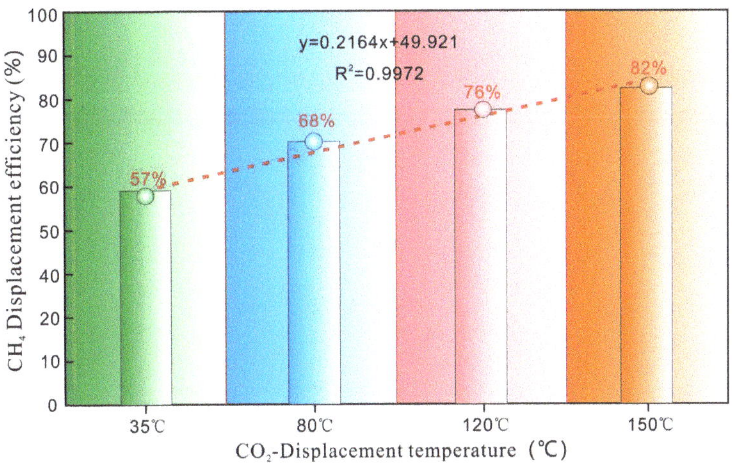

Figure 17. CH_4 desorption efficiency in coal under different CO_2 thermal flooding temperatures.

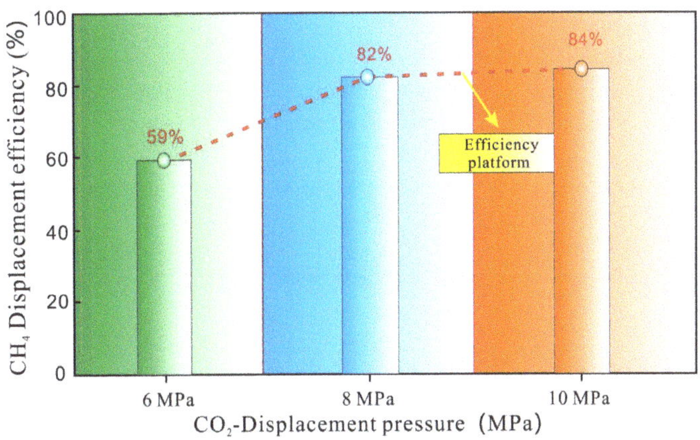

Figure 18. CH_4 desorption efficiency in coal under different CO_2 thermal flooding pressures.

Although increasing injection pressure enhances macroscopic permeability and creates fracture networks for gas transport, its displacement efficiency exhibits a distinct physical limitation: beyond 8 MPa, efficiency plateaus with minimal further improvement. This bottleneck stems from the strong capillary forces and adsorption potentials in isolated micropores, which tightly retain CH_4 molecules. Macroscopic fluid pressure neither transmits effectively into such nanoscale pores nor provides sufficient energy to release methane from its adsorbed state.

The limited efficacy of pressure in displacing residual methane from isolated micropores during CO_2 flooding arises from two physical constraints. First, the nanoscale confinement in micropores induces strong capillary forces and adsorption potentials that firmly trap CH_4 molecules, creating an energy barrier that fluid pressure alone cannot overcome. Second, pressure transmission becomes inefficient at the nanoscale: while CO_2 under high pressure can propagate through macroscopic fractures, it fails to effectively penetrate or fill isolated micropores, leaving the trapped methane unaffected by the applied pressure (Figure 19).

https://doi.org/10.3390/en19020496

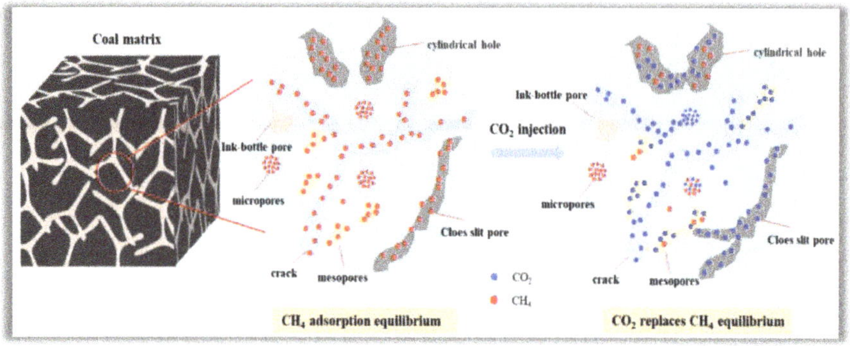

Figure 19. Schematic of the mechanism for ECBM in coal through CO_2 flooding.

Heating effectively overcomes the limitations of pressure-driven displacement. Elevated temperature not only improves CO_2 diffusion into isolated micropores but also supplies adsorbed CH_4 molecules with adequate thermal energy to facilitate desorption, thereby mobilizing residual gas inaccessible to pressure alone. The synergy is further enhanced by optimizing pressure application. Cyclic injection—alternating between injection and equilibration—ensures uniform reservoir pressurization and promotes complete gas penetration into micropores. This strategy increases the efficiency of pressure utilization at the microscopic scale, working in concert with thermal effects.

In summary, the TAPM (Thermally Augmented, Pressure-Managed) approach establishes a synergistic framework for enhanced coalbed methane recovery. This methodology employs cyclic pressurization to ensure macroscopic sweep and efficient transport, while thermal activation releases gas from micropores through enhanced desorption. The integrated process systematically improves overall extraction efficiency.

5. Conclusions

Integrated NMR and CT scanning revealed the dynamic evolution of coal's fracture network under supercritical CO_2 conditions. Temperature–pressure decoupling demonstrated their distinct roles in fracture propagation and connectivity, elucidating a mechanistic link between permeability enhancement and methane displacement based on the current experimental findings. These insights address fundamental knowledge gaps and support optimized deep CBM extraction with CO_2 storage.

(1) Based on the experimental data, a temperature-dependent shift in the dominant modification mechanism during thermal CO_2 displacement was identified. Below 80 °C, pore restructuring dominates, characterized by a reduction in micropores and an expansion of macropores. Beyond this temperature, thermal stress becomes predominant, converting micropores to macropores and generating thermal fractures, thereby shifting the system from a pore-adjusted to a fracture-dominated regime. Consequently, the permeability trend exhibits a distinct transition near 80 °C under the tested conditions, marking the change from pore-adjusted to fracture-dominated flow.

(2) NMR and CT scans demonstrate that elevated injection pressure promotes the expansion and interconnection of micro-fractures, forming dominant macro-fracture networks. Fracture proportion increased from 31.7% to 47.4%, with corresponding micropore reduction and macropore growth. This structural reorganization enhanced permeability from 0.54 mD to 2.37 mD (339% growth). The multi-scale analysis reveals the transition from localized propagation to global connectivity under pressure, providing key theoretical support for deep CBM extraction.

https://doi.org/10.3390/en19020496

(3) Based on the experimental data, this investigation supports a "pressure-dominated, temperature-assisted" synergistic mechanism for permeability enhancement under the tested conditions. The experimental data demonstrate that the absolute increase in permeability driven by elevated injection pressure (e.g., from 0.54 to 2.37 mD between 6 and 10 MPa at 150 °C) was substantially greater than that induced by temperature increase alone (e.g., from 0.49 to 1.18 mD between 80 and 150 °C at 8 MPa). This supports the dominant role of pressure in creating and interconnecting macro-fractures, establishing efficient percolation pathways. Temperature primarily assists by promoting gas desorption and facilitating the initiation of micro-fractures through thermal stress.

(4) Experimental results demonstrate a clear temperature–pressure synergy during CO_2 displacement. At 150 °C, permeability increased by 339% as pressure rose from 6 to 10 MPa, while displacement efficiency improved by only 2.4% beyond 8 MPa, indicating a pressure-driven limitation. Thermal activation effectively overcomes this constraint by enhancing CO_2 diffusion and providing activation energy for CH_4 desorption from isolated micropores, thereby mobilizing residual gas unreachable by pressure alone. This finding establishes a new theoretical framework for efficient coalbed methane extraction, challenging conventional pressure-centric approaches.

Limitations and Future Work: The conclusions of this study are drawn within the scope of the presented experimental system. It should be noted that this work did not incorporate numerical modeling, comparative analysis with other enhancement techniques (e.g., cyclic injection, N_2 flooding), or an energy balance assessment. These aspects represent important avenues for future research to further generalize and optimize the proposed mechanism.

Author Contributions: Conceptualization, X.X. and E.G.; Methodology, T.G.; Formal analysis, K.W. and A.W.; Investigation, E.G. and A.W.; Data curation, K.W.; Writing—original draft, X.X.; Writing—review and editing, S.L. and Y.L.; Visualization, Y.L.; Supervision, X.X., T.G., and S.L.; Project administration, T.G., E.G., and S.L. All authors have read and agreed to the published version of the manuscript.

Funding: This study is jointly supported by the PetroChina Coalbed Methane Co., Ltd. Science and Technology Project "Research on Supercritical Carbon Dioxide Displacement, Pyrolysis and Storage Technology" (2024-KJ-17); China United Coalbed Methane National Engineering Research Center Co., Ltd. Science and Technology Project "Basic Research on the Effects of Thermal Stimulation for Enhanced Gas Production in Medium-Shallow Coal Reservoirs" (2024-ZXKJ-07).

Data Availability Statement: The original contributions presented in this study are included in the article. Further inquiries can be directed to the corresponding author.

Conflicts of Interest: Authors X.X., T.G., E.G., and S.L. were employed by China United Coalbed Methane National Engineering Research Center Co., Ltd., and PetroChina Coalbed Methane Co., Ltd. The remaining authors declare that the research was conducted in the absence of any commercial or financial relationships that could be construed as potential conflicts of interest. The authors declare that this study received funding from the Science and Technology Project of PetroChina Coalbed Methane Co., Ltd. and the Science and Technology Project of China United Coalbed Methane National Engineering Research Center Co., Ltd. The funders were not involved in the study design; collection, analysis, or interpretation of data; the writing of this article; or the decision to submit it for publication.

References

1. Gu, J.Y.; Zhang, B.; Guo, M.Q. Deep coalbed methane enrichment rules and its exploration and development prospect in Linxing Block. *J. China Coal Soc.* **2016**, *41*, 72–79. [CrossRef]
2. Hou, S.H.; Wang, X.M.; Wang, X.J.; Yuan, Y.; Zhuang, X. Geological controls on gas saturation in the yanchuan nan coalbed methane field, southeastern Ordos basin, China. *Mar. Pet. Geol.* **2016**, *78*, 254–270. [CrossRef]

https://doi.org/10.3390/en19020496

3. Hou, X.; Liu, S.; Zhu, Y.; Yang, Y. Evaluation of gas contents for a multi-seam deep coalbed methane reservoir and their geological controls: In situ direct method versus indirect method. *Fuel* **2020**, *265*, 116917. [CrossRef]
4. Kolak, J.J.; Burruss, R.C. Geochemical investigation of the potential for mobilizing non-methane hydrocarbons during carbon dioxide storage in deep coal beds. *Energy Fuels* **2006**, *20*, 566–574. [CrossRef]
5. Pan, Z.; Ye, J.; Zhou, F.; Tan, Y.; Connell, L.D.; Fan, J. CO_2 storage in coal to enhance coalbed methane recovery: A review of field experiments in China. *Int. Geol. Rev.* **2017**, *60*, 754–776. [CrossRef]
6. Zhou, F.; Hussain, F.; Cinar, Y. Injecting pure N_2 and CO_2 to coal for enhanced coalbed methane: Experimental observations and numerical simulation. *Int. J. Coal Geol.* **2013**, *116*, 53–62. [CrossRef]
7. Guo, X.L.; Yan, X.Q.; Yu, J.L.; Zhang, Y.; Chen, S.; Mahgerefteh, H.; Martynov, S.; Collard, A.; Proust, C. Pressure response and phase transition in supercritical CO_2 releases from a large-scale pipeline. *Appl. Energy* **2016**, *178*, 189–197. [CrossRef]
8. Cao, Y.X.; Zhang, J.S.; Zhai, H.; Fu, G.; Tian, L.; Liu, S. CO_2 gas fracturing: A novel reservoir simulation technology in low permeability gassy coal seams. *Fuel* **2017**, *203*, 197–207. [CrossRef]
9. Fathi, E.; Akkutlu, I.Y. Multi-component gas transport and adsorption effects during CO_2 injection and enhanced shale gas recovery. *Int. J. Coal Geol.* **2014**, *123*, 52–61. [CrossRef]
10. Zhang, X.G.; Ranjith, P.G.; Li, D.Y.; Perera, M.S.A.; Ranathunga, A.S.; Zhang, B.N. CO_2 enhanced flow characteristics of naturally-fractured bituminous coals with N_2 injection at different reservoir depths. *J. CO2 Util.* **2018**, *28*, 393–402. [CrossRef]
11. Wang, Q.; Ye, J.B.; Yang, H.Y.; Liu, Q. Chemical composition and structural characteristics of oil shales and their kerogens using Fourier Transform Infrared (FTIR) Spectroscopy and Solid-State 13C Nuclear Magnetic Resonance (NMR). *Energy Fuels* **2016**, *30*, 6271–6280. [CrossRef]
12. Fang, H.H.; Sang, S.X.; Liu, S.Q. The coupling mechanism of the thermal-hydraulic-mechanical fields in CH_4-bearing coal and its application in the CO_2-enhanced coalbed methane recovery. *J. Pet. Sci. Eng.* **2019**, *181*, 106177. [CrossRef]
13. Chai, D.; Yang, G.; Fan, Z.; Li, X. Gas transport in shale matrix coupling multilayer adsorption and pore confinement effect. *Chem. Eng. J.* **2019**, *370*, 1534–1549. [CrossRef]
14. Yu, H.G.; Zhou, G.Z.; Fan, W.T.; Ye, J. Predicted CO_2 enhanced coalbed methane recovery and CO_2 sequestration in China. *Int. J. Coal Geol.* **2007**, *71*, 345–357. [CrossRef]
15. Yang, Z.Z.; Yuan, J.F.; Zhu, J.Y.; Li, X.; Li, Y.; Wang, H. Thermal injection stimulation to enhance coalbed methane recovery. *Pet. Reserv. Eval. Dev.* **2022**, *12*, 617–625. [CrossRef]
16. Xu, J.; Zhai, C.; Liu, S. Investigation of temperature effects from LCO_2 with different cycle parameters on the coal pore variation based on infrared thermal imagery and low-field nuclear magnetic resonance. *Fuel* **2018**, *215*, 528–540. [CrossRef]
17. Han, S.; Sang, S.; Liang, J.; Zhang, J. Supercritical CO_2 adsorption in a simulated deep coal reservoir environment, implications for geological storage of CO_2 in deep coals in the southern Qinshui Basin, China. *Energy Sci. Eng.* **2019**, *7*, 488–500. [CrossRef]
18. Hao, Z.Y.; Yue, L.X.; Sun, K.M.; Wang, S. Experiment study on the porosity and permeability of low permeability coal by supercritical CO_2 temperature variation. *Coal Geol. Explor.* **2018**, *46*, 64–71. [CrossRef]
19. Zhang, S.S. Seepage Characteristics of Coal and Pyrolysis Model Optimization in the Context of Underground Coal Gasification. Ph.D. Thesis, China University of Mining and Technology, Beijing, China, 2020. [CrossRef]
20. Li, B.B.; Gao, Z.; Yang, K.; Li, J.H.; Ren, C.H.; Xu, J.; Cao, J. Study on coal adsorption-permeability model under the coupling of temperature and pore pressure. *Chin. J. Rock Mech. Eng.* **2020**, *39*, 668–681. [CrossRef]
21. Liu, J.J.; Nie, Z.S.; Yu, B.Z.; Yang, D. Analysis of the mechanism and influencing factors of supercritical carbon dioxide on coal permeability enhancement. *Coal Sci. Technol.* **2023**, *51*, 204–216. [CrossRef]
22. Su, E.; Liang, Y.; Chang, X. Effects of cyclic saturation of supercritical CO_2 on the pore structures and mechanical properties of bituminous coal: An experimental study. *J. CO2 Util.* **2020**, *40*, 101208. [CrossRef]
23. Lu, J.; Hawthorne, S.; Sorensen, J.; Pekot, L.; Kurz, B.; Smith, S. Advancing CO_2 enhanced oil recovery and storage in unconventional oil play—Experimental studies on Bakken shales. *Appl. Energy* **2017**, *208*, 171–183. [CrossRef]
24. Middleton, R.S.; Carey, J.W.; Currier, R.P.; Hyman, J.D.; Kang, Q.; Karra, S.; Jiménez-Martínez, J.; Porter, M.L.; Viswanathan, H.S. Shale gas and non-aqueous fracturing fluids: Opportunities and challenges for supercritical CO_2. *Appl. Energy* **2015**, *147*, 500–509. [CrossRef]
25. Liu, S.Q.; Sang, S.X.; Ma, J.S.; Wang, T.; Du, Y.; Fang, H.H. Effects of supercritical CO_2 on micropores in bituminous and anthracite coal. *Fuel* **2019**, *242*, 96–108. [CrossRef]
26. Bai, G.; Zhou, Z.J.; Li, X.M.; Cheng, Y.T.; Hu, K.; Chen, Y.; Zhou, X.H. Quantitative analysis of carbon dioxide replacement of adsorbed methane in different coal ranks using low-field NMR technique. *Fuel* **2022**, *326*, 124980. [CrossRef]
27. Zhang, K.Z.; Cheng, Y.P.; Li, W.; Wu, D.M.; Liu, Z.D. Influence of supercritical CO_2 on pore structure and functional groups of coal: Implications for CO_2 sequestration. *J. Nat. Gas Sci. Eng.* **2017**, *40*, 288–298. [CrossRef]
28. Liu, S.Q.; Ma, J.S.; Sang, S.X.; Wang, T.; Du, Y.; Fang, H.H. The effects of supercritical CO_2 on mesopore and macropore structure in bituminous and anthracite coal. *Fuel* **2018**, *223*, 32–43. [CrossRef]

https://doi.org/10.3390/en19020496

29. Li, L.; Liang, W.G.; Li, Z.G.; He, W. Experimental investigation on enhancing coalbed methane recovery by injecting high-temperature CO_2. *J. China Coal Soc.* **2017**, *42*, 2044–2050. [CrossRef]
30. Zheng, S.J.; Yao, Y.B.; Sang, S.X.; Liu, D.M.; Wang, M.; Liu, S.Q. Dynamic characterization of multiphase methane during CO_2-ECBM: An NMR relaxation method. *Fuel* **2022**, *324*, 124526. [CrossRef]
31. Mazumder, S.; van Hemert, P.; Busch, A.; Wolf, K.A.A.; Tejera-Cuesta, P. Flue gas and pure CO_2 sorption properties of coal: A comparative study. *Int. J. Coal Geol.* **2006**, *67*, 267–279. [CrossRef]
32. Huang, Y.; Saipeng, H.; Jian, H.; Liu, X.; Huang, L.; Fang, C. Numerical Simulation of the Effect of Injected CO_2 Temperature and Pressure on CO_2-Enhanced Coalbed Methane. *Appl. Sci.* **2020**, *10*, 1385. [CrossRef]
33. Liu, J.; Yao, Y.B.; Liu, D.M.; Elsworth, D. Experimental evaluation of CO_2 enhanced recovery of adsorbed-gas from shale. *Int. J. Coal Geol.* **2017**, *179*, 211–218. [CrossRef]
34. Meng, M.; Qiu, Z.; Zhong, R.; Liu, Z.; Liu, Y.; Chen, P. Adsorption characteristics of supercritical CO_2/CH_4 on different types of coal and a machine learning approach. *Chem. Eng. J.* **2019**, *368*, 847–864. [CrossRef]
35. Fan, C.; Yang, L.; Wang, G.; Huang, Q.; Fu, X.; Wen, H. Investigation on coal skeleton deformation in CO_2 injection enhanced CH_4 drainage from underground coal seam. *Front. Earth Sci.* **2021**, *9*, 766011. [CrossRef]
36. Liu, Y.L.; Xu, H.; Tang, D.Z.; Xu, F.Y.; Jonathan, P.M.; Hou, W.; Yan, X.; Ding, F.F. Coalbed methane production of a heterogeneous reservoir in the Ordos Basin, China. *J. Nat. Gas Sci. Eng.* **2020**, *82*, 103502. [CrossRef]
37. Hodot, B.B. *Coal and Gas Outburst;* China Industry Publishing House: Beijing, China, 1966.

https://doi.org/10.3390/en19020496

energies

MDPI

Article

Thermo-Mechanical Controls on Permeability in Deep Fractured-Porous Carbonates During Underground Gas Storage

Zhen Zhai [1,2], Quan Gan [1,2], Yan Wang [3], Saipeng Huang [1,2], Yuchao Zhao [4], Limin Li [3], Mingnan Xu [5], Junlei Wang [6] and Sida Jia [1,2,*]

1 State Key Laboratory of Coal Mine Disaster Dynamics and Control, Chongqing University, Chongqing 400044, China
2 School of Resources and Safety Engineering, Chongqing University, Chongqing 400044, China
3 Chongqing Xiangguosi Underground Gas Storage Company Limited, Chongqing 401120, China
4 Exploration and Development Research Institute, PetroChina Southwest Oil & Gasfield Company, Chengdu 610051, China
5 State Key Laboratory of Geomechanics and Geotechnical Engineering, Institute of Rock and Soil Mechanics, Chinese Academy of Sciences, Wuhan 430071, China
6 PetroChina Research Institute of Petroleum Exploration & Development, Beijing 100083, China
* Correspondence: jsd@cqu.edu.cn

Abstract

Deep fractured-porous carbonate reservoirs used for underground gas storage (UGS) experience simultaneous changes in temperature and effective stress during cyclic injection and withdrawal, so predicting permeability evolution is essential for evaluating long-term injectivity and deliverability. Using the Xiangguosi UGS as the engineering background, we measured steady-state gas permeability of three fractured-porous carbonate cores under representative conditions (20–80 °C; 15–35 MPa). Permeability decreases nonlinearly under coupled loading: changing temperature or effective stress alone typically reduces permeability by 30–70%, while the maximum reduction under concurrent increases in both variables exceeds 80% relative to the reference condition. An exponential model was fitted to quantify the decay parameter of permeability with effective stress (0.038–0.046 MPa^{-1}) and with temperature (0.016–0.020 $°C^{-1}$). In addition, the temperature-related exponential decay parameter decreases with increasing effective stress, because compliant fractures and larger pores are progressively pre-closed, weakening the permeability response to temperature. Finally, we propose a parsimonious separable exponential model that reproduces the measurements with a mean relative error below 12%, providing a practical constitutive relation for multiphysics simulations of UGS in fractured-porous carbonates.

Keywords: fractured-porous carbonates; temperature and stress coupling; permeability evolution; underground gas storage; exponential model

1. Introduction

Converting depleted gas reservoirs or salt caverns into underground gas storage (UGS) facilities helps meet seasonal peak-shaving demand and enhance natural gas supply security [1–4]. Among the available storage types, gas reservoir UGS has become a major focus because its geology is well constrained, and existing well patterns and surface facilities can be reused [5,6]. In southwestern China, the construction of large-scale UGS facilities has increasingly focused on deeply buried marine fractured-porous carbonate reservoirs [7,8]. These reservoirs exhibit complex THMC (thermal-hydro-mechanical-chemical) coupling behaviors during frequent, high-rate injection-withdrawal cycles [9,10]. Specifically, the synergy between thermal fluctuations and effective stress perturbations directly dictates fracture-pore connectivity and reservoir deliverability [11,12]. While previous studies have addressed independent stress sensitivity [13–17], characterizing the joint permeability response through parsimonious coupling models remains a critical prerequisite for accurate field-scale numerical implementation [18,19].

During UGS operation, pore pressure and temperature in the vicinity of the wellbore and within the reservoir vary significantly in both time and space [9], whereas the regional in-situ stress field remains essentially stable over the engineering timescale. As a result, the effective stress borne by the reservoir skeleton fluctuates cyclically with each injection and withdrawal cycle. At the same time, the temperature difference between the injected gas and the formation induces thermoelastic and thermochemical responses in the rock-fluid system [10,20]. The combined action of effective stress and temperature exerts a direct control over a number of key processes, including fracture opening and closure, pore-structure evolution, temporal deterioration, and partial recovery of permeability. This makes it a pivotal factor for describing the evolution of reservoir injectivity and deliverability, as well as for assessing the long-term operational safety of UGS facilities [11,12,17,21]. The stress-related component of this permeability evolution has consequently been a central topic in previous studies.

With respect to permeability stress sensitivity, early experimental work mainly focused on sandstones and other complex lithologic reservoirs. These studies generally concluded that permeability decreases with increasing effective stress and can be described by exponential or power-law relationships [14–17]. In the context of carbonate reservoirs, Yan et al. [13] have reported that the permeability of fractured-porous cores exhibits a rapid and non-linear decrease at low effective stress levels. Conversely, at higher stress levels, the permeability undergoes a gradual decline. Distinct sensitivity coefficients are exhibited by different pore structures. Sun et al. [6] and Gou et al. [17] showed that systems with coexisting fractures and matrix pores commonly display moderate to strong stress sensitivity, which can further reduce injectivity and deliverability under multi-cycle injection and production operations. With advances in experimental testing and imaging techniques, quantitative characterization of permeability stress sensitivity in carbonates has shifted from simple empirical functions to frameworks that explicitly link pore structure, permeability, and effective stress. Sun et al. [11] showed that fracture morphology and skeletal stiffness can lead to noticeable differences in stress sensitivity. Yang et al. [18] further pointed out that fractured carbonate rocks under high-stiffness conditions are particularly sensitive to stress and tend to exhibit stronger nonlinear behavior. Mi et al. [16], Zivar et al. [15], and Haghi et al. [14] successively proposed empirical relations between permeability and effective stress that are suitable for numerical simulations. Pu et al. [19] used digital rock physics to simulate the evolution of porosity and permeability in fractured-porous networks under multiaxial stress states, providing additional insight into the mechanisms of stress-dependent flow. Overall, these studies clarify how permeability responds to effective stress and provide a range of empirical stress-permeability relations. However, because most experiments were conducted under isothermal or single-temperature conditions, they offer only limited insight into temperature-related effects.

Temperature changes not only modify fluid properties but also significantly affect the mechanical and flow behavior of rocks through thermal expansion, thermally induced cracking, and mineral reactions [22,23]. Tomás et al. [22] indicated that the initiation and evolution of high-temperature thermal fractures is one of the key mechanisms governing mechanical deterioration and permeability variation in rocks. Wang et al. [23] found that moderate heating can induce microfracture opening and increase permeability, whereas at higher temperatures and over longer durations, heating may cause structural collapse and

https://doi.org/10.3390/en19020553

pore clogging, leading to permeability reduction. Zhai et al. [24] observed that the relative permeability of sandstone is sensitive to temperature and that heating can substantially enhance oil-phase flow capacity. In the context of UGS, previous studies have shown that permeability generally decreases during heating and increases during cooling, and exhibits hysteresis under repeated heating-cooling cycles [20]. Even moderate cyclic temperature variations on the order of several tens of degrees Celsius can cumulatively alter reservoir flow properties by changing rock-skeleton stiffness, inducing thermal fractures, and triggering local structural rearrangement [25]. These observations highlight a key gap: under realistic UGS conditions, where temperature and effective stress vary simultaneously, the coupled microstructural evolution of deep fractured-porous carbonate reservoirs and its impact on permeability remain poorly constrained.

Operational conditions that truly reflect UGS behavior involve simultaneous changes in temperature and effective stress. Against this background, some studies have begun to incorporate coupled temperature and stress effects into experiments and theoretical models. Gao et al. [12] demonstrated that multi-cycle effective stress perturbations cause continuous reductions in permeability and porosity in carbonate reservoirs, leading to substantial and partly irreversible damage to UGS injectivity and deliverability. Sun et al. [11] and Luo et al. [26] investigated the coupled responses of fractured-porous marine carbonates and deep lacustrine carbonates under combined temperature and stress loading and proposed empirical models. Using micro-imaging and fractal theory, Huang et al. [27] and Yang et al. [18] showed that the cooperative deformation of the pore skeleton and the vuggy pore network under stress is a key factor controlling the intensity and nonlinearity of macroscopic stress sensitivity. Temperature influences this process mainly indirectly, by modifying skeleton stiffness and fracture evolution. Nevertheless, studies on permeability evolution under coupled temperature and stress conditions still have several limitations. First, most experiments use sandstones, granites, or generic carbonate blocks, and systematic tests on deep fractured-porous carbonate reservoir cores under in-situ stress and temperature conditions remain scarce. Second, part of the existing work is conducted in high-temperature geothermal settings (typically > 200 °C), which differ markedly from the moderate 20–80 °C thermal disturbances experienced by UGS facilities. Third, structurally clear, easily calibrated permeability models that couple temperature and effective stress for field-scale numerical implementation are still lacking.

To bridge these gaps, this study presents a targeted investigation focusing on the Carboniferous fractured-porous carbonates of the Xiangguosi UGS (XGS-UGS). By conducting experiments on natural reservoir cores under a complete $T - \sigma_{eff}$ matrix (20–80 °C; 15–35 MPa), we aim to systematically quantify the coupled thermo-mechanical response of permeability and develop a parsimonious constitutive model that can be readily utilized for field-scale numerical simulations.

2. Materials and Methods

2.1. Experimental Samples

The reservoir samples tested in this study were collected from XGS-UGS in southwestern China. The storage reservoir consists of Carboniferous carbonates and is classified as a fractured-porous reservoir [8]. The structural trap of XGS-UGS is an approximately north-south-trending, belt-like feature, and reservoir permeability shows clear spatial heterogeneity (Figure 1). To compare seepage responses under coupled temperature and effective stress conditions, representative core intervals were selected from the northern, central, and southern parts of the reservoir based on well-log interpretation. These intervals are hereafter referred to as North, Middle, and South, respectively.

https://doi.org/10.3390/en19020553

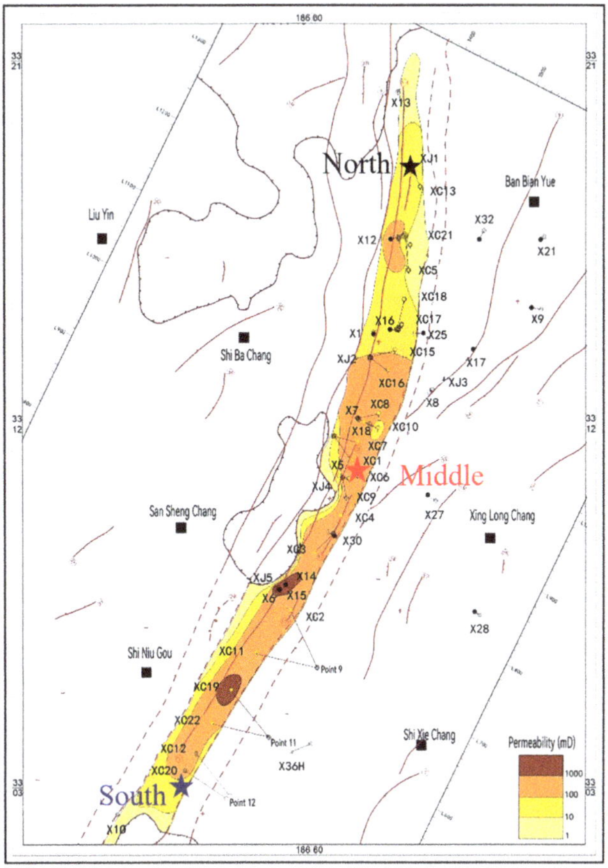

Figure 1. Plan-view permeability distribution of the Carboniferous carbonate reservoir in XGS-UGS and locations of the sampled core intervals (North, Middle, and South).

Sample preparation followed the Specifications for rock tests in water conservancy and hydropower engineering (SL/T 264-2020) [28] and the Standard for test methods of engineering rock masses (GB/T 50266-2013) [29]. Full-diameter cores were machined into cylindrical specimens with a diameter of 25 mm and a length of 50 mm. Dimensional tolerances and end-face flatness met the requirements of the standards. To provide a consistent preferential flow path among specimens, a throughgoing artificial fracture was introduced along the axial direction of each core using a diamond wire saw (Figure 2), which provides a simplified representation of a connected fracture pathway at the specimen scale. This treatment was designed to control the primary flow pathway and improve inter-specimen comparability, while acknowledging that it simplifies natural fracture complexity.

Before permeability tests, basic physical parameters of each specimen were measured (Table 1). Mineral composition was determined by powder X-ray diffraction (XRD) using a D8 ADVANCE diffractometer (Bruker, Karlsruhe, Germany). The XRD results show that the specimens are dolomite-dominated (65.0–100.0%). Quartz and calcite are mainly observed in the Middle specimen, whereas they are below the detection limit in the North and South specimens (Table 2), which is consistent with typical carbonate reservoirs.

https://doi.org/10.3390/en19020553

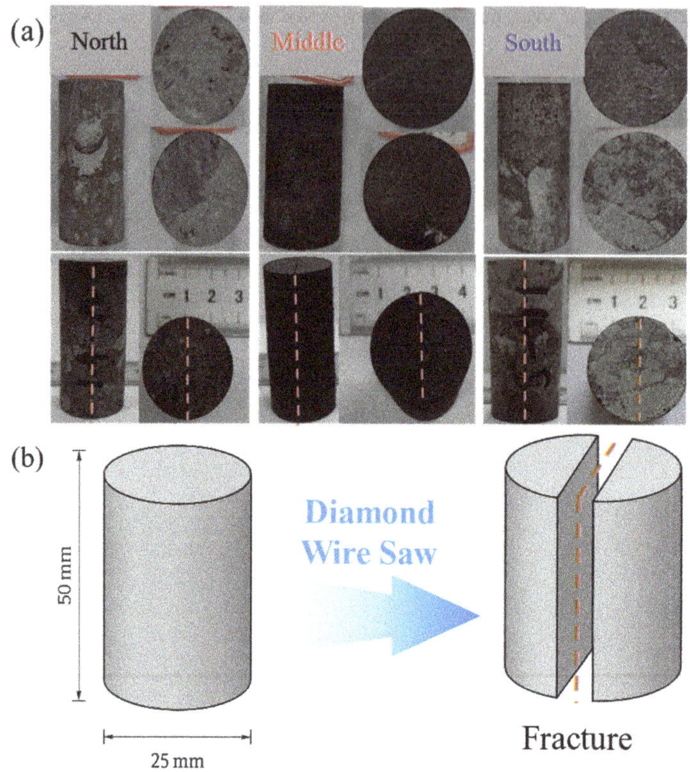

Figure 2. Photographs and schematic of specimen preparation: (**a**) representative core specimens from the North, Middle, and South intervals; (**b**) axial splitting using a diamond wire saw to create a throughgoing fracture.

Table 1. Basic information of the tested specimens.

Sample ID	Sampling Horizon	Sampling Depth (m)	Length (mm)	Diameter (mm)	Mass (g)
North	Carboniferous	2540.94–2541.11	50.00	25.10	64.566
Middle	Carboniferous	2352.11–2352.23	50.05	25.25	65.397
South	Carboniferous	2574.38–2574.54	50.00	25.15	63.473

Table 2. Mineral composition determined by powder XRD.

Sample ID	Quartz (%)	Dolomite (%)	Calcite (%)
North	BDL	100.0	BDL
Middle	21.4	65.0	13.7
South	BDL	100.0	BDL

Note: "BDL" indicates below the detection limit or trace amounts.

2.2. Experimental Apparatus and Permeability Calculation

Permeability tests were conducted using a high-temperature and high-pressure gas-flow system (Figure 3). The system consists of a core holder, a confining pressure control unit, a gas injection unit, a temperature control unit, and a flow measurement unit (Figure 3a,b). The core holder (316L stainless steel) has an inner diameter of 25 mm and an adjustable length of 25–100 mm; it is rated to 50 MPa confining pressure and 180 °C. Gas injection was provided by a TC-300D constant-rate/constant-pressure pump (maximum

https://doi.org/10.3390/en19020553

working pressure 70 MPa; accuracy ±1% full scale; flow rate range 0.0001–90 mL/min; cylinder volume 300 mL). To ensure that the specimen reached the preset temperature, five temperature sensors were installed at both ends and the middle of the specimen to monitor the core surface temperature in real time (Figure 3c).

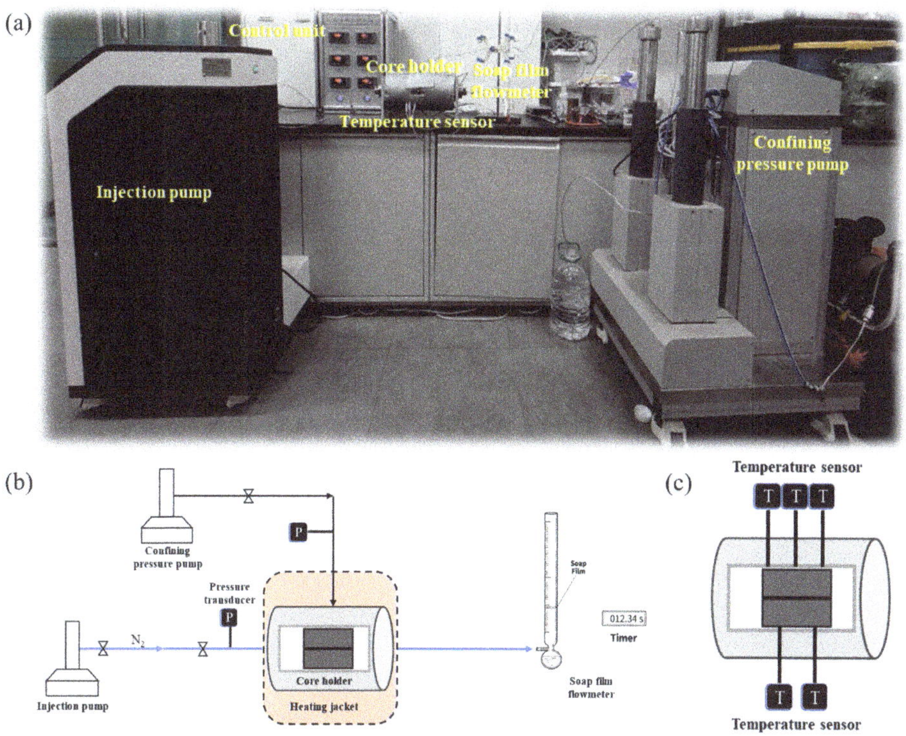

Figure 3. High-temperature and high-pressure gas-permeability apparatus: (**a**) photograph of the experimental setup; (**b**) schematic of the flow system; (**c**) layout of temperature sensors on the core holder.

The gas permeability measurements were conducted in strict accordance with the API RP 40 (Recommended Practices for Core Analysis) [30] and GB/T 29172-2012 (Practices for Core Analysis) [31] standards. A steady-state method was employed using high-purity nitrogen as the pore fluid. To account for the compressibility of gas, the permeability was calculated using the following integrated form of Darcy's Law for compressible flow [32–35]:

$$K = \frac{200 P_0 Q_0 \mu L}{A\left(P_1^2 - P_2^2\right)} \tag{1}$$

where Q_0 is the gas flow rate measured at the reference pressure P_0 (atmospheric pressure); μ is the gas viscosity at the test temperature [20,36,37]; L and A are the length and cross-sectional area of the core specimen, respectively; and P_1 and P_2 are the upstream (inlet) and downstream (outlet) pressures. By maintaining a constant pressure gradient across the specimen until the flow rate stabilized, the steady-state permeability was determined for each coupled temperature and stress condition.

Because this study focuses on the relative evolution of permeability with temperature and effective stress rather than absolute values, no Klinkenberg correction was applied. For fractured-porous specimens with permeabilities in the mD range, the slip-flow effect is sec-

https://doi.org/10.3390/en19020553

ondary to the 30–80% reductions driven by thermo-mechanical loading. Furthermore, the use of normalized permeability (K') largely mitigates systematic biases associated with gas slippage, as the effect remains relatively constant across the measured and reference states.

According to Biot's effective stress principle, the effective stress during the tests can be expressed as [15,38,39]:

$$\sigma_{eff} = \sigma_{tol} - \alpha P_p \tag{2}$$

where σ_{eff} is effective stress (MPa); σ_{tol} is total stress, which equals the confining pressure in this study (MPa); α is Biot's coefficient (0–1); and P_p is pore pressure (MPa). In these experiments, the pore pressure ranged from 0.2 to 0.7 MPa. Within the confining pressure range of 15–35 MPa, the term αP_p contributes less than 5% of σ_{eff} even when α = 1. While this approximation ($\sigma_{eff} \approx \sigma$) is numerically robust for calibrating parameters C and D under current laboratory conditions, it should be noted that in field-scale UGS operations, where pore pressure fluctuates significantly, the full effective stress term must be utilized to avoid systematic bias in predicting permeability evolution.

2.3. Experimental Procedures

Steady-state gas-flow tests were performed on three specimens at 20, 40, 60, and 80 °C. At each temperature, confining pressure was applied following predefined loading paths (Table 3). The overall workflow is shown in Figure 4.

Table 3. Test matrix and confining pressure loading paths.

Sample	Pore Pressure (MPa)	Temperature (°C)	Confining Pressure Path (MPa)
North, Middle, South	0.2–0.7	20	15→20→25→30→35
	0.2–0.7	40	15→25→35
	0.2–0.7	60	15→25→35
	0.2–0.7	80	15→25→35

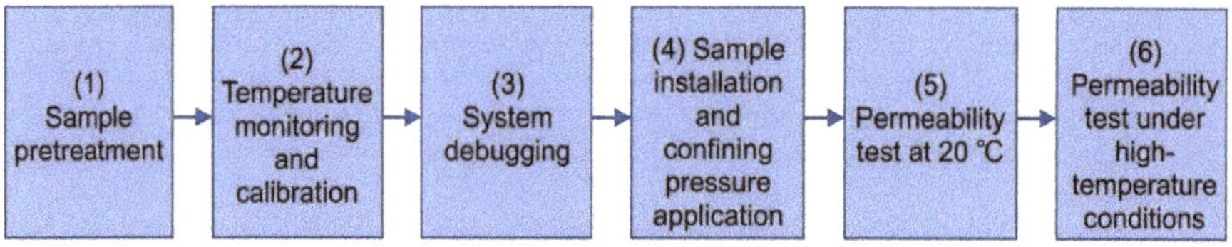

Figure 4. Workflow of the steady-state gas-permeability tests.

Tests at 20 °C. The gas-bottle pressure was adjusted, and the constant-rate/constant-pressure pump was started to supply N_2. The inlet pressure was regulated to maintain pore pressure within 0.2–0.7 MPa. Inlet pressure and volumetric flow rate were recorded every 10 min. Steady state was assumed when both pressure and flow rate varied by less than 5% over three consecutive measurements. After steady state was reached, three additional measurements were collected at 10 min intervals and averaged to calculate permeability using Equation (1). The confining pressure was then changed, and the above procedure was repeated to obtain permeability for the same specimen at effective stress levels of 15, 20, 25, 30, and 35 MPa.

Tests at elevated temperatures. For tests at 40, 60, and 80 °C, the target temperature was set, and the system was heated gradually while the core surface temperature was monitored in real time. Permeability testing started after the temperature stabilized at the target temperature for 1 h, with fluctuations not exceeding ±0.5 °C. At each elevated temperature, tests were performed at confining pressures of 15, 25, and 35 MPa, following the same steady-state criterion as at 20 °C.

https://doi.org/10.3390/en19020553

3. Experimental Results

3.1. Permeability Test Data

The steady-state gas permeability results of the three specimens under all test conditions are summarized in Table 4. Under the reference condition (T = 20 °C and σ_{eff} = 15 MPa), the initial permeabilities of the North, Middle, and South specimens are 2.01, 2.95, and 1.13 mD, respectively, differing by a factor of approximately 2–3.

Table 4. Steady-state permeability of the three specimens under different temperatures and effective stresses.

Temperature (°C)	Effective Stress (MPa)	Permeability of North (mD)	Permeability of Middle (mD)	Permeability of South (mD)
20	15	2.01	2.95	1.13
20	20	1.45	2.33	0.91
20	25	1.16	1.88	0.73
20	30	0.97	1.56	0.60
20	35	0.80	1.32	0.50
40	15	1.08	1.56	0.64
40	25	0.83	1.27	0.51
40	35	0.66	1.08	0.42
60	15	0.86	1.27	0.53
60	25	0.58	0.94	0.43
60	35	0.46	0.79	0.36
80	15	0.61	0.98	0.46
80	25	0.47	0.82	0.38
80	35	0.39	0.70	0.33

Note: For T = 40–80 °C, tests were conducted at σ_{eff} = 15, 25, and 35 MPa.

Such an initial difference is common for fractured carbonate cores and may reflect reservoir-scale heterogeneity as well as specimen-scale differences in fracture geometry. Previous studies suggest that flow in fractured-porous carbonates can be influenced by the connectivity between dissolution cavities (vug pores) and the fracture network, and that variations in fracture morphology or roughness may lead to substantial changes in measured permeability [18,27,40]. In this study, artificial fractures were prepared using the same wire-cutting procedure. However, variations in matrix fabric and cut surface micromorphology may still result in differences in fracture aperture and pore-fracture connectivity among specimens.

While these three representative cores (North, Middle, and South) capture the primary spatial heterogeneity of the XGS-UGS reservoir, they represent a localized population within a statistically larger and more complex geological system. Consequently, the measured permeabilities and subsequent sensitivity analyses should be viewed as indicative of the reservoir's behavior rather than an exhaustive characterization of all lithofacies present in the Carboniferous carbonate formation.

3.2. Permeability Variation with Effective Stress

Figure 5a shows the relationship between permeability and effective stress at T = 20 °C for the three specimens. For all specimens, permeability decreases nonlinearly with increasing σ_{eff}, and the rate of decrease diminishes at higher stress. For example, the permeability of the North specimen decreases from 2.01 mD at 15 MPa to 0.80 mD at 35 MPa, corresponding to a reduction of approximately 60%. The corresponding reductions are approximately 55% for the Middle specimen (2.95 to 1.32 mD) and 56% for the South specimen (1.13 to 0.50 mD). These results demonstrate a marked dependence of permeability on effective stress within the tested range (15–35 MPa).

https://doi.org/10.3390/en19020553

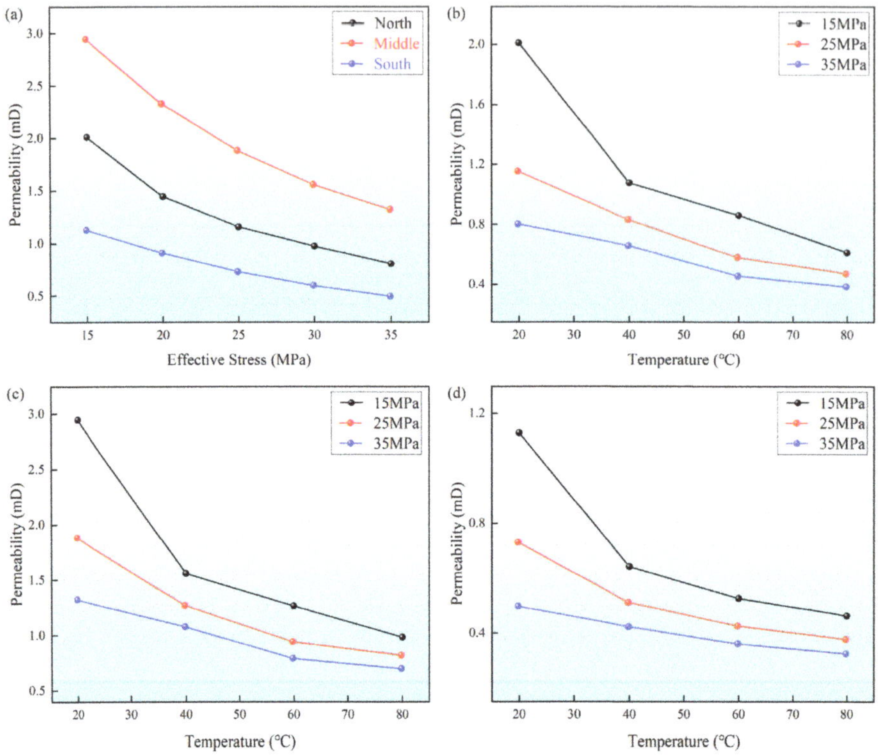

Figure 5. Permeability variations with effective stress and temperature. (**a**) Permeability versus effective stress at T = 20 °C for the three specimens. (**b–d**) Permeability versus temperature at σ_{eff} = 15, 25, and 35 MPa for the North, Middle, and South specimens.

3.3. Permeability Variation with Temperature

Figure 5b–d show permeability as a function of temperature at different effective-stress levels for the North, Middle, and South specimens. At a given σ_{eff}, permeability decreases with increasing temperature for all specimens. The trend is consistent across the tested stress levels, although the magnitude of reduction becomes smaller at higher σ_{eff}.

At σ_{eff} = 15 MPa, increasing temperature from 20 to 80 °C reduces permeability from 2.01 to 0.61 mD for the North specimen (approximately 70%), from 2.95 to 0.98 mD for the Middle specimen (approximately 67%), and from 1.13 to 0.46 mD for the South specimen (approximately 59%). Considering all specimens together, the relative reductions over 20–80 °C are approximately 59–70% at 15 MPa, 48–59% at 25 MPa, and 34–51% at 35 MPa. These results indicate that permeability is clearly temperature-dependent within the investigated range, and that its temperature sensitivity is stronger at lower effective stress.

When temperature and effective stress vary simultaneously from the reference condition (20 °C, 15 MPa) to the combined condition (80 °C, 35 MPa), permeability decreases from 2.01 to 0.39 mD (North), from 2.95 to 0.70 mD (Middle), and from 1.13 to 0.33 mD (South), corresponding to reductions of approximately 81%, 76%, and 71%, respectively. This joint reduction highlights that both temperature and effective stress should be considered when evaluating permeability evolution under UGS-relevant operating conditions.

https://doi.org/10.3390/en19020553

3.4. Normalized Permeability Representation

To reduce the influence of inter-specimen differences in initial permeability and enable a consistent comparison of permeability variations under different testing conditions, dimensionless normalized permeability variables are introduced using a common reference state (T_0 = 20 °C, σ_0 = 15 MPa) Hereafter, the superscript prime (′) denotes normalized permeability.

To characterize the relative change in permeability with effective stress at the reference temperature T_0, the stress-normalized permeability is defined as:

$$K'_{\sigma} = \frac{K\left(\sigma_{eff}, T_0\right)}{K(\sigma_0, T_0)} \tag{3}$$

where K'_{σ} is the dimensionless normalized permeability; $K\left(\sigma_{eff}, T_0\right)$ is the measured permeability at T_0 under a given effective stress σ_{eff}; and $K(\sigma_0, T_0)$ is the permeability at the reference condition (σ_0, T_0). This normalization approach has been widely used in studies of stress sensitivity and permeability evolution [41,42], as it can substantially reduce the impact of inter-specimen variability in initial permeability and highlight the effect of effective stress on permeability. The resulting normalized trends are illustrated in Figure 6a. K'_{σ} decreases monotonically and nonlinearly with increasing σ_{eff} for all specimens.

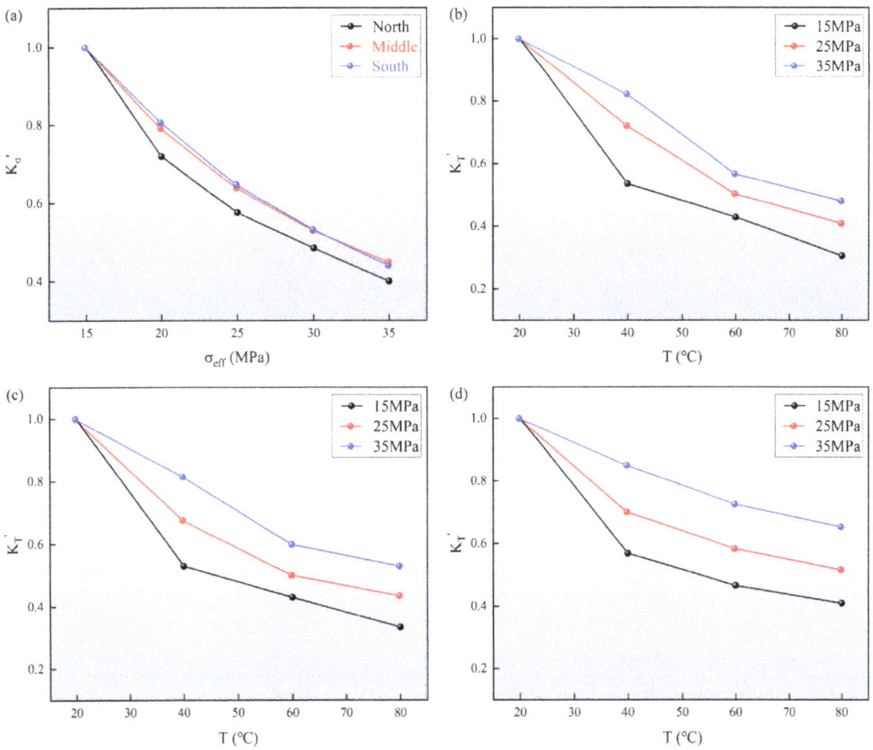

Figure 6. Normalized permeability results. (**a**) K'_{σ} versus σ_{eff} at T_0 = 20 °C for the three specimens. (**b–d**) K'_T versus T at σ_{eff} = 15, 25, 35 MPa for (**b**) North, (**c**) Middle, and (**d**) South specimens.

https://doi.org/10.3390/en19020553

To minimize the influence of different initial permeabilities on three specimens when examining temperature effects at a fixed effective stress, permeability is normalized by its value at the reference temperature T_0:

$$K'_T = \frac{K\left(\sigma_{eff}, T\right)}{K\left(\sigma_{eff}, T_0\right)} \tag{4}$$

where K'_T is the dimensionless normalized permeability, $K\left(\sigma_{eff}, T\right)$ is the measured permeability at effective stress σ_{eff} and temperature T, and $K\left(\sigma_{eff}, T_0\right)$ is the permeability at the same effective stress at T_0. The normalized results (Figure 6b–d) indicate that K'_T decreases with increasing temperature for all specimens, while the magnitude of reduction becomes less pronounced at higher σ_{eff}.

For a unified representation of permeability data obtained under combined variations of effective stress and temperature, a coupled normalized permeability is further defined with respect to the same reference state:

$$K' = \frac{K\left(\sigma_{eff}, T\right)}{K(\sigma_0, T_0)} \tag{5}$$

where K' is the dimensionless normalized permeability at a given $\left(\sigma_{eff}, T\right)$; $K\left(\sigma_{eff}, T\right)$ is the measured permeability; and $K(\sigma_0, T_0)$ is the permeability at the reference condition $T_0 = 20$ °C, $\sigma_0 = 15$ MPa. This coupled normalization enables permeability measurements acquired under different $\left(\sigma_{eff}, T\right)$ conditions to be compared within a unified dimensionless framework and will be used in the subsequent model development and validation.

4. Model Development

4.1. Model Scope, Assumptions, and Formulation Strategy

This chapter develops an exponential, engineering-oriented permeability model for fractured-porous carbonate specimens under coupled variations of effective stress and temperature. To enable cross-specimen comparison and to reduce the influence of inter-specimen variability in initial permeability, we use the dimensionless normalized permeability variables defined in Section 3.4, including the stress-normalized permeability K'_σ, the temperature-normalized permeability K'_T, and the coupled normalized permeability K'. The model is intended to represent the experimental conditions investigated in this study (i.e., the effective stress and temperature ranges in Section 3), and its performance is assessed through systematic validation and error analyses.

The exponential model is constructed under the following assumptions, which are consistent with the experimental design and the intended use of the model within the tested range:

(1) Reference-state normalization. Differences among specimens in initial permeability are accounted for by normalization to (σ_0, T_0) (Section 3.4), so that the model targets relative permeability changes rather than absolute values.

(2) Quasi-steady response at each loading step. The measured permeability at each imposed (σ_{eff}, T) condition is treated as representative of an equilibrated state for modeling purposes; time-dependent effects are not explicitly modeled.

(3) No explicit microstructural evolution law. Changes in fracture roughness, asperity damage, and other microstructural evolutions are not parameterized explicitly; their

https://doi.org/10.3390/en19020553

net influence is implicitly absorbed into fitted exponential parameters within the tested range.

(4) Separable stress and temperature effects (within the tested conditions). The coupled normalized permeability is approximated by a separable form,

$$K'\left(\sigma_{eff}, T\right) = f\left(\sigma_{eff}\right)g(T) \tag{6}$$

so that explicit cross-terms (e.g., $\left(\sigma_{eff} - \sigma_0\right)(T - T_0)$) are not included in the model. This separability assumption is grounded in the fundamental theory of fluid mechanics, where the hydraulic transmissivity of fracture-pore networks can often be decomposed into independent mechanical and thermal response functions under moderate loading conditions [43]. Similar simplifications have been successfully adopted in thermo-elastic-hydraulic modeling of tight sandstones and granites to balance computational efficiency with predictive accuracy [14,15,44]. This parsimonious form captures the primary thermo-mechanical evolution while avoiding the computational complexity associated with higher-order cross-terms. The separability assumption (Equation (6)) implies that the coupled permeability response can be approximated by the product of independent stress and temperature functions without explicit cross-terms. This approach is quantitatively justified by the subsequent error analysis (Section 4.5), which reveals a mean relative error of approximately 11–12% for the tested range. Within the moderate operating envelope of 20–80 °C and 15–35 MPa, this parsimonious form captures the primary thermo-mechanical evolution while avoiding the computational complexity and calibration challenges associated with higher-order coupling terms.

(5) Moderate-temperature regime. The investigated temperature range (20–80 °C) is treated as the applicability range of the proposed exponential model.

Within this framework, exponential expressions are adopted because they are widely used in permeability-stress or temperature studies and provide strictly positive predictions with a minimal number of parameters, facilitating stable calibration and sensitivity quantification. Based on this strategy, we first calibrate a stress model to obtain the stress-related exponential decay parameter C (Section 4.2), then calibrate a temperature model to obtain the temperature-related exponential decay parameter D (Section 4.3), and finally combine them into a coupled separable expression for prediction and validation (Sections 4.4 and 4.5).

4.2. Fitting of the Stress-Related Exponential Decay Parameter

As shown in Figure 6a, K'_σ decreases nonlinearly with increasing σ_{eff} for all three specimens. This curve shape is consistent with the exponential type trends commonly reported for tight and fractured rock [45,46]. Therefore, an exponential model is adopted to fit the normalized data:

$$K'_\sigma = \exp\left[-C\left(\sigma_{eff} - \sigma_0\right)\right] \tag{7}$$

where σ_0 = 15 MPa is the reference effective stress, and C (MPa^{-1}) is the stress-related exponential decay parameter. The parameter C directly reflects how sensitive the permeability is to changes in effective stress and has a clear physical meaning [47,48].

The fitting results are presented in Figure 7a–c and Table 5. For all three cores, the goodness-of-fit R^2 is greater than 0.97, suggesting that Equation (7) provides a reasonable description of the stress-dependent decay in permeability over 15–35 MPa. The fitted C values for the North, Middle and South cores are 0.0506, 0.0424 and 0.0420 MPa^{-1}.

https://doi.org/10.3390/en19020553

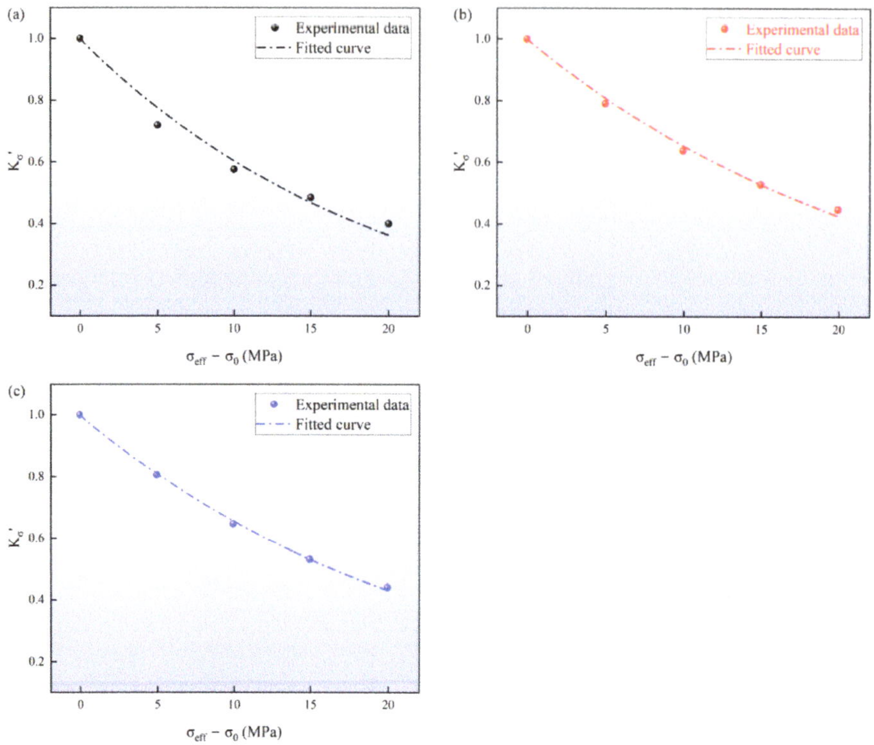

Figure 7. Normalized permeability exponential fitting results. (**a**–**c**) Experimental data and fitted curves for the North, Middle, and South cores.

Table 5. Fitted parameters of the exponential stress-sensitivity model (Equation (7)) at T = 20 °C.

Sample	Temperature (°C)	Model Form	C (MPa^{-1})	R^2
North	20	$K'_\sigma = \exp\left[-C\left(\sigma_{eff} - \sigma_0\right)\right]$	0.0506 ± 0.0030	0.9746
Middle	20		0.0424 ± 0.0011	0.9948
South	20		0.0420 ± 0.0005	0.9990

4.3. Fitting of the Temperature-Related Exponential Decay Parameter

As shown in Figure 6b–d, K'_T decreases monotonically and nonlinearly with increasing temperature for all three cores at each effective stress. Previous work has shown that, within the mid to low temperature range, the permeability-temperature relationship can often be captured by an exponential-type model [22]. In this sense, the parameter D can be interpreted as an experimental indicator that integrates the combined effects of thermoelastic expansion of the pore skeleton and microstructural adjustment. Similar formulations have also been widely adopted to characterize temperature-dependent permeability in low-permeability sandstones [44,49]. Referring to the exponential functions commonly used in studies of rock temperature-permeability coupling, the relationship between K'_T and T is fitted as

$$K'_T = \exp[-D(T - T_0)] \quad (8)$$

where T_0 = 20 °C is the reference temperature and $D > 0$ (°C^{-1}) is the temperature-related exponential decay parameter, representing the relative decay rate of permeability per Celsius increase with respect to T_0.

https://doi.org/10.3390/en19020553

Here, D is fitted separately at each σ_{eff} to evaluate the influence of effective stress on the temperature sensitivity of permeability. The coupled model in Section 4.4 adopts constant C and D to couple temperature and effective-stress effects via a parsimonious separable exponential form.

The fitted curves (Figures 8–10) and Table 6 show that the exponential decay model provides a satisfactory description of the data. For all test conditions, the goodness-of-fit R^2 is greater than 0.90 and, in most cases, exceeds 0.95, confirming the applicability of Equation (8) to the 20–80 °C temperature conditions relevant to UGS engineering. The fitted temperature-related exponential decay parameter D mainly falls in the range of 0.0075–0.023 $°C^{-1}$. As illustrated in Figures 8–10, the exponential decay model accurately captures the nonlinear reduction of permeability with temperature across all tested stress levels. A key feature of these plots is the systematic change in the curvature: the curves are steepest at the lowest effective stress (15 MPa) and become significantly flatter as the stress triples to 35 MPa. This visual 'flattening' directly reflects the decline in the temperature-related decay parameter D (Table 6).

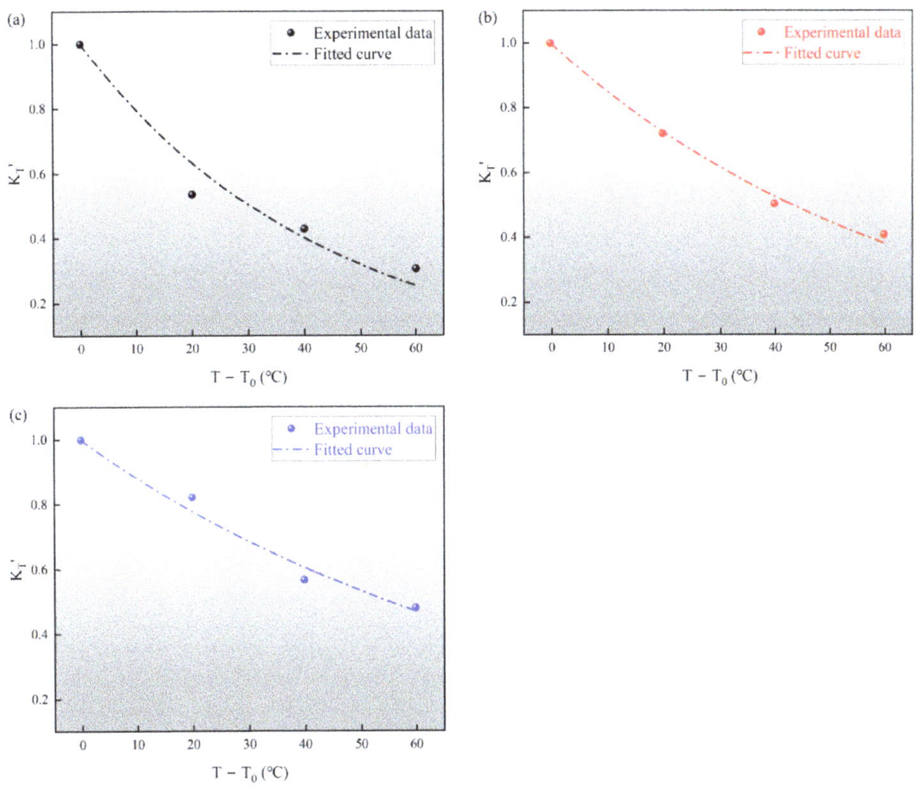

Figure 8. Exponential fitting of normalized permeability (K'_T) versus temperature for the North core under different effective stresses: (**a**) 15 MPa; (**b**) 25 MPa; and (**c**) 35 MPa. The progressive flattening of the fitted curves at higher stress levels illustrates the mechanical 'passivation' of the thermal response as compliant conduits are pre-closed by effective stress.

https://doi.org/10.3390/en19020553

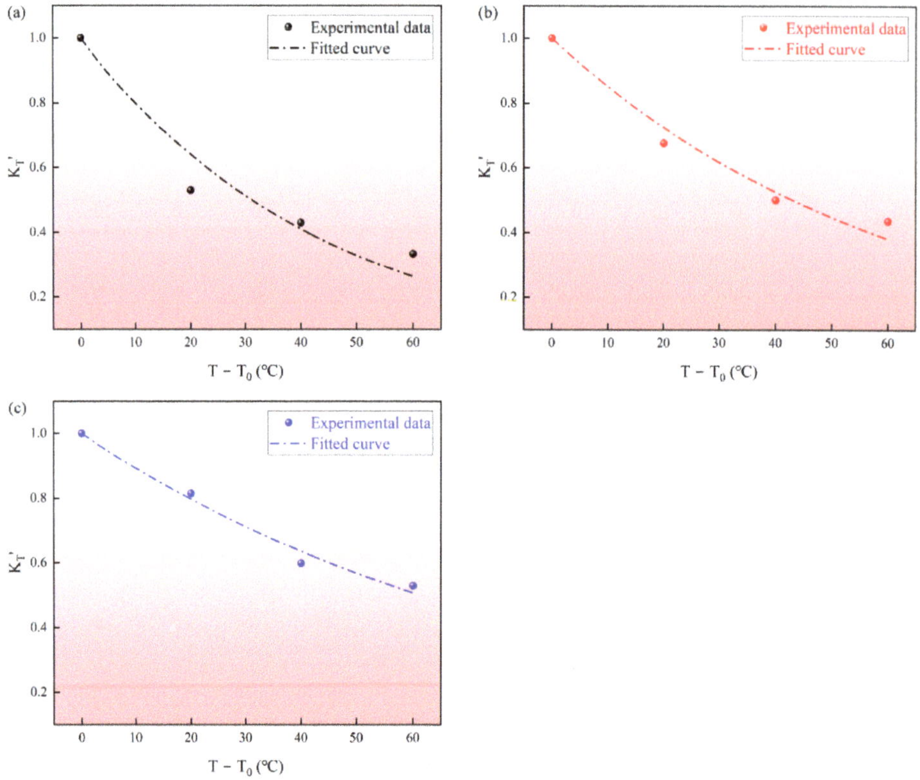

Figure 9. Exponential fitting of normalized permeability (K'_T) versus temperature for the Middle core under different effective stresses: (**a**) 15 MPa; (**b**) 25 MPa; and (**c**) 35 MPa. The progressive flattening of the fitted curves at higher stress levels illustrates the mechanical 'passivation' of the thermal response as compliant conduits are pre-closed by effective stress.

Table 6. Fitted parameters of the exponential temperature-sensitivity model at different effective stresses.

Sample	Effective Stress (MPa)	Model Form	D (°C^{-1})	R^2
North	15		0.0229 ± 0.0026	0.9534
	25		0.0160 ± 0.0006	0.9938
	35		0.0125 ± 0.0008	0.9797
Middle	15	$K'_T = \exp[-D(T - T_0)]$	0.0223 ± 0.0030	0.9323
	25		0.0161 ± 0.0013	0.9685
	35		0.0113 ± 0.0006	0.9842
South	15		0.0189 ± 0.0028	0.9004
	25		0.0128 ± 0.0013	0.9383
	35		0.0075 ± 0.0003	0.9915

https://doi.org/10.3390/en19020553

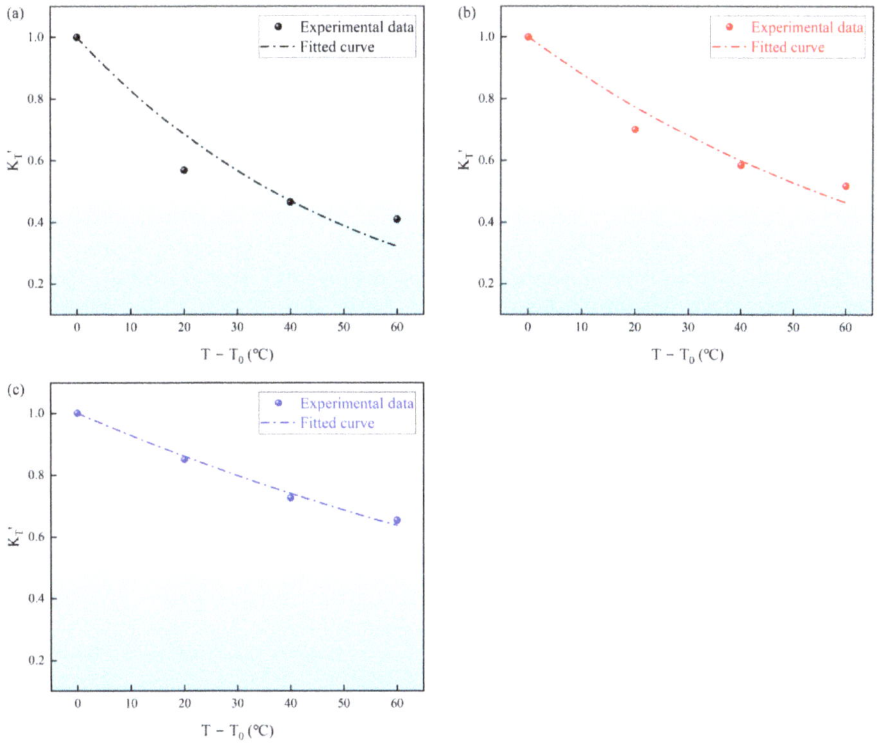

Figure 10. Exponential fitting of normalized permeability (K'_T) versus temperature for the South core under different effective stresses: (**a**) 15 MPa; (**b**) 25 MPa; and (**c**) 35 MPa. The progressive flattening of the fitted curves at higher stress levels illustrates the mechanical 'passivation' of the thermal response as compliant conduits are pre-closed by effective stress.

4.4. Exponential Permeability Model Under Coupled Temperature and Stress Conditions

To represent the combined effects of effective stress and temperature within a unified framework, combining Equations (7) and (8) under the separability assumption gives:

$$K' = \exp[-C(\sigma - \sigma_0) - D(T - T_0)] \quad (9)$$

where C (MPa^{-1}) and D ($°\text{C}^{-1}$) are the stress-related and temperature-related exponential decay parameters. Equation (9) follows the separability assumption stated in Section 4.1 and does not include an explicit cross-term such as $\left(\sigma_{eff} - \sigma_0\right)(T - T_0)$.

The fitted parameters of the coupled model for the three specimens are listed in Table 7. The fitted C values are approximately 0.038–0.046 MPa^{-1}, and the fitted D values are approximately 0.016–0.020 $°\text{C}^{-1}$, with R^2 ranging from 0.878 to 0.929. This indicates that, within the effective stress range of this study (15–35 MPa), a 10 MPa increase in effective stress leads to a permeability reduction of roughly 32–37%. Within 20–80 °C, if effective stress is held constant, a 10 °C increase in temperature reduces permeability by about 15–18%.

Table 7. Fitted parameters of the coupled model for the three specimens based on Equation (9).

Sample	Model Form	C (MPa^{-1})	D ($°\text{C}^{-1}$)	R^2
North		0.0459 ± 0.0039	0.0203 ± 0.0017	0.9289
Middle	$K' = \exp[-C(\sigma - \sigma_0) - D(T - T_0)]$	0.0384 ± 0.0037	0.0194 ± 0.0017	0.9152
South		0.0376 ± 0.0039	0.0160 ± 0.0017	0.8776

https://doi.org/10.3390/en19020553

4.5. Model Validation and Error Analysis

Using the fitted exponential model Equation (9), the normalized permeability of each specimen under different $\left(\sigma_{eff}, T\right)$ is predicted. To evaluate the model performance, the relative error is introduced:

$$\delta = \frac{\left|K'_{pred} - K'\right|}{K'} \times 100\% \tag{10}$$

where K'_{pred} and K' are the modeled and measured normalized permeability. For each specimen, the relative errors δ, the mean relative error δ_{mean}, the maximum relative error $\delta_{\max}$ and the 80th percentile of the relative error δ_{80} are summarized in Tables 8–10.

The error statistics in Tables 8–10 provide insight into the model's reliability. The mean relative error of approximately 11–12% for all specimens suggests that the parsimonious separable form is sufficient for capturing the primary trends of permeability evolution. However, it is noted that the maximum relative errors (up to 39%) typically occur at the lowest permeability values (e.g., 80 °C and 35 MPa). This is because, as absolute permeability approaches the reservoir's lower limit, even small numerical residuals result in high relative percentages. Despite these localized deviations, the 80th percentile of errors remains within a narrow band (22–24%), confirming the model's overall consistency for engineering applications.

Table 8. Error statistics of permeability model predictions for the North specimens.

T (°C)	σ_{eff} (MPa)	K'	K'_{pred}	δ (%)	δ_{mean} (%)	δ_{max} (%)	δ_{80} (%)
20	15	1.00	1.00	0.00	11.45	38.67	22
20	20	0.72	0.79	10.50			
20	25	0.58	0.63	9.80			
20	30	0.48	0.50	3.68			
20	35	0.40	0.40	0.27			
40	15	0.54	0.67	24.38			
40	25	0.41	0.42	1.50			
40	35	0.33	0.27	19.10			
60	15	0.43	0.44	3.64			
60	25	0.29	0.28	3.13			
60	35	0.23	0.18	21.94			
80	15	0.31	0.30	3.05			
80	25	0.24	0.19	20.59			
80	35	0.19	0.12	38.67			

Table 9. Error statistics of permeability model predictions for the Middle specimens.

T (°C)	σ_{eff} (MPa)	K'	K'_{pred}	δ (%)	δ_{mean} (%)	δ_{max} (%)	δ_{80} (%)
20	15	1.00	1.00	0.00	12.03	39.11	24
20	20	0.79	0.83	4.32			
20	25	0.64	0.68	6.61			
20	30	0.53	0.56	6.13			
20	35	0.45	0.46	3.26			
40	15	0.53	0.68	28.18			
40	25	0.43	0.46	7.11			
40	35	0.37	0.31	14.00			
60	15	0.43	0.46	7.14			
60	25	0.32	0.31	1.94			
60	35	0.27	0.21	20.63			
80	15	0.33	0.31	6.41			
80	25	0.28	0.21	23.52			
80	35	0.24	0.14	39.11			

https://doi.org/10.3390/en19020553

Table 10. Error statistics of permeability model predictions for the South specimens.

T (°C)	σ_{eff} (MPa)	K'	K'_{pred}	δ (%)	δ_{mean} (%)	δ_{max} (%)	δ_{80} (%)
20	15	1.00	1.00	0.00	12.44	37.34	22
20	20	0.81	0.83	2.77			
20	25	0.65	0.69	6.10			
20	30	0.53	0.57	6.95			
20	35	0.44	0.47	6.91			
40	15	0.57	0.73	27.77			
40	25	0.45	0.50	9.96			
40	35	0.37	0.34	8.69			
60	15	0.47	0.53	13.22			
60	25	0.38	0.36	4.16			
60	35	0.32	0.25	22.30			
80	15	0.41	0.38	6.53			
80	25	0.33	0.26	21.43			
80	35	0.29	0.18	37.34			

5. Discussion

This chapter interprets the observed permeability evolution of fractured-porous carbonate rocks under coupled temperature and effective stress conditions. We first explain the stress-controlled exponential-type permeability decay and its microphysical origin, and then discuss why the temperature sensitivity is progressively attenuated by increasing effective stress within the investigated 20–80 °C conditions. Based on these observations, two full-path conceptual models are proposed for permeability evolution under effective-stress loading or temperature loading. Finally, we discuss the engineering relevance of the proposed exponential model and clarify its current applicability and limitations.

5.1. Stress-Controlled Permeability Decay and Full-Path Conceptual Model

Table 7 shows that the fitted stress-related exponential decay parameter C for the North, Middle, and South cores is 0.0459, 0.0384, and 0.0376 MPa^{-1}. This level of stress sensitivity is comparable to that reported for carbonate cores from the Malm geothermal reservoir in Germany [50] and is also consistent with observations in tight gas sandstones, where permeability decreases markedly as effective stress increases [51].

The exponential-type decay of permeability with increasing effective stress (Figure 7) can be interpreted as a non-uniform progressive closure of the fracture-pore network. Farahani et al. [52] suggested that flow pathways in fractured-porous rocks can be conceptualized as a network composed of highly compliant fractures (soft hydraulic units) connected with less compliant matrix pores (stiff hydraulic units). At relatively low effective stress, wide fractures and large-aperture channels accommodate most deformation. Their rapid closure and geometric rearrangement lead to a steep reduction in flow capacity. As effective stress increases further, the dominant fracture pathways become increasingly restricted, and the remaining flow is progressively carried by smaller pores and stiff contact areas. Consequently, additional stress increments cause smaller permeability changes, and the curve transitions to a gentler decay regime. This interpretation is consistent with the theoretical analysis of Zimmerman and Bodvarsson [43], who showed that the transmissivity of a single fracture decreases approximately with the cube of its mechanical aperture, revealing the microphysical origin of the macroscopic nonlinear stress sensitivity observed in this study.

Building on this interpretation, a conceptual model is proposed to describe permeability evolution along a broader effective-stress path (Figure 11). Two competing stages can be distinguished:

Stage I (mechanical compaction zone): When the effective stress has not yet reached the dilatancy threshold of the rock, the dominant mechanism is the mechanical closure of

https://doi.org/10.3390/en19020553

pre-existing fractures and pores. Macroscopically, permeability decreases monotonically and rapidly with increasing effective stress. The stress range covered by the present experiments falls within this stage.

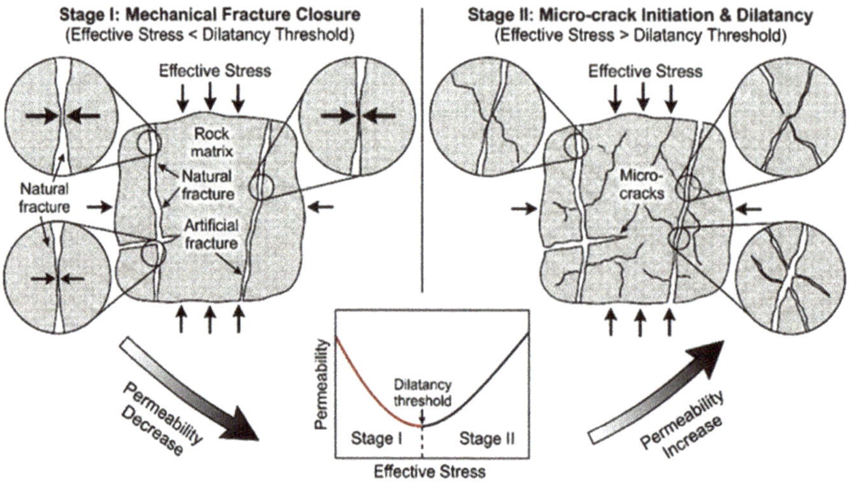

Figure 11. Conceptual model of permeability evolution in fractured-porous rocks under increasing effective stress, illustrating a fracture closure-dominated stage (Stage I) and a potential damage-induced micro-cracks stage (Stage II). The black arrows represent the application of effective stress, and the grey arrows indicate the overall trend of permeability evolution (decrease or increase).

Stage II (damage-induced dilatancy zone): If the effective stress continues to increase beyond the rock's dilatancy threshold, micro-crack initiation and propagation are triggered. As experimentally demonstrated by Zoback and Byerlee [53], microcrack dilatancy in the vicinity of the failure stress can cause a substantial increase in permeability, potentially exceeding initial values by several times, due to the formation of new flow channels. While our current data reside strictly within the mechanical compaction regime (Stage I), this inclusion of Stage II provides a comprehensive roadmap for reservoir behavior under extreme stress loading.

This model captures the entire nonlinear competitive evolution from "mechanical fracture closure" to "micro-crack initiation and dilatancy" and provides a theoretical framework for understanding reservoir behavior under more extreme loading conditions.

5.2. Temperature Sensitivity of Permeability Under Effective-Stress Regulation and Full-Path Conceptual Model

From a mechanistic perspective, the maximum temperature in this study is 80 °C, which is well below the threshold for extensive thermal cracking commonly reported in heating-induced fracturing experiments (typically 200–300 °C) [22,54]. Therefore, within 20–80 °C, the observed permeability reduction is mainly attributed to thermoelastic deformation of the rock skeleton and microstructural rearrangement driven by mismatched thermal expansion among mineral constituents. When pre-existing fractures are not fully closed, heating tends to impose additional compression on relatively wide fracture segments, whereas the geometric adjustment of pore throats and smaller fractures is comparatively limited; as a result, the effective flow channels shrink without generating new fracture surfaces. Consistent behaviors have been reported in temperature-stress-permeability experiments on sandstones and granites, where permeability commonly decreases slowly with temperature at low to intermediate temperatures and increases sharply only after

https://doi.org/10.3390/en19020553

pervasive thermal cracking develops at higher temperatures [23,49,54,55]. This indicates that the present experiments primarily capture the stage dominated by thermoelastic expansion-compression rather than thermal cracking.

As shown in Figure 6b–d, the temperature-caused permeability reduction is stronger at low effective stress and becomes progressively weaker at higher effective stress. For example, at σ_{eff} = 15 MPa, heating from 20 to 80 °C reduces permeability by roughly 60% for the North, Middle, and South cores, whereas at σ_{eff} = 35 MPa the reduction is only about 30%. The experimental results provide a clear quantitative link between mechanical confinement and the attenuation of temperature sensitivity. Specifically, as shown in Table 6 and Figure 12, the temperature-related decay parameter D drops by approximately 45.4% to 60.3% across the three specimens as the effective stress increases from 15 to 35 MPa. This systematic reduction in D quantifies the passivation effect: at 15 MPa, the unclosed and compliant fractures maximize the permeability response to thermoelastic expansion, leading to higher sensitivity. However, as the effective stress triples to 35 MPa, the mechanical pre-closure of these primary conduits, which accounts for an initial 55–60% permeability loss (Section 3.2), leaves a stiffer skeleton dominated by less-compressible matrix pores. This transition from fracture-dominated to matrix-dominated flow significantly weakens the capacity of mineral expansion to further restrict the seepage pathways, thereby numerically reducing the temperature sensitivity of the bulk rock. A comparable "passivation" of temperature sensitivity at higher effective stresses has also been reported for Beishan granite and several Chinese sandstones [23,55].

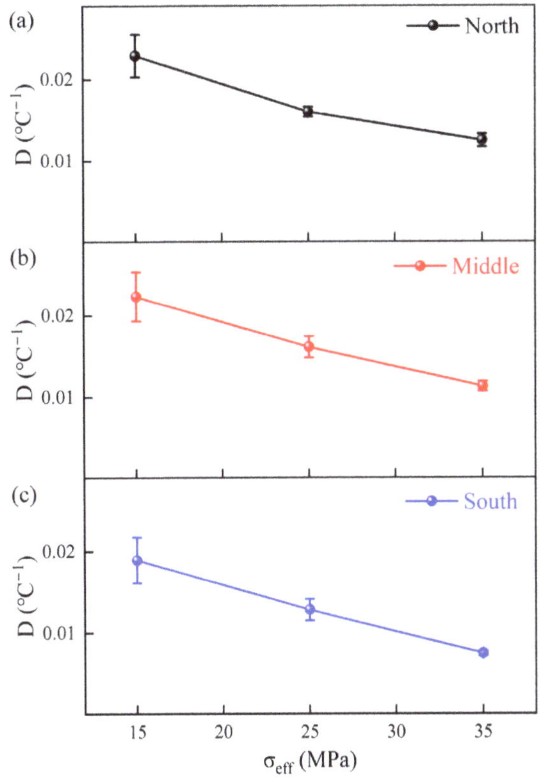

Figure 12. Fitted temperature-related exponential decay parameter D under different effective stress for the (**a**) North, (**b**) Middle, and (**c**) South cores. Error bars denote the standard error of D obtained from exponential fitting.

https://doi.org/10.3390/en19020553

Based on the experimental results and rock thermophysical theory, we propose a full path mechanistic model for permeability evolution under temperature loading (Figure 13). The model comprises two competing stages.

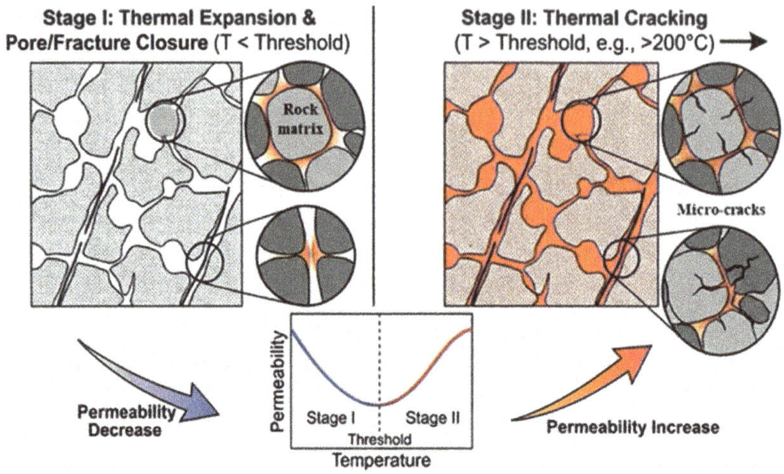

Figure 13. Conceptual model for the full path permeability evolution under temperature loading. Stage I (20–80 °C in this study) is dominated by thermoelastic expansion and progressive closure of pores and fractures. Stage II (typically > 200 °C) is dominated by thermal cracking, which may enhance permeability and reverse the overall trend.

Stage I (thermal expansion induced permeability reduction zone): Within the operational temperature conditions of this study (20–80 °C), permeability decreases mainly because matrix minerals undergo thermoelastic expansion. Owing to mismatched thermal expansion among mineral constituents and the confinement imposed by external stresses, the expanding grains progressively encroach on pre-existing pore and fracture space, narrowing effective flow paths, especially relatively wide fractures. This mechanism provides a microphysical explanation for the exponential type permeability decay with increasing temperature observed here, and it represents the dominant response for XGS-UGS operating conditions.

Stage II (thermal cracking induced permeability enhancement zone): If temperature continues to rise beyond the thermal cracking threshold (typically > 200 °C) [22,54], thermally induced stresses may exceed intergranular cohesive strength, triggering intergranular microcrack initiation and the propagation of pre-existing fractures [49,54,55]. According to the study on carbonates by Yang et al. [56], increasing temperature leads to the progressive development of internal micro-fractures and a decrease in the elastic modulus. This real-time thermal cracking can enhance network connectivity and potentially reverse the permeability reduction trend. Although the temperatures in this study (20–80 °C) remain within the Stage I closure regime, the model reflects the well-documented thermo-mechanical evolution of carbonate reservoirs under higher thermal loads.

5.3. Limitations

The 1:1 comparison between modeled and measured values, together with the 80% error band, are presented in Figure 14, which indicates a systematic bias of the model: in the range $K' = 0$–0.4 (usually corresponding to high temperatures and high effective stresses), most predicted values fall below the 1:1 line (slight underestimation), whereas in the range $K' = 0.4$–1.0 (corresponding to low temperatures and low effective stresses), the model tends

https://doi.org/10.3390/en19020553

to overestimate permeability. This suggests that the actual temperature-stress coupling does not strictly follow a single exponential law with constant parameters, but exhibits piecewise nonlinearity: at low temperatures and low effective stresses, fractures are only partially closed and highly compliant; as temperature and effective stress increase, most fractures become nearly closed, additional load is mainly borne by matrix pores, and permeability changes become less pronounced. Therefore, an exponential model with constant C and D cannot fully capture the "initially steep then gradually gentle" curve shape. To further mitigate this systematic bias, future model development could benefit from adopting a piecewise parameterization or a stress-dependent formulation (e.g., $C\left(\sigma_{eff}\right)$ and $D\left(\sigma_{eff}\right)$). Such an approach would allow the decay parameters to evolve with the mechanical state of the rock: using higher sensitivity values in the low-stress regime to represent compliant fracture closure and lower values at high stresses to reflect the dominance of stiffer matrix pores. By explicitly accounting for the varying stiffness of the hydraulic units, such refinements could effectively reduce the current overestimation at high normalized permeability values and the underestimation at low values. In addition, Equation (9) assumes separability and does not include higher-order coupling terms; at higher temperatures and larger stresses, or in scenarios involving pronounced dissolution-precipitation and thermally induced fracture development [57], a separable exponential form may be insufficient to describe strongly nonlinear coupling. With broader experimental data covering wider conditions and controlling factors, more flexible constitutive forms incorporating explicit cross terms could be developed to further refine and generalize the present exponential model. While more complex constitutive forms incorporating explicit cross-terms could potentially reduce residuals further, the current separable model provides a robust balance between mathematical simplicity and predictive accuracy. The consistency shown in the 1:1 comparison (Figure 14) confirms that for fractured-porous carbonates in XGS-UGS, the independent parameterization of C and D is sufficient to reproduce the overall permeability trends across diverse temperature and stress conditions.

Another limitation is the reliance on macroscopic seepage measurements to infer micro-scale deformation processes. While the current flow-based results accurately quantify the hydraulic aperture response, which is essential for evaluating UGS injectivity and deliverability, they do not provide direct visual evidence of the internal geometric evolution. High-resolution micro-CT imaging would be invaluable for visualizing localized fracture closure, asperity damage, and thermally induced micro-cracking under loading. However, performing high-resolution in-situ scans at the investigated moderate temperatures and high stresses (80 °C; 35 MPa) remains a significant technical challenge. Moreover, ex-situ measurements may introduce stress-release artifacts that alter the fracture contact state. Consequently, microstructural changes are currently absorbed into the fitted exponential parameters rather than being parameterized explicitly. Future research incorporating specialized in-situ CT scanning systems will be prioritized to provide more rigorous geometric validation for the proposed thermo-mechanical coupling mechanisms.

Furthermore, the use of a single throughgoing artificial fracture to represent the flow path (Section 2.1) introduces certain scale-dependent simplifications. While this approach ensures experimental repeatability and allows for clear quantification of thermo-mechanical coupling mechanisms, it simplifies the multi-scale nature of natural fracture networks. In natural reservoirs, fractures exhibit diverse orientations relative to the core axis, varying surface roughness, and heterogeneous mineral infillings (e.g., calcite or clay). These features not only define the initial seepage capacity but also significantly influence the local mechanical stiffness and thermal expansion response of the rock skeleton. Additionally, natural fractured specimens are often susceptible to structural damage during the high-precision machining process, making it technically challenging to prepare high-quality

https://doi.org/10.3390/en19020553

samples that meet strict laboratory standards. Furthermore, while the axial orientation of the artificial fractures maximizes the initial seepage capacity of the core plugs, it does not account for the tortuosity and connectivity losses inherent in multi-scale fracture networks. Consequently, when extrapolating these findings to field-scale simulations, the proposed model should be viewed as a fundamental characterization of the 'hydraulic unit' response, which must be coupled with reservoir-scale geological models to account for the structural complexity and network-scale amplification of thermo-mechanical effects. Despite these simplifications, the proposed exponential parameters provide a consistent baseline for characterizing the response of primary seepage conduits in the XGS-UGS, serving as a critical input for upscale multiphysics simulations.

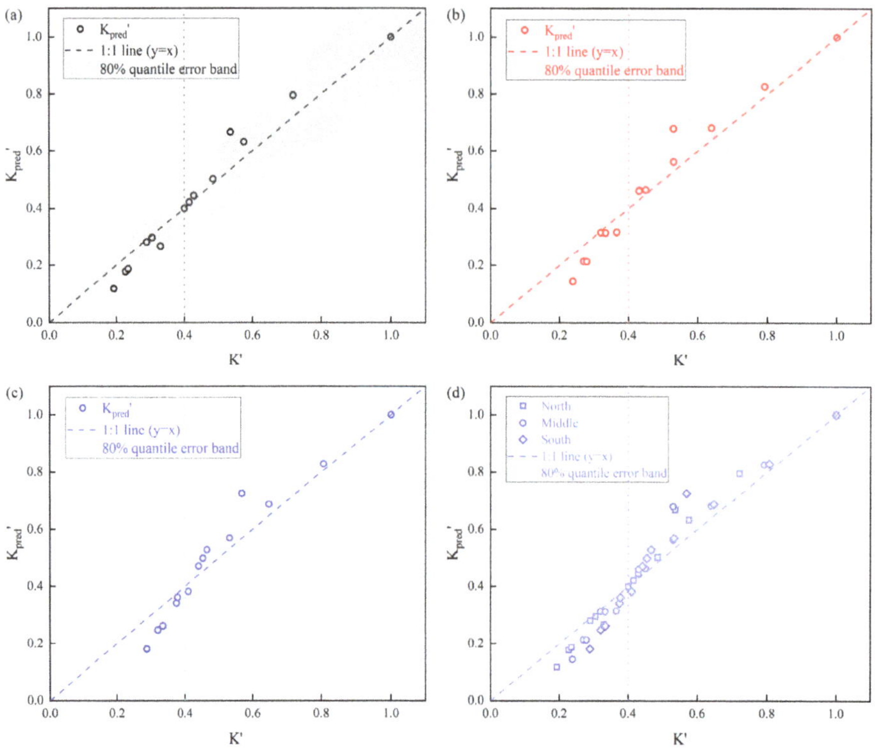

Figure 14. 1:1 comparison between modeled and measured normalized permeability under coupled temperature and stress conditions. The dashed line indicates $K'_{pred} = K'$; the shaded region denotes the 80th percentile error band. (**a**) North core; (**b**) Middle core; (**c**) South core; (**d**) combined data from all three cores.

The transition from a single laboratory-scale fracture to a complex reservoir-scale network can be achieved through the equivalent permeability (k_{eq}) concept. In this framework, the laboratory-derived parameters C and D serve as the fundamental constitutive response for the individual hydraulic conduits. When integrated into Discrete Fracture Network (DFN) or equivalent continuum models, the macroscopic permeability of a reservoir grid block can be expressed as an assembly of these responsive units. As the effective stress and temperature evolve during UGS operations, the aperture of each fracture within the network follows the calibrated exponential decay, collectively driving the evolution of the field-scale equivalent permeability. However, it is important to note that network-scale connectivity and the cooperative deformation of fractured-vuggy clusters may introduce additional nonlinearities. Therefore, reservoir-scale simulations should utilize these laboratory results as a calibrated baseline while accounting for broader statistical variations in the coefficients to address the impact of complex heterogeneity.

The sensitivity parameters C and D are intrinsically linked to the mechanical state of the rock skeleton. In actual UGS facilities, gas injection increases pore pressure and reduces effective stress, potentially enhancing the reactivity of compliant fractures to both mechanical loading and thermal expansion. If the pore-pressure term in Equation (2) is neglected under high-pressure injection conditions, the model would likely underestimate the rate of permeability recovery. Therefore, for engineering implementation, the exponential model (Equation (9)) should be coupled with a dynamic effective stress calculation that explicitly accounts for localized pore pressure variations.

It is also worth noting that while apparent gas permeability was used without Klinkenberg correction, the identified sensitivity trends remain robust. In the low-mD range of XGS-UGS carbonates, the magnitude of permeability decay caused by fracture closure and mineral thermal expansion far outweighs potential slip-flow deviations. This ensures that the derived exponential parameters (C and D) accurately reflect the relative seepage response of the reservoir skeleton to operational fluctuations.

Another limitation of the current study is the use of monotonic loading paths to calibrate the exponential parameters C and D. While these parameters successfully quantify the sensitivity of the reservoir skeleton to initial stress and temperature perturbations, actual UGS operations involve high-frequency cycles of gas injection and withdrawal. Previous studies have indicated that multi-cycle effective stress perturbations can cause irreversible structural damage to the pore-fracture network, leading to permeability hysteresis where the unloading path deviates from the loading path. Such hysteresis often leads to cumulative permeability loss as fracture asperities undergo non-elastic deformation or mechanical fatigue over time. Therefore, the values of C and D reported here (Table 7) specifically represent the loading-path behavior. While they provide a robust upper-bound estimate for permeability decay during initial reservoir adjustment, they may not fully capture the long-term rock fatigue or the partial recovery of flow channels under cyclic conditions. Future research incorporating long-term cyclic testing is needed to extend the exponential model into a full-cycle constitutive relation for UGS performance prediction.

Finally, it is essential to acknowledge the indicative nature of the derived exponential parameters. The ranges identified for the stress-related decay parameter C (0.038–0.046 MPa^{-1}) and the temperature-related decay parameter D (0.016–0.020 $°C^{-1}$) provide a calibrated baseline for the fractured-porous units within the XGS-UGS under the investigated conditions (20–80 °C; 15–35 MPa). These parameters are representative of the coupled thermo-mechanical response of the primary flow conduits; however, reservoir-scale simulations should account for broader statistical variations in these coefficients to fully address the impact of complex heterogeneity on long-term UGS deliverability.

5.4. Engineering Significance

Within the investigated range (20–80 °C and the tested effective-stress levels), the proposed exponential model offers a concise engineering-scale tool to quantify coupled temperature and stress effects and to compare sensitivity among cores or reservoir intervals using fitted parameters (C and D) in a consistent manner. The identified sensitivity provides a mechanistic basis for understanding field-scale permeability fluctuations in UGS facilities. During the gas injection phase, the rising pore pressure reduces effective stress, and the injection of relatively cold gas typically leads to localized cooling near the wellbore. Our experimental results confirm that both of these thermo-mechanical shifts favor the recovery of flow conduits, which is consistent with field observations where reservoir permeability

https://doi.org/10.3390/en19020553

is generally higher during injection than during the production phase. However, the macroscopic impact on well performance is governed by the synergistic coupling of the sensitivity parameters (C and D) and the spatial extent (radius) of the stress and temperature disturbance zones. While extreme permeability variations are primarily concentrated in the near-wellbore region, they act as a dynamic skin factor that dictates the Injectivity Index. Incorporating these localized and cyclic effects into reservoir simulations allows for a more accurate prediction of deliverability evolution and operational optimization over the entire UGS service life, preventing over-optimistic forecasts that fail to account for localized thermo-mechanical seepage restriction.

To illustrate the practical impact of the proposed model on UGS operations, a representative case study was conducted based on the operating parameters of the XGS-UGS. This facility typically operates with pore pressures between 12 and 30 MPa and formation temperatures ranging from 30 to 60 °C. During a withdrawal phase, as the pore pressure decreases from 30 MPa to 12 MPa, the effective stress acting on the rock skeleton increases by approximately 18 MPa. Concurrently, geothermal heat transfer can cause the localized near-wellbore temperature to rise from the injection state (30 °C) to the formation equilibrium state (60 °C). According to the calibrated exponential model and the sensitivity parameters derived in this study ($C \approx 0.042$ MPa^{-1} and $D \approx 0.019$ °C^{-1}), this coupled thermo-mechanical loading leads to a localized permeability reduction of approximately 73% within the primary seepage conduits during a single withdrawal cycle. Such a substantial decay acts as a dynamic skin effect that significantly constrains the well's deliverability index.

From an engineering perspective, the proposed constitutive model $K(\sigma, T)$ can be integrated into reservoir simulators through two primary pathways:

Direct Analytical Embedding: The exponential relation can be implemented as a dynamic multiplier within the simulator's constitutive subroutines. This allows the solver to update the permeability of each grid block at every time step based on the instantaneous local stress and temperature fields.

Multi-dimensional Lookup Tables: For commercial simulators with restricted constitutive flexibility, the model can generate a 2D lookup table (K' vs σ_{eff} and T). The simulator then utilizes bilinear interpolation to determine the effective permeability multiplier during predictive calculations.

By accounting for these localized cyclic variations, the proposed model provides a quantitative tool for optimizing injection-production schedules and ensuring the long-term deliverability of fractured-porous carbonate reservoirs.

6. Conclusions

Steady-state gas permeability tests were conducted on three fractured-porous carbonate cores from the XGS-UGS under coupled temperature and effective stress conditions (20–80 °C; 15–35 MPa). The main conclusions are as follows:

(1) Overall permeability reduction under coupled loading. Permeability decreases markedly with increasing effective stress and temperature within the investigated conditions. Varying a single factor (either temperature or effective stress) typically leads to an approximately 30–70% reduction, whereas their combined increase can produce a maximum reduction exceeding 80% for the tested cores.

(2) Hierarchy of controlling factors and stress regulation of temperature sensitivity. Exponential-type fittings indicate that permeability is primarily governed by effective stress, while temperature plays a secondary, regulating role. The stress-related exponential decay parameter C (0.038–0.046 MPa^{-1}) is consistently higher than the temperature-related exponential decay parameter D (0.016–0.020 °C^{-1}) for all three

cores. In addition, D decreases systematically with increasing effective stress, demonstrating that temperature sensitivity is strongly modulated by the stress state because compliant fractures and larger pores are progressively pre-closed.

(3) Mechanistic interpretation and conceptual regimes. The stress-induced permeability decay can be interpreted as a non-uniform progressive closure of the fracture-pore network: rapid closure of compliant flow paths dominates at relatively low effective stress, followed by a gentler decay as flow becomes increasingly controlled by stiffer pores and contact areas. Within 20–80 °C, the observed temperature effect is mainly attributed to thermoelastic deformation and microstructural rearrangement that narrow existing flow channels, rather than pervasive thermal cracking. Conceptually, under more extreme effective-stress or temperature loading, microcracks may propagate and induce dilatancy, potentially enhancing permeability; however, the operating-relevant conditions examined here remain within a closure-dominated regime.

(4) Exponential model and applicability. A concise, physically interpretable exponential model was proposed to describe permeability evolution under coupled temperature-effective stress loading, showing good predictive performance within the investigated ranges (mean relative error on the order of 10%). The model can be implemented as a constitutive permeability module in multiphysics simulations to support injectivity/deliverability evaluation and operational optimization for XGS-UGS and other UGS sites with similar fractured-porous carbonate settings.

Author Contributions: Z.Z.: Conceptualization, Methodology, Formal analysis, Investigation, Data curation, Writing—original draft—review & editing. Q.G.: Conceptualization, Methodology, Validation, Resources, Formal analysis, Investigation, Data curation, Writing—review & editing, Funding acquisition. Y.W.: Conceptualization, Methodology, Validation, Formal analysis, Writing—review & editing. S.H.: Conceptualization, Methodology, Investigation, Data curation, Writing—review & editing. Y.Z.: Conceptualization, Methodology, Validation, Formal analysis, Writing—review & editing. L.L.: Conceptualization, Methodology, Validation, Formal analysis, Writing—review & editing. M.X.: Conceptualization, Methodology, Validation, Formal analysis, Writing—review & editing. J.W.: Conceptualization, Methodology, Validation, Formal analysis, Writing—review & editing. S.J.: Conceptualization, Methodology, Validation, Resources, Formal analysis, Investigation, Data curation, Writing—review & editing, Funding acquisition. All authors have read and agreed to the published version of the manuscript.

Funding: National Natural Science Foundation of China (Project Nos. U23B20160, 5217041034, and 42402248), CO_2 Enhanced Shale Gas Recovery and Sequestration by Numerical Thermal-Hydro-Mechanical-Chemical modelling (Grant No. 2023DQ02-0206) funded by PetroChina Research Institute of Petroleum Exploration & Development, and Chongqing Natural Science Foundation (Project No. CSTB2025NSCQ-GPX0718).

Data Availability Statement: The original contributions presented in this study are included in the article. Further inquiries can be directed to the corresponding author(s).

Acknowledgments: This research was funded by the National Natural Science Foundation of China (Project Nos. U23B20160, 5217041034, and 42402248), CO_2 Enhanced Shale Gas Recovery and Sequestration by Numerical Thermal-Hydro-Mechanical-Chemical modelling (Grant No. 2023DQ02-0206), and the Chongqing Natural Science Foundation (Project No. CSTB2025NSCQ-GPX0718).

Conflicts of Interest: Author Yan Wang and Limin Li were employed by the company Chongqing Xiangguosi Underground Gas Storage Company Limited. Author Yuchao Zhao was employed by the company Exploration and Development Research Institute, PetroChina Southwest Oil & Gasfield Company. The remaining authors declare that the research was conducted in the absence of any commercial or financial relationships that could be construed as a potential conflict of interest.

https://doi.org/10.3390/en19020553

References

1. Ma, X.H.; Zheng, D.W.; Wei, G.Q.; Ding, G.S.; Zheng, S.J. Development Directions of Major Scientific Theories and Technologies for Underground Gas Storage. *Nat. Gas Ind.* **2022**, *42*, 93–99. (In Chinese) [CrossRef]
2. Hu, Y.; Wang, M.Y.; Luo, L.; Li, L.X.; Peng, X.; Zhao, Z.H. Technical Basis and Development Idea for the Construction of "Gas Storage Capacity in Southwest China". *Nat. Gas Ind.* **2024**, *44*, 1–13. (In Chinese) [CrossRef]
3. Zheng, D.W.; Wanyan, Q.Q.; Zhao, K. Development Status and Prospect of Underground Natural Gas Storage in China. *Int. Pet. Econ.* **2025**, *33*, 37–45. (In Chinese) [CrossRef]
4. Liu, J.; Pei, J.; Wei, J.; Yang, J.; Xu, H. Development Status and Prospect of Salt Cavern Energy Storage Technology. *Earth Energy Sci.* **2025**, *1*, 159–179. [CrossRef]
5. Wei, G.Q.; Zheng, Y.L.; Qiu, X.S.; Sun, J.C.; Shi, L.; Lei, X. Geological Theory and Application of Underground Gas Storage in China. *Acta Pet. Sin.* **2019**, *40*, 1519–1530. (In Chinese) [CrossRef]
6. Sun, J.C.; Xu, H.C.; Wang, J.M.; Shi, L.; Li, C.; Tang, L.; Zhong, R. Injection–Production Mechanisms and Key Evaluation Technologies for Underground Gas Storages Rebuilt from Gas Reservoirs. *Nat. Gas Ind.* **2018**, *38*, 138–144. (In Chinese) [CrossRef]
7. Zheng, X.; Zhao, Y.C.; Zhao, Z.H.; Tang, H.Y.; Zhao, Y.L. Mechanism Investigation on In-Situ Stress Characteristics and Mechanical Integrity of Fracture-Cavity Carbonate Underground Gas Storage Reservoir. *Pet. Reserv. Eval. Dev.* **2024**, *14*, 814–824.
8. Zhao, Y.C.; Luo, Y.; Li, L.X.; Zhou, Y.; Li, L.M.; Wang, X. In-Situ Stress Simulation and Integrity Evaluation of Underground Gas Storage: A Case Study of the Xiangguosi Underground Gas Storage, Sichuan, SW China. *J. Geomech.* **2022**, *28*, 523–536. (In Chinese)
9. Gan, Q.; Song, H.; Elsworth, D.; Jia, S.; Chen, J.; Ma, F.; Li, Q.; Yang, Y.; Wang, X.; Dai, Z. Deep Learning-Enhanced Global Sensitivity Analysis for Uncertainty Quantification in THMC Coupled $scCO_2$-EGS. *Energy* **2025**, *335*, 138086. [CrossRef]
10. Fan, Y.; Gao, X.P.; Ning, F. Seepage Law of Gas-Reservoir UGSs during Multi-Cycle Injection and Production. *Nat. Gas Ind.* **2023**, *43*, 103–111. (In Chinese) [CrossRef]
11. Sun, K.; Liu, H.; Leung, J.Y.; Wang, J.; Feng, Y.; Liu, R.; Kang, Z.; Zhang, Y. Impact of Effective Stress on Permeability for Carbonate Fractured-Vuggy Rocks. *J. Rock Mech. Geotech. Eng.* **2024**, *16*, 942–960. [CrossRef]
12. Gao, X.; Yang, J.; Han, H.; Peng, J.; Peng, H.; Yang, Z. Experimental Research on the Effect Caused by Effective Stress Change in Injection Production Capacity of Carbonate Reservoir. *Front. Earth Sci.* **2023**, *11*, 1101056. [CrossRef]
13. Yan, F.M.; Kang, Y.L.; Li, S.; Du, C.C.; Li, D.M. Simulated Experiment on Stress Sensitivity in Fractured-Vuggy Reservoir. *Nat. Gas Geosci.* **2010**, *21*, 489–493+507. (In Chinese)
14. Haghi, A.H.; Chalaturnyk, R.; Talman, S. Stress-Dependent Pore Deformation Effects on Multiphase Flow Properties of Porous Media. *Sci. Rep.* **2019**, *9*, 15004. [CrossRef]
15. Zivar, D.; Foroozesh, J.; Pourafshary, P.; Salmanpour, S. Stress Dependency of Permeability, Porosity and Flow Channels in Anhydrite and Carbonate Rocks. *J. Nat. Gas Sci. Eng.* **2019**, *70*, 102949. [CrossRef]
16. Mi, O.; Wang, Z.; Zhang, R.; Zhang, Z.; Zhang, B. Permeability Stress Sensitivity Effect and Evolution Model of Dual Medium in Tight Sandstone Reservoirs. *Measurement* **2026**, *259*, 119627. [CrossRef]
17. Gou, Y.; Sun, J.C.; Yang, Z.M.; Zhou, X.M. Permeability-Stress Sensitivity of Complex Lithology Reservoir and Its Effect on Gas Well Productivity. *Rock Soil Mech.* **2014**, *35*, 2535–2542. (In Chinese) [CrossRef]
18. Yang, Y.; Tao, L.; Yang, H.; Iglauer, S.; Wang, X.; Askari, R.; Yao, J.; Zhang, K.; Zhang, L.; Sun, H. Stress Sensitivity of Fractured and Vuggy Carbonate: An X-Ray Computed Tomography Analysis. *J. Geophys. Res. Solid Earth* **2020**, *125*, e2019JB018759. [CrossRef]
19. Pu, Z.; Tian, Y.; Fu, J.; Yang, Y.; Satea, A.; Xiao, Z.; Zhao, Y.; Zhang, L. Stress-Sensitive Porosity and Permeability in Carbonate Rocks for Underground Hydrogen Storage: A Digital Rock Simulation Study. *Int. J. Coal Sci. Technol.* **2025**, *12*, 55–71. [CrossRef]
20. Zheng, S.J.; Zheng, D.W.; Sun, J.C.; Li, C.; Wu, Z.D.; Zhu, Z.S.; Liu, X.S. Temperature-sensitivity of Underground Gas Reservoir Storage and Its Effect on Well Deliverability. *Pet. Geol. Exp.* **2022**, *44*, 365–372. (In Chinese) [CrossRef]
21. Zhu, S.N.; Zeng, D.Q.; Zhang, G.Q.; Jia, Y.W.; Yang, X.S.; Wang, D.D. Evaluation Method and Evolutionary Characteristics of Damaged Cap Rock Sealing Capacity in Gas Reservoir Type Underground Gas Storage under Multi-Cycle Injection-Production Conditions. *Nat. Gas Geosci.* **2024**, *35*, 2078–2090. (In Chinese)
22. Tomás, R.; Benavente, D.; Martínez-Ibáñez, V.; Garrido, M.E. How Do High Temperatures Affect Rock Properties? A Comprehensive Review of Experimental Thermal Effects and Underlying Mechanisms. *Eng. Geol.* **2025**, *357*, 108323. [CrossRef]
23. Wang, L.; Wu, Y.; Huang, Z.; Lin, J.; Wang, Y.; Zhang, W. Effects of Temperature and Confining Pressure on the Permeability of Beishan Granite from High-Level Radioactive Waste Disposal Repository. *Case Stud. Therm. Eng.* **2024**, *55*, 104111. [CrossRef]
24. Zhai, Z.; Li, K.; Bao, X.; Tong, J.; Yang, H.; Han, B. Experimental Studies on Influence of Temperature on Relative Permeability Curves in Sandstone Reservoirs. *Energy Explor. Exploit.* **2022**, *40*, 206–223. [CrossRef]
25. Wang, H.J. New Approach to the Thermal Properties of Carbonate Rocks in Geothermal Reservoirs: Molecular Dynamics Calculation and Case Studies. *Renew. Energy* **2022**, *198*, 861–871. [CrossRef]

https://doi.org/10.3390/en19020553

26. Luo, X.; Zheng, Y.; Huang, P.; Yang, L.; Chen, X.; Ren, X.; Wang, G.; Wang, Y.; Wang, Z.; Li, X. Experimental Study of the Flow Mechanism of Deep Lacustrine Carbonate Reservoirs in the Yingxi Area of the Qaidam Basin. *Geofluids* **2022**, *2022*, 8328764. [CrossRef]
27. Huang, Z.; Zhou, X.; Wang, H.; Wang, Q.; Wu, Y.-S. Stress Sensitivity Analysis of Vuggy Porous Media Based on Two-Scale Fractal Theory. *Sci. Rep.* **2024**, *14*, 20710. [CrossRef]
28. *SL/T 264-2020*; Specifications for Rock Tests in Water Conservancy and Hydropower Engineering. China Water & Power Press: Beijing, China, 2020.
29. *GB/T 50266-2013*; Standard for Test Methods of Engineering Rock Masses. China Planning Press: Beijing, China, 2013.
30. *API RP 40*; Recommended Practices for Core Analysis. American Petroleum Institute: Washington, DC, USA, 1998.
31. *GB/T 29172-2012*; Practices for Core Analysis. Standards Press of China: Beijing, China, 2012.
32. Li, X.C.; Feng, Z.C.; Guo, J.Z.; Liang, Y.K.; Bai, J.; Wang, C. Study on Influence Laws of Temperature and Stress on Sandstone Permeability. *Coal Sci. Technol.* **2019**, *47*, 96–100. (In Chinese)
33. Bu, Y.S.; Yang, S.Q.; Huang, Y.H. Experimental Study on the Influence of Temperature and Damage Degree on the Permeability of Sandstone. *Eng. Mech.* **2021**, *38*, 122–130. (In Chinese)
34. Liu, J.F.; Ni, H.Y.; Pu, H.; Huang, B.X.; Yao, Q.L.; Mao, X.B. Test Theory, Method and Device of Gas Permeability of Porous Media and the Application. *Chin. J. Rock Mech. Eng.* **2021**, *40*, 137–146. (In Chinese)
35. Jia, C.J.; Wang, H.L.; Xu, W.Y.; Zuo, J.; Skoczylas, F. Study on Gas Permeation and Slippage Effects in Rock Mass of Underground Water-Sealed Cave. *Eng. Mech.* **2015**, *32*, 50–57. (In Chinese)
36. Hibi, Y.; Fujinawa, K.; Nishizaki, S.; Okamura, K.; Tasaki, M. Investigation for Necessity of Dispersivity and Tortuosity in the Dusty Gas Model for a Binary Gas System in Soil. *Soils Found.* **2010**, *50*, 143–159. [CrossRef]
37. Huber, M.L.; Perkins, R.A.; Lemmon, E.W. Reference Correlation for the Viscosity of Nitrogen from the Triple Point to 1000 K and Pressures up to 2200 MPa. *Int. J. Thermophys.* **2024**, *45*, 146. [CrossRef] [PubMed]
38. Sun, Y.Z.; Xie, L.Z.; He, B.; Gao, C.; Wang, J. Effects of Effective Stress and Temperature on Permeability of Sandstone from CO_2-Plume Geothermal Reservoir. *J. Rock Mech. Geotech. Eng.* **2016**, *8*, 819–827. [CrossRef]
39. Zhang, J.; Zhou, X.; Liu, X.; Fang, L.; Liu, Y.; Wang, Y. Deformation and Permeability of Fractured Rocks Using Fluid-Solid Coupling under Loading-Unloading Conditions. *J. Rock Mech. Geotech. Eng.* **2025**, *17*, 4889–4907. [CrossRef]
40. Jouini, M.S.; Alabere, A.O.; Alsuwaidi, M.; Morad, S.; Bouchaala, F.; Al Jallad, O.A. Experimental and Digital Investigations of Heterogeneity in Lower Cretaceous Carbonate Reservoir Using Fractal and Multifractal Concepts. *Sci. Rep.* **2023**, *13*, 20306. [CrossRef]
41. Shi, R.; Liu, J.; Wei, M.; Elsworth, D.; Wang, X. Mechanistic Analysis of Coal Permeability Evolution Data under Stress-Controlled Conditions. *Int. J. Rock Mech. Min. Sci.* **2018**, *110*, 36–47. [CrossRef]
42. Eggers, C.G.; Berli, M.; Accorsi, M.L.; Or, D. Deformation and Permeability of Aggregated Soft Earth Materials. *J. Geophys. Res. Solid Earth* **2006**, *111*, 2005JB004123. [CrossRef]
43. Zimmerman, R.W.; Bodvarsson, G.S. Hydraulic Conductivity of Rock Fractures. *Transp. Porous Media* **1996**, *23*, 1–30. [CrossRef]
44. Guo, X.; Zou, G.; Wang, Y.; Wang, Y.; Gao, T. Investigation of the Temperature Effect on Rock Permeability Sensitivity. *J. Pet. Sci. Eng.* **2017**, *156*, 616–622. [CrossRef]
45. Shi, J.; Durucan, S. Near-Exponential Relationship between Effective Stress and Permeability of Porous Rocks Revealed in Gangi's Phenomenological Models and Application to Gas Shales. *Int. J. Coal Geol.* **2016**, *154–155*, 111–122. [CrossRef]
46. Liang, Z.; Jiang, Z.; Wu, W.; Guo, J.; Wang, M.; Nie, Z.; Li, Z.; Xu, D.; Xue, Z.; Chen, R.; et al. Study and Classification of Porosity Stress Sensitivity in Shale Gas Reservoirs Based on Experiments and Optimized Support Vector Machine Algorithm for the Silurian Longmaxi Shale in the Southern Sichuan Basin, China. *ACS Omega* **2022**, *7*, 33167–33185. [CrossRef]
47. Liu, K.; Yin, D.Y.; Sun, Y.H. The Mathematical Model of Stress Sensitivities on Tight Reservoirs of Different Sedimentary Rocks and Its Application. *J. Pet. Sci. Eng.* **2020**, *193*, 107372. [CrossRef]
48. Wu, J.-B.; Yang, S.-L.; Li, Q.; Yang, K.; Huang, C.; Lv, D.-P.; Zhou, W. Micromechanism and Mathematical Model of Stress Sensitivity in Tight Reservoirs of Binary Granular Medium. *Pet. Sci.* **2024**, *21*, 1780–1795. [CrossRef]
49. Xu, Y.; Xiao, J.; Li, X.; Xia, K.; Peng, J. Experimental Study on the Permeability Evolution of Argillaceous Sandstone under Elevated Temperatures. *Eng. Geol.* **2023**, *313*, 106974. [CrossRef]
50. Bohnsack, D.; Potten, M.; Freitag, S.; Einsiedl, F.; Zosseder, K. Stress Sensitivity of Porosity and Permeability under Varying Hydrostatic Stress Conditions for Different Carbonate Rock Types of the Geothermal Malm Reservoir in Southern Germany. *Geotherm. Energy* **2021**, *9*, 15–74. [CrossRef]
51. Shar, A.M.; Mahesar, A.A.; Chandio, A.D.; Memon, K.R. Impact of Confining Stress on Permeability of Tight Gas Sands: An Experimental Study. *J. Petrol. Explor. Prod. Technol.* **2017**, *7*, 717–726. [CrossRef]
52. Farahani, M.; Aghaei, H.; Masoumi, H. Effect of Pore Type on Porosity, Permeability and Pore Volume Compressibility of Geological Formations Due to in-Situ Stress Change. *J. Pet. Sci. Eng.* **2022**, *218*, 110986. [CrossRef]

53. Zoback, M.D.; Byerlee, J.D. The Effect of Microcrack Dilatancy on the Permeability of Westerly Granite. *J. Geophys. Res.* **1975**, *80*, 752–755. [CrossRef]
54. Hu, J.; Xie, H.; Li, C.; Liu, G. Evolution Mechanism of Permeability of Hot Dry Rock under Coupled Effect of Thermal Fatigue and Seawater Interaction during Coastal Geothermal Development. *Renew. Sustain. Energy Rev.* **2024**, *189*, 114061. [CrossRef]
55. Ge, Z. Temperature and Pressure Effect on Permeability of Chinese Sandstone: A Review. *Acta Geodyn. Geomater.* **2018**, *15*, 289–296. [CrossRef]
56. Yang, W.D.; Wang, B.Q.; Yao, J.; Jing, W.J.; Zhang, X. Experimental study on mechanical properties of carbonate rocks under real-time high temperature and heat treatment under triaxial compression. *Chin. J. Rock Mech. Eng.* **2024**, *43*, 1347–1358. (In Chinese)
57. Jia, S.; Ma, F.; Yang, Z.; Zhou, Z.; Ling, H.; Lan, T.; Wang, W.; Tian, Y.; Thanh, H.V.; Soltanian, M.R.; et al. Retardation Factor Scaling for Contaminant Transport in Fractured Media. *Adv. Water Resour.* **2024**, *188*, 104709. [CrossRef]

https://doi.org/10.3390/en19020553

MDPI AG
Grosspeteranlage 5
4052 Basel
Switzerland
Tel.: +41 61 683 77 34

Energies Editorial Office
E-mail: energies@mdpi.com
www.mdpi.com/journal/energies

www.ingramcontent.com/pod-product-compliance
Lightning Source LLC
Chambersburg PA
CBHW061808210926
51850CB00006B/644
* 9 7 8 3 7 2 5 8 7 4 0 4 0 *